Lecture Notes in Mathematics 2121

ι

More information about this series at http://www.springer.com/series/304

Daniel Robertz

Formal Algorithmic Elimination for PDEs

 Springer

Daniel Robertz
Plymouth University
School of Computing & Mathematics
Plymouth, UK

ISBN 978-3-319-11444-6   ISBN 978-3-319-11445-3 (eBook)
DOI 10.1007/978-3-319-11445-3
Springer Cham Heidelberg New York Dordrecht London

Lecture Notes in Mathematics ISSN print edition: 0075-8434
                              ISSN electronic edition: 1617-9692

Library of Congress Control Number: 2014952682

Mathematics Subject Classification (2010): 12H05, 13N10, 16S32, 35C05

Printed on acid-free paper

Springer is part of Springer Science+Business Media (www.springer.com)

Preface

This monograph originates in my habilitation treatise. The theme is to understand the correspondence between systems of partial differential equations and their analytic solutions from a formal viewpoint. We consider on the one hand the problem of determining the set of analytic solutions to such a system and on the other hand that of finding differential equations whose set of solutions coincides with a given set of analytic functions. Insight into one of these problems may advance the understanding of the other one. In particular, as shown in this monograph, symbolic solving of partial differential equations profits from the study of the implicitization problem for certain parametrized sets of analytic functions.

Two algorithms are fundamental for this work. Janet's and Thomas' algorithms transform a given system of linear or nonlinear partial differential equations into an equivalent form allowing to determine all formal power series solutions in a straightforward way. Moreover, these effective differential algebra techniques solve certain elimination problems when applied in an appropriate way. The first part of this monograph discusses these methods and related constructions in detail. Efficient implementations of both algorithms in the computer algebra system Maple have been developed by the author of this monograph and his colleagues.

The second part of this monograph addresses the problem of finding an implicit description in terms of differential equations of a set of analytic functions which is given by a parametrization of a certain kind. Effective methods of different generality are developed to solve the differential elimination problems that arise in this context. As a prominent application of these results it is demonstrated how some family of exact solutions of the Navier-Stokes equations can be computed.

I would like to thank several persons for many fruitful discussions about algebra, about PDEs, and their connection, and for supporting my work. In particular, I would like to express my gratitude to Prof. Dr. W. Plesken, Prof. Dr. J. Bemelmans, Prof. Dr. C. Melcher, Prof. Dr. S. Walcher at RWTH Aachen and Priv.-Doz. Dr. M. Barakat at TU Kaiserslautern. Thanks go to Prof. Dr. V. P. Gerdt (JINR Dubna) and Dr. A. Quadrat (Inria Saclay), not only because I have been enjoying our collaborations, but also for reading parts of a first version of this monograph.

In the context of work on Thomas' algorithm, I am grateful to my colleagues Dr. T. Bächler and Dr. M. Lange-Hegermann for their help. Moreover, we profited very much from discussions with Prof. Dr. D. Wang (Paris 6), Prof. Dr. F. Boulier, and Dr. F. Lemaire (both at Université de Lille) on this topic.

Furthermore, I would like to thank all colleagues at Lehrstuhl B für Mathematik, RWTH Aachen, for a very nice atmosphere, and, in particular, Mr. W. Krass for linguistic advice.

Finally, I am indebted to two anonymous referees for excellent suggestions on an earlier version of this monograph.

Plymouth, England   *Daniel Robertz*
July 2014

Contents

Chapter 1
Introduction

Differential equations are ubiquitous in the modeling of phenomena studied in the natural sciences, in their simulation, and in engineering applications. A lot of mathematical research is devoted to understanding many different aspects of differential equations and to developing methods which determine (exact or approximate) solutions of differential equations.

This monograph approaches partial differential equations (PDEs) from the viewpoint of algebra and contributes algorithmic methods which allow to investigate effectively the relationship of systems of PDEs and their sets of solutions. Employing formal techniques, the focus is on polynomial differential equations and their analytic solutions.

We borrow quite a few concepts from algebraic geometry, which is a theory that evolved over centuries from the continued quest to solve algebraic equations. The contemporary understanding of geometric facts in this context is formulated in terms of polynomial ideals, the precise connection between geometry and algebra being established by Hilbert's Nullstellensatz.

Whenever a set of points is given by a polynomial or rational parametrization, an elimination of the parameters from the equations which express the coordinates of the points yields equations that are satisfied by the coordinates of every point of the set. If there exists an implicit description of this set as solution set of a system of polynomial equations, then elimination constructs such a description. Since not every such solution set admits a rational parametrization, the scope of the inverse problem has to be restricted in some sense, but methods are known which construct rational parametrizations for special varieties.

This monograph develops algorithmic methods which accomplish the analogous elimination task for systems of polynomial partial differential equations and their (complex) analytic solutions. It builds on work by Charles Riquier, Maurice Janet, Joseph Miller Thomas, Joseph Fels Ritt, Ellis Robert Kolchin, and others, who laid the foundation of differential algebra.

A given multivariate polynomial, whose coefficients are analytic functions, is interpreted as a parametrization of a set of analytic functions, i.e., every element of this

© Springer International Publishing Switzerland 2014
D. Robertz, *Formal Algorithmic Elimination for PDEs*,
Lecture Notes in Mathematics 2121, DOI 10.1007/978-3-319-11445-3_1

set arises from substitution of appropriate analytic functions for the indeterminates of the polynomial. Moreover, the substitution of functions for the indeterminates also involves the composition with prescribed analytic functions. If the polynomial is linear, then the resulting set is a vector space over the field of constants. In general, however, the parametrized set is rarely closed under addition.

As a simple example we mention that the analytic functions of the form

$$F(x-t) + G(x+t)$$

admit an implicit description as solution set of the wave equation

$$\frac{\partial^2 u}{\partial t^2} = \frac{\partial^2 u}{\partial x^2},$$

a well-known fact when considered in the reverse direction.

An implicit description of an algebraic variety admits a straightforward check whether or not a given point belongs to the variety, namely by verifying if it is a solution of the defining equations. Similarly, if a set of analytic functions is the solution set of a system of differential equations, deciding membership to this set reduces to checking whether or not a given analytic function is annihilated by the corresponding differential operators.

Depending on the representation of a function at hand, it may not be obvious at all whether the function has another representation of a special form, e.g., as sum of functions depending on a smaller number of arguments, etc. A prominent example, which is only loosely related to the contents of this monograph, is Vladimir I. Arnold's solution of Hilbert's 13th problem showing that every continuous function of several variables can be represented as a composition of finitely many continuous functions of two variables. We restrict our attention to decomposition problems for analytic functions which may be answered by constructing a system of partial differential equations and inequations, whose set of solutions coincides with the set of decomposable functions.

The algorithmic methods developed in this monograph allow to improve symbolic solving of PDEs. On the one hand, membership of a solution to a family of solutions, which is implicitly described by PDEs, is decidable, so that questions regarding completeness of the family of solutions can be addressed. On the other hand, adding a PDE system characterizing analytic functions of a special form to a PDE system to be investigated may allow to extract explicit solutions of the prescribed type (cf., e.g., Subsect. 3.3.5, which records only a small family of explicit solutions computed in this way for the Navier-Stokes equations). This approach generalizes the well-known method of separation of variables.

Among the techniques discussed and used in this monograph are formal methods for solving systems of linear and polynomially nonlinear partial differential equations (Chap. 2). In the linear case, an algorithm by Maurice Janet is recalled and generalized to Ore algebras (Sect. 2.1), which include algebras of differential operators, of difference operators, combinations of them, and many more. Applied to a system of linear PDEs, it computes an equivalent system, called a Janet basis, from which it is possible to read off a partition of the set of Taylor coefficients of a formal power series solution into those whose values can be chosen arbitrarily and those whose values are then uniquely determined by these choices. Since the highest derivative may cancel in a linear combination of derivatives of two equations, this information is not apparent from the original system in general.

The case of nonlinear PDEs is dealt with using an algorithm by Joseph Miller Thomas (Sect. 2.2). It transforms a PDE system into a finite collection of so-called simple differential systems, whose solution sets form a partition of the solution set of the original system. Each of the simple systems allows to determine its formal power series solutions in a way analogous to the linear case, which is achieved by combining Thomas' splitting procedure with the ideas of Janet. Simple systems in the sense of Thomas consist of equations and inequations in general, the latter ensuring, e.g., non-vanishing of highest coefficients of equations for polynomial division and pairwise disjointness of the solutions sets of the Thomas decomposition.

For systems of linear PDEs we recall in Subsect. 2.1.5 the notion of generalized Hilbert series (cf. also [PR05, Rob06]) and present applications of this important combinatorial invariant, e.g., for computing a Noether normalization of a finitely generated commutative algebra over a field (cf. also [Rob09]). We introduce the corresponding notion for simple differential systems in Subsect. 2.2.5 and use it for elimination purposes in Sect. 3.3.

If Thomas' algorithm is applied to a system of PDEs whose left hand sides generate a prime differential ideal, then there exists a unique simple system in the resulting decomposition which is the most generic system in a precise sense. This is shown in Subsect. 2.2.3 and applied for solving the implicitization problem in Sect. 3.3.

Emphasizing that the algorithms presented in this monograph are of practical interest, we summarize work by the author of this monograph and his colleagues at Lehrstuhl B für Mathematik, RWTH Aachen, on implementations of Janet's and Thomas' algorithm in the computer algebra system Maple (Subsects. 2.1.6 and 2.2.6).

Chapter 3 is subdivided into three parts. The first section recalls classical elimination techniques using Janet bases and generalizes a method that is more efficient in practice (called "degree steering" in [PR08]) to Ore algebras using Janet's algorithm (Subsects. 3.1.1 and 3.1.3) and to systems of nonlinear PDEs using Thomas' algorithm (Subsect. 3.1.4). The idea is to associate weights to all variables and to apply these algorithms repeatedly while gradually increasing the weights of the variables to be eliminated. For the case where the given polynomials have a special monomial structure, a more efficient elimination technique, adapted to this situation, is developed in Subsect. 3.1.2 and is used for solving implicitization problems later. Subsect. 3.1.5 deals with compatibility conditions for inhomogeneous linear systems of functional equations and PDEs in particular. Whereas determining all compatibility conditions is an elimination problem, in the sense that all consequences of the given equations are to be found that involve only the inhomogeneous parts, a Janet basis for the linear system almost provides a Janet basis for its compatibility conditions at the same time.

Section 3.2 solves the implicitization problem for families of analytic functions that are parametrized by linear polynomials. More precisely, such a family arises as the set of linear combinations of finitely many given analytic functions, the coefficients being arbitrary analytic functions with prescribed arguments. This section is based on [PR10], elaborates on details of alternative implicitization methods, and presents applications to questions concerning the separation of variables and symbolic solving of PDEs.

Section 3.3 addresses the implicitization problem for families of analytic functions that are parametrized by multilinear polynomials. Again substitution of functions for the indeterminates of such a polynomial is combined with the composition with given analytic functions. Following the pattern in Sect. 3.2, implicitization methods of different generality are presented, ways to determine parameters which realize a given function as member of a parametrized family are explained, and applications to the problem of computing explicit solutions of PDE systems are given.

Both Sect. 3.2 and Sect. 3.3 build in a crucial way on the elimination procedures discussed in Sect. 3.1 and on Janet's and Thomas' algorithm (Sects. 2.1 and 2.2), respectively. New methods of differential algebra dealing with composite functions are developed.

An appendix reviews basic principles of module theory and differential algebra. We also derive the chain rule for higher derivatives, which is used in Sect. 3.2.

Chapter 2
Formal Methods for PDE Systems

Abstract This chapter discusses formal methods which transform a system of partial differential equations (PDEs) into an equivalent form that allows to determine its power series solutions. Janet's algorithm deals with the case of systems of linear PDEs. The first section presents a generalization of this algorithm to linear functional equations defined over Ore algebras. As a byproduct of a Janet basis computation, a generalized Hilbert series enumerates either a vector space basis for the linear equations that are consequences of the given system or those Taylor coefficients of a power series solution of the PDE system which can be chosen arbitrarily. Systems of polynomially nonlinear PDEs are treated in the second section from the same point of view. A Thomas decomposition of such a system consists of finitely many so-called simple differential systems whose sets of solutions form a partition of the solution set of the given system. Each simple differential system admits a straightforward method of determining its power series solutions. If the given PDE system generates a prime differential ideal, then exactly one of the simple differential systems is the most generic one in a precise sense. Both Janet's and Thomas' algorithm also solve certain elimination problems as described and employed in the following chapter.

2.1 Janet's Algorithm

A *system of linear functional equations*

$$Ru = 0 \tag{2.1}$$

for a vector u of p unknown functions is given by a matrix $R \in D^{q \times p}$ with entries in a ring D of linear operators. We outline an algebraic approach of handling such a linear system. This approach assumes that the set $\mathscr{F}$ of functions which are candidates for solutions of (2.1) is chosen as a left D-module, the left action of D being the one used in (2.1). In particular, the assumption that the result of applying any

© Springer International Publishing Switzerland 2014
D. Robertz, *Formal Algorithmic Elimination for PDEs*,
Lecture Notes in Mathematics 2121, DOI 10.1007/978-3-319-11445-3_2

operator in D to any function in $\mathscr{F}$ is a function in $\mathscr{F}$ is unavoidable as soon as algebraic manipulations are performed on the equations of system (2.1).

In this section the focus is on the case of linear partial differential equations (PDEs). Then we choose D as a ring of differential operators with left action on smooth functions by partial differentiation. (We will concentrate on analytic functions.)

Every solution of (2.1) satisfies all consequences of (2.1); we restrict our attention here to consequences which are obtained from (2.1) by multiplying a matrix with q columns and entries in D from the left. The condition that a vector of p functions solves (2.1) can be restated as follows. Let $(e_1, \ldots, e_p)$ be the standard basis of the free left D-module $D^{1 \times p}$. Then every homomorphism $\varphi \colon D^{1 \times p} \to \mathscr{F}$ of left D-modules is uniquely determined by its values $u_1, \ldots, u_p$ for $e_1, \ldots, e_p$, and every choice of values for $e_1, \ldots, e_p$ defines such a homomorphism. Now, $(u_1, \ldots, u_p)$ solves (2.1) if and only if the corresponding homomorphism φ factors over $D^{1 \times p}/D^{1 \times q} R$, i.e., is well-defined on residue classes modulo $D^{1 \times q} R$. In other words, we have

$$\hom_D(D^{1 \times p}/D^{1 \times q} R, \mathscr{F}) \cong \{u \in \mathscr{F}^{p \times 1} \mid Ru = 0\} \tag{2.2}$$

as abelian groups. (We attribute this remark to B. Malgrange [Mal62, Subsect. 3.2]; it is a basic principle of algebraic analysis, cf., e.g., [Kas03]. For recent work combining algebraic analysis with systems and control theory, cf., e.g., [PQ99], [Pom01], [CQR05], [CQ08], [Qua10b], [Rob14], and the references therein.)

Understanding the structure of the solution set of (2.1) therefore requires at least being able to compute in the residue class module

$$M := D^{1 \times p}/D^{1 \times q} R.$$

Moreover, the left D-module M is an intrinsic description of the given system of linear functional equations in the following sense. Let $Sv = 0$ with $S \in D^{s \times r}$ be another system of linear functional equations (defined over the same ring) and assume that $Ru = 0$ and $Sv = 0$ are equivalent, i.e., there exist $T \in D^{r \times p}$ and $U \in D^{p \times r}$ such that the homomorphisms of abelian groups

$$\mathscr{F}^{p \times 1} \longrightarrow \mathscr{F}^{r \times 1} \colon u \longmapsto T u, \qquad \mathscr{F}^{r \times 1} \longrightarrow \mathscr{F}^{p \times 1} \colon v \longmapsto U v$$

induce isomorphisms between $\{u \in \mathscr{F}^{p \times 1} \mid Ru = 0\}$ and $\{v \in \mathscr{F}^{r \times 1} \mid Sv = 0\}$ which are inverse to each other. If the set $\mathscr{F}$ is chosen appropriately (viz. an injective cogenerator for the category of left D-modules, cf. Remark 3.1.52, p. 154), then this implies that the homomorphisms of left D-modules

$$D^{1 \times p} \longrightarrow D^{1 \times r} \colon a \longmapsto aU, \qquad D^{1 \times r} \longrightarrow D^{1 \times p} \colon b \longmapsto bT$$

induce isomorphisms between $D^{1 \times p}/D^{1 \times q} R$ and $D^{1 \times r}/D^{1 \times s} S$ which are inverse to each other. Hence, the left D-modules which are associated with two equivalent systems of linear functional equations are isomorphic.

Computation in $M = D^{1 \times p}/D^{1 \times q} R$ for certain rings D of linear operators is made possible by (a generalization of) an algorithm named after the French mathematician Maurice Janet (1888–1983). Having computed a special generating set for the left D-module $D^{1 \times q} R$, called *Janet basis*, a (unique) normal form for the representatives of each residue class in M is defined and can be computed effectively.

The origin of Janet bases can roughly be described as follows. In work of C. Méray [Mér80] and C. Riquier [Riq10] in the second half of the 19th century the analytic solvability of systems of PDEs was investigated and a generalization of the Cauchy-Kovalevskaya Theorem was obtained. A typical formulation of this classical theorem (cf., e.g., [RR04, Sect. 2.2], [Eva10, Thm. 2 in Subsect. 4.6.3]) assumes a Cauchy problem for unknown functions $u_1, \ldots, u_m$ of $x_1, \ldots, x_n$ in a neighborhood of the origin of the following form. The given PDEs are solved for the partial derivatives of $u_1, \ldots, u_m$ with respect to the first argument x_1, say, their right hand sides being linear in the other (first order) partial derivatives of $u_1, \ldots, u_m$ with coefficients which are analytic in $x_2, \ldots, x_m$ and $u_1, \ldots, u_m$, and boundary data for $u_i(0, x_2, \ldots, x_n)$, $i = 1, \ldots, m$, are given by analytic functions. Then this Cauchy problem has a unique analytic solution. Other common formulations of this theorem allow higher order of differentiation (which can be reduced to the above situation by introducing further unknown functions). Analytic coordinate changes may be used to transform boundary data on an analytic hypersurface which is non-characteristic for the first order PDE system to the hypersurface $x_1 = 0$.

Riquier's Existence Theorem asserts the existence of analytic solutions to systems of PDEs of a certain class (cf. also [Tho28, Tho34], [Rit34, Chap. IX], [Rit50, Chap. VIII]). The equations are solved for certain distinct partial derivatives and their right hand sides are analytic functions of $x_1, \ldots, x_n$ and of partial derivatives of $u_1, \ldots, u_m$ which are less than the ones on the respective left hand side with respect to some total ordering[1]. Moreover, the system is assumed to incorporate all integrability conditions in some sense (i.e., to be passive as in Definition 2.1.40, p. 28, or Definition 2.2.48, p. 94).

First the existence of formal power series solutions is investigated (formal integrability). Given appropriate boundary conditions, convergence is considered as a second step. Confining ourselves, for the moment, to systems of linear PDEs, the first problem can be solved by transforming any given system into an equivalent one whose formal power series solutions can readily be determined. More precisely, the resulting system allows to partition the set of Taylor coefficients of a power series solution into two sets: coefficients which can be chosen arbitrarily and coefficients which are then uniquely determined by these choices. M. Janet developed an effective procedure which accomplishes such a transformation into a formally integrable system of PDEs (cf. [Jan29, Jan20]; certainly Janet was influenced by work of D. Hilbert [Hil90] as well). The result is now called a Janet basis.

More details on Riquier's Existence Theorem and, in particular, a version for differential regular chains can be found in [PG97, Sect. I.2], [Lem02, Chap. 3]. For

---

[1] The ordering is assumed to be a Riquier ranking as discussed in Remark 3.1.39, p. 142, and is assumed to respect the differentiation order; a PDE system of this form is called orthonomic.

applications of the theory of Riquier and Janet to the study of Bäcklund transformations, symmetries of differential equations, and related questions, we refer, e.g., to [Sch84], [Sch08a], [RWB96], [MRC98], [Dra01].

Around 1990 the similarity of Janet bases and Gröbner bases became evident to several researchers (cf., e.g., [Wu91], [Pom94, pp. 16–17], [ZB96]). In the case of a commutative polynomial algebra D, the result of Janet's algorithm is actually a Gröbner basis for the ideal of D which is generated by the input. The development of the notion of *involutive division* and *involutive basis* by V. P. Gerdt, Y. A. Blinkov, A. Y. Zharkov, and others (cf. [Ger05] and the references therein) turned Janet's algorithm into an efficient alternative to Buchberger's algorithm [Buc06] for computing Gröbner bases, cf. also Remark 2.1.49 below. In fact, the decomposition of multiple-closed sets of monomials into disjoint cones in a computation of an involutive basis (cf. Subsect. 2.1.1) allows to neglect many S-polynomials that are dealt with by Buchberger's original algorithm (cf. also [Ger05, Sect. 5]).

Janet's and Buchberger's algorithms solve the problem of constructing a convergent (i.e., confluent and terminating) rewriting system for the representatives of residue classes of a multivariate polynomial ring modulo an ideal. In other words, given a representative of a residue class, reduction modulo a Janet or Gröbner basis constructs the unique irreducible representative of the same residue class in finitely many steps. In fact, a unification of Buchberger's algorithm and the Knuth-Bendix completion procedure [KB70] can be achieved (cf. [Buc87, pp. 24–25], [BG94]), e.g., by incorporating constraints for coefficients into term rewriting (the inverse of a non-zero element of the ground field being the solution of an equation). In contrast to the general Knuth-Bendix completion procedure, Janet's and Buchberger's algorithms always terminate. For a study of rewriting systems for free associative algebras over commutative rings, we refer to [Ber78].

Generalizations of Gröbner bases to non-commutative algebras have been studied since a couple of decades, cf., e.g., [KRW90], [Kre93], [Mor94], [Lev05], [GL11]; for rings of differential operators, cf., e.g., [CJ84], [Gal85], [IP98], [SST00]. Buchberger's algorithm was adapted to Ore algebras by F. Chyzak (cf. [Chy98], [CS98], where it is also applied to the study of special functions and combinatorial sequences). Involutive divisions were studied for the Weyl algebra in [HSS02] and were extended to non-commutative rings in [EW07]. However, we follow a more direct approach below, in order to develop Janet's algorithm for Ore algebras.

The following presentation generalizes earlier descriptions that were given in [PR05, Rob06, Rob07]. In Subsect. 2.1.1 we discuss the combinatorics on which Janet's algorithm is based. In each non-zero polynomial a unique term is selected as the most significant one in a certain sense, and it is the technique of forming a partition of the set of monomials arising in this way which directs Janet's algorithm to new polynomials to be included in the resulting Janet basis. The same technique will be used in Sect. 2.2 for the computation of Thomas decompositions of systems of nonlinear partial differential equations and inequations.

After recalling the concept of Ore algebra in Subsect. 2.1.2, Janet's algorithm is adapted to a certain class of Ore algebras in Subsect. 2.1.3. The relation between

Janet bases and Gröbner bases is described in Subsect. 2.1.4, where we also comment on the complexity of their computation.

Subsection 2.1.5 develops the notion of generalized Hilbert series and applies this combinatorial device to the construction of a Noether normalization of a finitely generated commutative algebra over a field and to the solution of systems of linear partial differential equations. Subsection 2.1.6 summarizes work by the author of this monograph which resulted in implementations of the involutive basis technique in Maple and C++ and refers to related software.

2.1.1 Combinatorics of Janet Division

Janet's algorithm constructs a distinguished generating set, called Janet basis, for an ideal of a commutative polynomial algebra, or for a left ideal of a ring of differential operators, or, more generally, for finitely generated left modules over certain Ore algebras. Following Maurice Janet [Jan29], this method examines in a precise sense the highest terms occurring in the generators and their divisibility relations. We therefore restrict our attention in this subsection to the combinatorial properties of certain sets of monomials which are relevant for Janet's algorithm.

Let $X := \{x_1, \ldots, x_n\}$ be a set of n symbols. For any subset $Y = \{y_1, \ldots, y_r\}$ of X we denote by

$$\text{Mon}(Y) := \left\{ \prod_{i=1}^{r} y_i^{\alpha_i} \,\middle|\, \alpha \in (\mathbb{Z}_{\geq 0})^r \right\}$$

the monoid of monomials in $y_1, \ldots, y_r$, which is the free commutative semigroup with identity element generated by $y_1, \ldots, y_r$ with the usual divisibility relation $|$. For $m = y_1^{\alpha_1} \cdot \ldots \cdot y_r^{\alpha_r}$ we define $\deg_{y_i}(m) := \alpha_i$, $i = 1, \ldots, r$. We will often write m as y^α, and we denote by $|\alpha|$ the length $\alpha_1 + \ldots + \alpha_r$ of the multi-index α.

Definition 2.1.1. A set $S \subseteq \text{Mon}(X)$ is said to be $\text{Mon}(X)$-*multiple-closed*, if

$$ms \in S \qquad \text{for all} \quad m \in \text{Mon}(X), \quad s \in S.$$

Every set $G \subseteq \text{Mon}(X)$ satisfying

$$\text{Mon}(X) \cdot G = \{ mg \mid m \in \text{Mon}(X), g \in G \} = S$$

is called a *generating set* for S.

Janet used the following lemma for his "calcul inverse de la dérivation" (cf. [Jan29]). It can be seen as the special case of Hilbert's Basis Theorem dealing with ideals generated by monomials, which amounts to the statement that every sequence of monomials in which no monomial has a divisor among the previous ones is finite. This combinatorial fact is also referred to as Dickson's Lemma and is proved by induction on n.

Lemma 2.1.2. *Every* $\mathrm{Mon}(X)$-*multiple-closed subset of* $\mathrm{Mon}(X)$ *has a finite generating set. Equivalently, every ascending chain of* $\mathrm{Mon}(X)$-*multiple-closed subsets of* $\mathrm{Mon}(X)$ *terminates.*

Remark 2.1.3. Every $\mathrm{Mon}(X)$-multiple-closed set has a unique minimal generating set, which is obtained from any generating set G by removing all elements which have a proper divisor in G.

We are going to partition multiple-closed sets (and, more importantly, their complements in $\mathrm{Mon}(X)$) into cones of monomials, one instrumental fact being that the latter are again $\mathrm{Mon}(\mu)$-multiple-closed sets for some $\mu \subseteq X$.

Definition 2.1.4. a) A set $C \subseteq \mathrm{Mon}(X)$ is called a *(monomial) cone* if there exist $m \in C$ and $\mu \subseteq \{x_1, \ldots, x_n\}$ such that $\mathrm{Mon}(\mu)m = C$. The monomial m is uniquely determined by C and is called the *generator* of the cone C, and the elements of μ (of $\overline{\mu} := \{x_1, \ldots, x_n\} - \mu$) are called the *multiplicative* (resp. *non-multiplicative*) *variables* for C. Geometrically speaking, the extremal rays of the cone are parallel to the coordinate axes corresponding to multiplicative variables when monomials are visualized as points in the positive orthant (cf. Ex. 2.1.7). We often refer to such a cone C by the pair (m, μ).

b) Let $S \subseteq \mathrm{Mon}(X)$ be a set of monomials. A *cone decomposition of* S is a finite set $\{(m_1, \mu_1), \ldots, (m_r, \mu_r)\}$ of monomial cones such that the sets $C_i := \mathrm{Mon}(\mu_i)m_i$, $i = 1, \ldots, r$, satisfy $C_1 \cup \ldots \cup C_r = S$ and $C_i \cap C_j = \emptyset$ for all $i \neq j$.

Given a finite set $M = \{m_1, \ldots, m_r\}$ of monomials, there may exist in general no or many ways of arranging sets of multiplicative variables $\mu_1, \ldots, \mu_r$ such that $\{(m_1, \mu_1), \ldots, (m_r, \mu_r)\}$ is a cone decomposition of the $\mathrm{Mon}(X)$-multiple-closed set S generated by M. After enlarging the set M by elements of S, cone decompositions of S of this form exist. The possible strategies generating such cone decompositions are addressed by the notion of *involutive division*, studied, e.g., by Gerdt, Blinkov [GB98a, GB98b], Apel [Ape98], Seiler [Sei10] and others; cf. [Ger05] for a survey[2]. We restrict our attention to the strategy developed by Janet:

Definition 2.1.5. [GB98a] Let $M \subset \mathrm{Mon}(X)$ be finite. For each $m \in M$, the *Janet division* defines the set μ of multiplicative variables for the cone with generator m as follows. Let $m = x^\alpha = x_1^{\alpha_1} \cdot \ldots \cdot x_n^{\alpha_n} \in M$. For $1 \leq i \leq n$, let

$$x_i \in \mu \quad :\Longleftrightarrow \quad \alpha_i = \max\{\beta_i \mid x^\beta \in M, \beta_j = \alpha_j \text{ for all } j < i\},$$

i.e., x_i is a multiplicative variable for the cone with generator m if and only if its exponent in m is maximal among the corresponding exponents of all monomials in M whose sequence of exponents of $x_1, x_2, \ldots, x_{i-1}$ coincides with that of m.

---

[2] At the time of this writing, computer experiments have been carried out by Y. A. Blinkov and V. P. Gerdt which indicate that an involutive division which is computationally superior to the one of Janet can be defined by determining the non-multiplicative variables for a generator m_i as the union of those necessary for separating each two cones $\mathrm{Mon}(X)m_i$, $\mathrm{Mon}(X)m_j$, $i \neq j$, and by deciding the latter using a suitable term ordering on $\mathrm{Mon}(X)$, cf. [GB11].

There are also other common involutive divisions. For instance, J. M. Thomas proposed to define x_i to be a multiplicative variable for the cone with generator m if and only if $\alpha_i = \max\{\beta_i \mid x^\beta \in M\}$ (cf. [Tho37, § 36]). *Pommaret division* defines the set of multiplicative variables for the cone with generator $m \neq 1$ to be $\{x_1, \ldots, x_k\}$, where $k = \min\{j \mid \alpha_j \neq 0\}$ is the *class* of m (cf. [Pom94, p. 90], [Jan29, no. 58]).

We mention that, in the context of combinatorics, cone decompositions as defined above are referred to as Stanley decompositions, cf., e.g., [SW91].

Given a finite generating set for a $\mathrm{Mon}(X)$-multiple-closed set S, the following algorithm constructs a cone decomposition of S using the strategy proposed by Janet division. A given total ordering $>$ on X determines the order in which exponents of monomials are compared. (In Definition 2.1.5 we have $x_1 > x_2 > \ldots > x_n$.)

Algorithm 2.1.6 (*Decompose*).

Input: A finite subset G of $\mathrm{Mon}(X)$, a subset η of X, and a total ordering $>$ on
$\quad X = \{x_1, \ldots, x_n\}$

Output: A cone decomposition of the $\mathrm{Mon}(\eta)$-multiple-closed subset of $\mathrm{Mon}(X)$
$\quad$ generated by G

Algorithm:

1: **if** $|G| \leq 1$ **or** $\eta = \emptyset$ **then**
2: $\quad$ **return** $\{(g, \eta) \mid g \in G\}$
3: **else**
4: $\quad$ let y be the maximal element of η with respect to $>$
5: $\quad$ $d \leftarrow \max\{\deg_y(g) \mid g \in G\}$
6: $\quad$ **for** $i = 0, \ldots, d$ **do**
7: $\quad\quad$ $C^{(i)} \leftarrow Decompose(\bigcup_{j=0}^{i} \{y^{i-j} g \mid g \in G, \deg_y(g) = j\}, \eta - \{y\}, >)$
8: $\quad$ **end for**
9: $\quad$ replace each (m, μ) in $C^{(d)}$ with $(m, \mu \cup \{y\})$
10: $\quad$ **return** $\bigcup_{i=0}^{d} C^{(i)}$
11: **end if**

Proof. Termination follows from the fact that the cardinality of η decreases in recursive calls of the algorithm.

We show the correctness by induction on $|\eta|$. First of all, a $\mathrm{Mon}(\eta)$-multiple-closed set which is generated by a single element or is empty admits a trivial cone decomposition. If $\eta = \emptyset$, then each element of G is the generator of a cone without multiplicative variables. In any other case, the sets of multiples of elements of G with a fixed degree in the maximal variable y in η are treated separately. Let us assume that Algorithm 2.1.6 is correct if the input is any $\mathrm{Mon}(\eta')$-multiple-closed subset of Mon(X), where $\eta' \subset X$ has cardinality less than $|\eta|$. Then we assert that the monomial cones in each $C^{(i)}$, $i = 0, \ldots, d$, in step 10 are mutually disjoint. This assertion holds for $i = d$ because mutually disjoint cones with generators of the same degree d in y for which y is a non-multiplicative variable are still mutually disjoint after y has been added as a multiplicative variable (in step 9). By the induction hypothesis, the assertion is true for $i = 0, \ldots, d - 1$. Since y is only chosen to be multiplicative for the cones in $C^{(d)}$, and since cones in different $C^{(i)}$ contain only monomials of distinct degrees in y, it is clear that the cones in $\bigcup_{i=0}^{d} C^{(i)}$ are mutually disjoint. Finally, we show that we have

$$\bigcup_{(m,\mu) \in \bigcup_{i=0}^{d} C^{(i)}} \mathrm{Mon}(\mu)m = \mathrm{Mon}(\eta)G$$

after step 9. The inclusion "$\subseteq$" is obvious. By the induction hypothesis, for each $i = 0, \ldots, d$, the cones in $C^{(i)}$ resulting from step 7 form a partition of

$$\mathrm{Mon}(\eta - \{y\}) \cdot \bigcup_{j=0}^{i} \{y^{i-j}g \mid g \in G, \deg_y(g) = j\}.$$

Every element s in $\mathrm{Mon}(\eta)G$ can be written as $my^k g$ for some $m \in \mathrm{Mon}(\eta - \{y\})$, $k \in \mathbb{Z}_{\geq 0}$, and $g \in G$. If the degree i of s in y is at most d, then s is an element of a unique cone in $C^{(i)}$. If i is greater than d, then s is an element of the cone in $C^{(d)}$ resulting from step 9 which contains $my^{k-(i-d)}g$. $\qquad\square$

Algorithm 2.1.6 will be applied both in Subsect. 2.1.3 and Subsect. 2.2.2 for the construction of Janet bases for Ore algebras and of Thomas decompositions of differential systems, respectively. Whenever possible, Algorithm 2.1.6 should be applied to the minimal generating set for the multiple-closed set under consideration (also in recursive calls). This is easily achieved by an additional preliminary step which removes all elements from G which have a proper divisor in G. The algorithms discussed in Subsects. 2.1.3 and 2.2.2 produce a more compact result when making use of this modification. It is not incorporated into Algorithm 2.1.6 because for the computation of Janet bases over the ring of integers, the numeric coefficients of highest terms of polynomials must be taken into account, so that an adaptation of (auto-) reduction of a generating set is required (cf. also Def. 2.1.33).

Example 2.1.7. Let R denote the commutative polynomial algebra $K[x_1, x_2, x_3]$ over a field K and define $X := \{x_1, x_2, x_3\}$. We are going to apply the previous algorithm with the total ordering $x_1 > x_2 > x_3$. Let $\eta = X$ and let $S \subset \mathrm{Mon}(X)$ be the $\mathrm{Mon}(X)$-multiple-closed set generated by $\{x_1x_2, x_1^3x_3\}$. Then Algorithm 2.1.6 sets $d = 3$ and is applied recursively to

$$(\emptyset, \{x_2, x_3\}), \quad (\{x_1x_2\}, \{x_2, x_3\}), \quad (\{x_1^2x_2\}, \{x_2, x_3\}), \quad (\{x_1^3x_2, x_1^3x_3\}, \{x_2, x_3\}),$$

where the first component in each pair is a generating set for a $\mathrm{Mon}(\{x_2, x_3\})$-multiple-closed set. Only the last recursive run starts new recursions; the respective (minimized) arguments are $(\{x_1^3 x_3\}, \{x_3\})$, $(\{x_1^3 x_2\}, \{x_3\})$. The final result is

$$\{(x_1^3 x_2, \{x_1, x_2, x_3\}), (x_1^3 x_3, \{x_1, x_3\}), (x_1^2 x_2, \{x_2, x_3\}), (x_1 x_2, \{x_2, x_3\})\}.$$

We also display this decomposition in the following form, where the symbol $*$ indicates a non-multiplicative variable and does not represent an element of the set of multiplicative variables:

$$x_1^3 x_2, \{x_1, x_2, x_3\},$$
$$x_1^3 x_3, \{x_1, *, x_3\},$$
$$x_1^2 x_2, \{*, x_2, x_3\},$$
$$x_1 x_2, \{*, x_2, x_3\}.$$

The cones of this decomposition may also be visualized in the positive orthant of a coordinate system whose axes specify the exponents of x_1, x_2, x_3 in monomials:

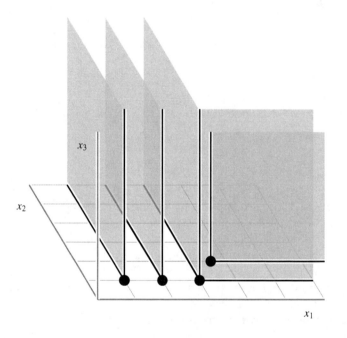

Fig. 2.1 A visualization of the cone decomposition in Example 2.1.7

Next we give a similar algorithm which produces a cone decomposition for the complement of a $\mathrm{Mon}(X)$-multiple-closed set S in $\mathrm{Mon}(X)$. Decompositions produced by this algorithm will be used later in the case of the set of leading monomials of a submodule of $D^{1 \times q}$, where D is an Ore algebra, viz. to get a partition of the set of "standard monomials", and in the case of the set of leaders of a system of polynomial differential equations (cf. also Remarks 2.1.67 and 2.2.79).

Algorithm 2.1.8 (*DecomposeComplement*).

Input: A finite subset G of $\mathrm{Mon}(X)$, a subset η of X, and $v \in \mathrm{Mon}(X)$ such that
   $G \subseteq \mathrm{Mon}(\eta)v$, and a total ordering $>$ on $X = \{x_1, \ldots, x_n\}$

Output: A cone decomposition of $\mathrm{Mon}(\eta)v - S$, where S is the $\mathrm{Mon}(\eta)$-multiple-
   closed subset of $\mathrm{Mon}(X)$ generated by G

Algorithm:

1: **if** $G = \emptyset$ **then**   // *the complement equals* $\mathrm{Mon}(\eta)v$, *which is a cone*

2:    **return** $\{(v, \eta)\}$

3: **else if** $\eta = \emptyset$ **then**   // *thus,* $G = S = \{v\}$

4:    **return** $\emptyset$

5: **else**

6:    let y be the maximal element of η with respect to $>$

7:    $d \leftarrow \max \{\deg_y(g) \mid g \in G\}$; $e \leftarrow \deg_y(v)$

8:    **for** $i = e, \ldots, d$ **do**

9:       $C^{(i)} \leftarrow DecomposeComplement(\bigcup_{j=e}^{i} \{y^{i-j}g \mid g \in G, \deg_y(g) = j\}, \eta - \{y\},$
         $y^{i-e}v, >)$

10:   **end for**

11:   replace each (m, μ) in $C^{(d)}$ with $(m, \mu \cup \{y\})$

12:   **return** $\bigcup_{i=e}^{d} C^{(i)}$

13: **end if**

Proof. It is clear that Algorithm 2.1.8 terminates. If G is empty, then $S = \emptyset$, and
$\{(v, \eta)\}$ is a trivial cone decomposition of $\mathrm{Mon}(\eta)v$. Otherwise, if η is empty, then
$S = \{v\}$, and $\mathrm{Mon}(\eta)v - S$ is empty. If $|\eta| = 1$, then the algorithm enumerates the
monomials in $\mathrm{Mon}(\eta)v - \mathrm{Mon}(\eta)G$, which are finitely many. These monomials are
generators of cones without multiplicative variables. The rest of Algorithm 2.1.8 is
similar to Algorithm 2.1.6. The only difference is the additional argument v, which
comprises the information in which set $\mathrm{Mon}(\eta)v$ the complement is to be taken.
The recursive treatment of the sets of multiples of elements of G with a fixed degree
in y needs to consider only monomials of degree at least $e = \deg_y(v)$. $\square$

Remark 2.1.9. The result of applying Algorithm 2.1.8 to a finite generating set G
for a $\mathrm{Mon}(X)$-multiple-closed subset S of $\mathrm{Mon}(X)$ and $v = 1$ is a cone decomposition of $\mathrm{Mon}(X) - S$. An additional preliminary step removing all elements from G
which have a proper divisor in G reduces the number of unnecessary recursive calls.

Example 2.1.10. Applying Algorithm 2.1.8 to the same data as in Example 2.1.7
and $v = 1$ leads again to $d = 3$ and the same recursive calls with additional arguments
$v = 1, x_1, x_1^2$, and x_1^3, respectively. After additional recursive runs, the results are

$$\{(1, \{x_2, x_3\})\}, \quad \{(x_1, \{x_3\})\}, \quad \{(x_1^2, \{x_3\})\}, \quad \text{and} \quad \{(x_1^3, \emptyset)\},$$

respectively. The final result is: $\{(1,\{x_2,x_3\}), (x_1,\{x_3\}), (x_1^2,\{x_3\}), (x_1^3,\{x_1\})\}$. An alternative representation of the result is the following, where, as in Example 2.1.7, the symbol $*$ replaces a non-multiplicative variable in the set of all variables and is not to be understood as an element of the set:

$$1, \{ * , x_2, x_3 \},$$
$$x_1, \{ * , * , x_3 \},$$
$$x_1^2, \{ * , * , x_3 \},$$
$$x_1^3, \{ x_1, * , * \}.$$

A visualization of the cone decompositions of both the multiple-closed set S and its complement in the same orthant is given as follows:

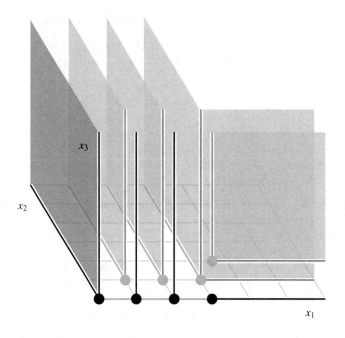

Fig. 2.2 A visualization of the cone decompositions in Examples 2.1.7 and 2.1.10

Definition 2.1.11. Let S be a $\mathrm{Mon}(X)$-multiple-closed subset of $\mathrm{Mon}(X)$ with finite generating set G, and let $>$ be a total ordering on X. We call the cone decomposition of S (of $\mathrm{Mon}(X) - S$) which is constructed by Algorithm 2.1.6 (resp. 2.1.8) a *Janet decomposition* of S (resp. of $\mathrm{Mon}(X) - S$). (If Algorithms 2.1.6 and 2.1.8 reduce G to the minimal generating set in the beginning, then this notion only depends on S and $>$.) The set of generators of the cones is called the *Janet completion* of G.

2.1.2 Ore Algebras

Ore algebras form a large class of algebras, many instances of which are encountered in applications as algebras of linear operators. The name refers to Ø. Ore, who studied non-commutative rings of polynomials under the assumption that the degree of a product of two non-zero polynomials is the sum of their degrees [Ore33]. Under the same assumption E. Noether and W. Schmeidler proved earlier that one-sided ideals of such rings are finitely generated and investigated the decompositions of such ideals as intersections of irreducible ones [NS20].

For instance, the Weyl algebra $A_1(\mathbb{R})$ consists of the polynomials in $\frac{d}{dt}$ whose coefficients are real polynomials in t, and the structure of $A_1(\mathbb{R})$ as a (non-commutative) algebra is defined in such a way that its elements represent ordinary differential operators with polynomial coefficients (i.e., the commutation rules in $A_1(\mathbb{R})$ are determined by the product rule of differentiation; cf. Ex. 2.1.18 a)). Many types of systems of linear equations can be analyzed structurally by viewing them as (left) modules over appropriate Ore algebras. The Ore algebra is chosen to contain all polynomials in the operators occurring in the system equations (cf. also the introduction to this section).

An Ore algebra is obtained as an iterated *Ore extension* of another algebra. An Ore extension forms a skew polynomial ring by adjoining one indeterminate, which does not necessarily commute with the specified algebra of coefficients. After giving the definition of skew polynomial rings and Ore algebras following [CS98], several examples of Ore algebras are discussed. At the end of this subsection important properties of Ore algebras are recalled.

In what follows, let K be a field (of any characteristic) or $K = \mathbb{Z}$, and let A be a (not necessarily commutative) K-algebra which is a domain, i.e., an associative and unital[3] algebra over K without zero divisors.

Definition 2.1.12 ([MR01], [Coh71]). Let ∂ be an indeterminate, $\sigma: A \to A$ a K-algebra endomorphism and $\delta: A \to A$ a σ-derivation, i.e., a K-linear map which satisfies

$$\delta(ab) = \sigma(a)\,\delta(b) + \delta(a)\,b \qquad \text{for all } a, b \in A.$$

The *skew polynomial ring* $A[\partial; \sigma, \delta]$ is the (not necessarily commutative) K-algebra generated by A and ∂ obeying the commutation rules

$$\partial a = \sigma(a)\,\partial + \delta(a) \qquad \text{for all } a \in A.$$

(K-linearity of both σ and δ implies that ∂ commutes with every element of K.)

Remark 2.1.13. If σ is injective, then $A[\partial; \sigma, \delta]$ is a domain because the maximum multiplicity of ∂ as a factor in the terms of an element p of $A[\partial; \sigma, \delta]$ is then referred

---

[3] All algebra homomorphisms are assumed to map the multiplicative identity element to the multiplicative identity element.

to as the degree of p, and the degree of a product of two non-zero elements of $\Lambda[\partial; \sigma, \delta]$ equals the sum of their degrees.

We recall the notion of Ore algebra as defined in [Chy98, CS98], which is an iterated skew polynomial ring with commuting indeterminates.

Definition 2.1.14. Let A be a K-algebra which is a domain and $\partial_1, \ldots, \partial_l$ indeterminates, $l \in \mathbb{Z}_{\geq 0}$. The *Ore algebra* $D = A[\partial_1; \sigma_1, \delta_1][\partial_2; \sigma_2, \delta_2] \ldots [\partial_l; \sigma_l, \delta_l]$ is the (not necessarily commutative) K-algebra generated by A and $\partial_1, \ldots, \partial_l$ subject to the relations

$$\partial_i d = \sigma_i(d)\partial_i + \delta_i(d), \quad d \in A[\partial_1; \sigma_1, \delta_1] \ldots [\partial_{i-1}; \sigma_{i-1}, \delta_{i-1}], \quad i = 1, \ldots, l, \quad (2.3)$$

where the map σ_i is a K-algebra monomorphism of $A[\partial_1; \sigma_1, \delta_1] \ldots [\partial_{i-1}; \sigma_{i-1}, \delta_{i-1}]$ and δ_i is a σ_i-derivation of $A[\partial_1; \sigma_1, \delta_1] \ldots [\partial_{i-1}; \sigma_{i-1}, \delta_{i-1}]$ (cf. Def. 2.1.12) satisfying for all $1 \leq j < i \leq l$

$$\begin{cases} \sigma_i(\partial_j) = \partial_j, \\ \delta_i(\partial_j) = 0 \end{cases} \quad (2.4)$$

and

$$\begin{cases} \sigma_i \circ \sigma_j = \sigma_j \circ \sigma_i, \\ \delta_i \circ \delta_j = \delta_j \circ \delta_i, \\ \sigma_i \circ \delta_j = \delta_j \circ \sigma_i, \\ \sigma_j \circ \delta_i = \delta_i \circ \sigma_j \end{cases} \quad (2.5)$$

as restrictions to $A[\partial_1; \sigma_1, \delta_1] \ldots [\partial_{i-1}; \sigma_{i-1}, \delta_{i-1}]$. Moreover, we require that (2.3) holds for all $d \in D$ by extending σ_i and δ_i to D as K-algebra monomorphism and σ_i-derivation, respectively, subject to $\sigma_i(\partial_j) = \partial_j$ and $\delta_i(\partial_j) = 0$ for all $1 \leq i < j \leq l$.

Remark 2.1.15. Conditions (2.4) imply that the indeterminates ∂_i and ∂_j commute in D for all $1 \leq i, j \leq l$, and conditions (2.5) ensure that this postulation is compatible with associativity of the multiplication in D. Indeed, for all $1 \leq j < i \leq l$ and all $d \in A[\partial_1; \sigma_1, \delta_1] \ldots [\partial_{j-1}; \sigma_{j-1}, \delta_{j-1}]$ we have

$$\begin{aligned}
\partial_i(\partial_j d) &= \partial_i(\sigma_j(d)\partial_j + \delta_j(d)) \\
&= \sigma_i(\sigma_j(d)\partial_j + \delta_j(d))\partial_i + \delta_i(\sigma_j(d)\partial_j + \delta_j(d)) \\
&= \sigma_i(\sigma_j(d))\partial_j\partial_i + \sigma_i(\delta_j(d))\partial_i + \delta_i(\sigma_j(d))\partial_j + \delta_i(\delta_j(d)) \\
&= \sigma_j(\sigma_i(d))\partial_i\partial_j + \sigma_j(\delta_i(d))\partial_j + \delta_j(\sigma_i(d))\partial_i + \delta_j(\delta_i(d)) \\
&= \sigma_j(\sigma_i(d)\partial_i + \delta_i(d))\partial_j + \delta_j(\sigma_i(d)\partial_i + \delta_i(d)) \\
&= \partial_j(\sigma_i(d)\partial_i + \delta_i(d)) \\
&= \partial_j(\partial_i d).
\end{aligned}$$

Moreover, since all maps σ_i and δ_j are K-linear, each indeterminate ∂_i commutes with every element of K. Extending Remark 2.1.13 we note that, since every σ_i is a K-algebra monomorphism, D is a domain.

We will concentrate on K-algebras A that are either fields (e.g., a field of rational functions over a field K or a field of meromorphic functions on a connected open subset of K^n, where $K = \mathbb{C}$) or commutative polynomial algebras over K (where K is a field or $\mathbb{Z}$) with finitely many indeterminates. The definition of a monomial in an Ore algebra depends on the type of the K-algebra A in this sense.

Definition 2.1.16. Let $D = A[\partial_1; \sigma_1, \delta_1] \ldots [\partial_l; \sigma_l, \delta_l]$ be an Ore algebra.

a) In case $A = K[z_1, \ldots, z_n]$ is a commutative polynomial algebra over a field K or over $K = \mathbb{Z}$, then the set of *indeterminates* of D is defined by

$$\mathrm{Indet}(D) := \{ z_1, \ldots, z_n, \partial_1, \ldots, \partial_l \}.$$

A *monomial* of D is then defined to be an element of the form $z^\alpha \partial^\beta$, where $z^\alpha := z_1^{\alpha_1} \cdot \ldots \cdot z_n^{\alpha_n}$, $\partial^\beta := \partial_1^{\beta_1} \cdot \ldots \cdot \partial_l^{\beta_l}$, $\alpha \in (\mathbb{Z}_{\geq 0})^n$, $\beta \in (\mathbb{Z}_{\geq 0})^l$, and we set

$$\mathrm{Mon}(D) := \{ z^\alpha \partial^\beta \mid \alpha \in (\mathbb{Z}_{\geq 0})^n, \beta \in (\mathbb{Z}_{\geq 0})^l \}.$$

The *total degree* of $z^\alpha \partial^\beta$ is defined to be $|\alpha| + |\beta| = \alpha_1 + \ldots + \alpha_n + \beta_1 + \ldots + \beta_l$.

b) If A is a field, then we define

$$\mathrm{Indet}(D) := \{ \partial_1, \ldots, \partial_l \}, \qquad \mathrm{Mon}(D) := \{ \partial^\beta \mid \beta \in (\mathbb{Z}_{\geq 0})^l \},$$

and the total degree of ∂^β is defined to be $|\beta| = \beta_1 + \ldots + \beta_l$.

We denote the total degree of a monomial $m \in \mathrm{Mon}(D)$ by $\deg(m)$.

For any subset Y of $\mathrm{Indet}(D)$, let $\mathrm{Mon}(Y)$ be the subset of elements of $\mathrm{Mon}(D)$ which do not involve any indeterminate in $\mathrm{Indet}(D) - Y$.

Let $q \in \mathbb{N}$ and denote by $e_1, \ldots, e_q$ the standard basis vectors of the free left D-module $D^{1 \times q}$. We set

$$\mathrm{Mon}(D^{1 \times q}) := \bigcup_{k=1}^{q} \mathrm{Mon}(D) e_k.$$

Remark 2.1.17. The definition of the commutation rules of D implies that $D^{1 \times q}$ is a free left A-module with basis

$$\{ \partial^\beta e_i \mid \beta \in (\mathbb{Z}_{\geq 0})^l, 1 \leq i \leq q \}. \tag{2.6}$$

Moreover, if $A = K[z_1, \ldots, z_n]$, then $\mathrm{Mon}(D^{1 \times q})$ is a basis of $D^{1 \times q}$ as a free left K-module. In other words, every $p \in D^{1 \times q}$ has a unique representation

$$p = \sum_{k=1}^{q} \sum_{m \in \mathrm{Mon}(D)} c_{k,m} \, m \, e_k \tag{2.7}$$

as linear combination of the elements of $\mathrm{Mon}(D^{1 \times q})$ with coefficients $c_{k,m} \in K$, where only finitely many $c_{k,m}$ are non-zero. In case A is a field the same holds true with $c_{k,m} \in A$ (because the basis in (2.6) equals $\mathrm{Mon}(D^{1 \times q})$).

Since D is a non-commutative ring in general, elements $p \in D^{1 \times q}$ may have more than one representation as sum of terms with unspecified order of the indeterminates. However, by the previous definition of monomials and the choice to write coefficients in A on the left, we distinguish a *normal form* (2.7) for the elements of $D^{1 \times q}$. For any $p \in D - \{0\}$, we define the *total degree* of p by

$$\deg(p) := \max \{\deg(m) \mid c_{m,k} \neq 0\},$$

using the representation (2.7). (Note that, if $\sigma_1, \ldots, \sigma_l$ are injective, then the maximum of the total degrees of monomials with non-zero coefficient is the same for any representation of p as sum of terms.)

We list important examples of Ore algebras.

Examples 2.1.18. a) If $A = K[z_1, \ldots, z_n]$, $\sigma_i = \mathrm{id}_D$ and $\delta_i = 0$ for all $i = 1, \ldots, l$, then the Ore algebra $D = K[z_1, \ldots, z_n][\partial_1; \sigma_1, \delta_1] \ldots [\partial_l; \sigma_l, \delta_l]$ is the commutative polynomial algebra over K in $n + l$ indeterminates.
b) For $n \in \mathbb{N}$, the *Weyl algebra*

$$A_n(K) := K[z_1, \ldots, z_n][\partial_1; \sigma_1, \delta_1] \ldots [\partial_n; \sigma_n, \delta_n]$$

over K is defined by

$$\sigma_i = \mathrm{id}_{A_n(K)}, \quad \delta_i = \left(a \mapsto \frac{\partial a}{\partial z_i}\right), \quad i = 1, \ldots, n.$$

In $A_n(K)$ the commutation rules

$$\partial_j z_i = z_i \partial_j + \delta_{i,j}, \quad 1 \leq i, j \leq n,$$

hold, where $\delta_{i,j}$ is the Kronecker symbol, i.e., $\delta_{i,j} = 1$ if $i = j$ and $\delta_{i,j} = 0$ otherwise. Let K be $\mathbb{R}$ or $\mathbb{C}$. We may interpret $z_1, \ldots, z_n$ as coordinates of the smooth manifold K^n. Then the indeterminate z_i in $A_n(K)$ can be understood as a name for the linear operator acting from the left on the K-vector space of smooth functions on K^n by multiplication with z_i, and the indeterminate ∂_j represents the partial differential operator with respect to z_j. (The indeterminates ∂_i and ∂_j commute, cf. Rem. 2.1.15, which is required by Schwarz' Theorem in this context.) Another variant of the Weyl algebra is the *algebra of differential operators with rational function coefficients*

$$B_n(K) := K(z_1, \ldots, z_n)[\partial_1; \sigma_1, \delta_1] \ldots [\partial_n; \sigma_n, \delta_n],$$

where σ_i and δ_i, $i = 1, \ldots, n$, are defined in the same way as above, but the elements of $A = K(z_1, \ldots, z_n)$ are rational functions in $z_1, \ldots, z_n$.

c) For $h \in \mathbb{R}$ let $S_h := \mathbb{R}[t][\delta_h; \sigma, \delta]$ be the *algebra of shift operators*, where

$$\sigma = (a(t, \delta_h) \mapsto a(t - h, \delta_h)), \quad \delta = (a \mapsto 0), \quad a = a(t, \delta_h) \in S_h.$$

This implies the commutation rule

$$\delta_h t = (t - h)\, \delta_h$$

in S_h. Hence, δ_h represents the linear operator which shifts the argument of a function of t by the amount h.

d) For $h \in \mathbb{R}$ define $D_h := \mathbb{R}[t][\partial; \sigma_1, \delta_1][\delta_h; \sigma_2, \delta_2]$, where $\sigma_1 = \mathrm{id}_{D_h}$ is the identity map, δ_1 is defined by formal differentiation with respect to t, and

$$\sigma_2 = (a(t, \partial, \delta_h) \mapsto a(t - h, \partial, \delta_h)), \quad \delta_2 = (a \mapsto 0), \quad a = a(t, \partial, \delta_h) \in D_h.$$

This algebra consists of linear operators which are relevant for differential time-delay systems.

e) Let $D = K[z_1, \ldots, z_n][\partial_1; \sigma_1, \delta_1] \ldots [\partial_n; \sigma_n, \delta_n]$, where

$$\sigma_i(a) = a(z_1, \ldots, z_{i-1},\, z_i - 1,\, z_{i+1}, \ldots, z_n, \partial_1, \ldots, \partial_n), \quad a \in D,$$

and $\delta_i = 0$, $i = 1, \ldots, n$. This algebra is used for the algebraic treatment of (multidimensional) discrete systems. Of course, the direction of the shifts can be reversed.

We only recall the essential property of Ore algebras, studied by Ore [Ore33], which ensures the existence of left skew fields of fractions. (All concepts dealing with left multiplication, left ideals, etc., can of course be translated into analogous concepts for right multiplication, right ideals and so on.)

Definition 2.1.19. A ring D is said to satisfy the *left Ore condition* if for all a_1, $a_2 \in D - \{0\}$ there exist $b_1, b_2 \in D - \{0\}$ such that $b_1 a_2 = b_2 a_1$.

If the left Ore condition is satisfied, then every right-fraction $a_1 \cdot \frac{1}{a_2}$ has a representation[4] as left-fraction $\frac{1}{b_2} \cdot b_1$. Thus, if $D - \{0\}$ is multiplicatively closed, noncommutative localization with set of denominators $D - \{0\}$ is made possible.

Proposition 2.1.20 ([MR01], Cor. 2.1.14). *Let D be a domain. A left skew field of fractions of D exists if and only if D satisfies the left Ore condition.*

In fact, if we confine ourselves to left Noetherian rings, i.e., rings for which every ascending chain of left ideals terminates, then every domain has this property.

Proposition 2.1.21 ([MR01], Thm. 2.1.15). *If D is a left Noetherian domain, then D satisfies the left Ore condition.*

---

[4] If Janet bases can be computed over D, as explained in the next subsection, then pairs (b_1, b_2) can be determined effectively as syzygies of (a_2, a_1), cf. Subsect. 3.1.5, p. 147.

Moreover, in analogy to Hilbert's Basis Theorem, we have the following important proposition.

Proposition 2.1.22 ([MR01], Thm. 1.2.9 (iv)). *If A is a left Noetherian domain and σ is an automorphism of A, then $A[\partial;\sigma,\delta]$ is also a left Noetherian domain.*

All Ore algebras in the Examples 2.1.18 are left Noetherian (with bijective twist).

2.1.3 Janet Bases for Ore Algebras

In this subsection we present Janet's algorithm for a certain class of Ore algebras D. Given a submodule M of the free left D-module $D^{1 \times q}$, $q \in \mathbb{N}$, in terms of a finite generating set, a distinguished generating set for M is constructed, which, in particular, allows to decide effectively whether a given element of $D^{1 \times q}$ is in M and to read off important invariants of M. In case D is a commutative polynomial algebra over a field, Janet's algorithm can be viewed as a simultaneous generalization of Euclid's algorithm (dealing with univariate polynomials) and Gaussian elimination (dealing with linear polynomials).

We deal at the same time with both cases of D being an iterated Ore extension of either a field K (whose elements do not necessarily commute with every element of D) or of a commutative polynomial algebra (over a field or over $\mathbb{Z}$) with finitely many indeterminates (cf. Def. 2.1.16).

In the former case we define

$$D = K[\partial_1;\sigma_1,\delta_1]\ldots[\partial_l;\sigma_l,\delta_l],$$

where for some subfield K_0 of K, each monomorphism σ_i is assumed to be a K_0-algebra automorphism and each δ_i is a K_0-linear σ_i-derivation as in Definition 2.1.14, $i = 1, \ldots, l$. We set $n := 0$ in this case. Note that K plays the role of the algebra A in the previous subsection, so that elements of K do not necessarily commute with the elements $\partial_1, \ldots, \partial_l$ in D. We also use the notation $K\langle\partial_1,\ldots,\partial_l\rangle$ for such a skew polynomial ring.

In the latter case we define

$$D = K[z_1,\ldots,z_n][\partial_1;\sigma_1,\delta_1]\ldots[\partial_l;\sigma_l,\delta_l], \qquad n \in \mathbb{Z}_{\geq 0},$$

where each monomorphism σ_i is assumed to be a K-algebra automorphism and each δ_i is a σ_i-derivation as in Definition 2.1.14, $i = 1, \ldots, l$, and where K is a field (of any characteristic) or $K = \mathbb{Z}$, and $K[z_1,\ldots,z_n]$ is the commutative polynomial algebra over K with standard grading. Moreover, in order to be able to develop Janet's algorithm for Ore algebras employing the notion of multiple-closed sets of monomials (as discussed in Subsect. 2.1.1), we restrict ourselves to the following class of Ore algebras. Let K^* denote the group of multiplicatively invertible elements of K. (In case $n = 0$ the next assumption is vacuous.)

Assumption 2.1.23. The automorphisms $\sigma_1, \ldots, \sigma_l$ are of the form

$$\sigma_i(z_j) = c_{ij} z_j + d_{ij}, \qquad c_{ij} \in K^*, \quad d_{ij} \in K, \quad j = 1, \ldots, n, \quad i = 1, \ldots, l,$$

and each σ_i-derivation δ_i satisfies

$$\delta_i(z_j) = 0 \quad \text{or} \quad \deg(\delta_i(z_j)) \leq 1, \qquad j = 1, \ldots, n, \quad i = 1, \ldots, l,$$

where $\deg(\delta_i(z_j))$ denotes the total degree of the polynomial $\delta_i(z_j) \in K[z_1, \ldots, z_n]$.

By Proposition 2.1.22, in both of the above cases D is a left Noetherian domain. We assume that the operations in D which are necessary for executing the algorithms described below can be carried out effectively, e.g., arithmetic in D and deciding equality of elements in D.

Let $q \in \mathbb{N}$, and denote by $e_1, \ldots, e_q$ the standard basis vectors of the free left D-module $D^{1 \times q}$. Recall from Remark 2.1.17 that every $p \in D^{1 \times q}$ has a unique representation

$$p = \sum_{k=1}^{q} \sum_{m \in \mathrm{Mon}(D)} c_{k,m} m e_k \tag{2.8}$$

as linear combination of monomials in $\mathrm{Mon}(D^{1 \times q})$ with coefficients $c_{k,m} \in K$, where only finitely many $c_{k,m}$ are non-zero.

Definition 2.1.24. A *term ordering* $>$ on $\mathrm{Mon}(D^{1 \times q})$ (or on $D^{1 \times q}$) is a total ordering on $\mathrm{Mon}(D^{1 \times q})$ which satisfies the following two conditions.

a) For all $1 \leq i \leq n$, $1 \leq j \leq l$, and $1 \leq k \leq q$, we have $z_i e_k > e_k$ and $\partial_j e_k > e_k$.
b) For all $m_1 e_k, m_2 e_l \in \mathrm{Mon}(D^{1 \times q})$ the following implications hold:

$$m_1 e_k > m_2 e_l \quad \Longrightarrow \quad z_i m_1 e_k > z_i m_2 e_l \qquad \text{for all } i = 1, \ldots, n$$

and

$$m_1 e_k > m_2 e_l \quad \Longrightarrow \quad m_1 \partial_j e_k > m_2 \partial_j e_l \qquad \text{for all } j = 1, \ldots, l.$$

If a term ordering $>$ on $\mathrm{Mon}(D^{1 \times q})$ is fixed, then for every non-zero $p \in D^{1 \times q}$ the $>$-greatest monomial occurring (with non-zero coefficient) in the representation (2.8) of p as left K-linear combination of monomials is uniquely determined and is called the *leading monomial* of p, denoted by $\mathrm{lm}(p)$. The coefficient of $\mathrm{lm}(p)$ in this representation of p is called the *leading coefficient* of p, denoted by $\mathrm{lc}(p)$. For any subset $S \subseteq D^{1 \times q}$ we define

$$\mathrm{lm}(S) := \{\, \mathrm{lm}(p) \mid 0 \neq p \in S \,\}.$$

Remark 2.1.25. By Lemma 2.1.2, every term ordering on $D^{1 \times q}$ is a well-ordering, i.e., every non-empty subset of $\mathrm{Mon}(D^{1 \times q})$ has a least element. Equivalently, every descending sequence of elements of $\mathrm{Mon}(D^{1 \times q})$ terminates.

Example 2.1.26. Let $\pi\colon \{1,\ldots,n+l\} \to \mathrm{Indet}(D)$ be a bijection. The *lexicographical ordering (lex)* on $\mathrm{Mon}(D)$ (which extends the total ordering $\pi(1) > \pi(2) > \ldots > \pi(n+l)$ of the indeterminates) is defined for monomials $m_1, m_2 \in \mathrm{Mon}(D)$ by

$$m_1 > m_2 \quad :\Longleftrightarrow \quad \begin{cases} m_1 \neq m_2 \quad \text{and} \quad \deg_{\pi(j)}(m_1) > \deg_{\pi(j)}(m_2) \quad \text{for} \\[2mm] j = \min\{\, 1 \leq i \leq n+l \mid \deg_{\pi(i)}(m_1) \neq \deg_{\pi(i)}(m_2) \,\}. \end{cases}$$

Example 2.1.27. Let $\pi\colon \{1,\ldots,n+l\} \to \mathrm{Indet}(D)$ be a bijection. The *degree-reverse lexicographical ordering (degrevlex)* on $\mathrm{Mon}(D)$ (extending the total ordering $\pi(1) > \pi(2) > \ldots > \pi(n+l)$ of the indeterminates) is defined for monomials $m_1, m_2 \in \mathrm{Mon}(D)$ by

$$m_1 > m_2 \quad :\Longleftrightarrow \quad \begin{cases} \deg(m_1) > \deg(m_2) \quad \text{or} \\[2mm] \big(\ \deg(m_1) = \deg(m_2) \quad \text{and} \quad m_1 \neq m_2 \quad \text{and} \\[2mm] \deg_{\pi(j)}(m_1) < \deg_{\pi(j)}(m_2) \quad \text{for} \\[2mm] j = \max\{\, 1 \leq i \leq n+l \mid \deg_{\pi(i)}(m_1) \neq \deg_{\pi(i)}(m_2) \,\}\ \big). \end{cases}$$

Example 2.1.28. Two ways of extending a given term ordering $>_1$ on $\mathrm{Mon}(D)$ to $\mathrm{Mon}(D^{1 \times q})$ for $q > 1$ are often used. The *term-over-position ordering* (extending $>_1$ and the total ordering $e_1 > \ldots > e_q$ of the standard basis vectors) is defined for $m_1, m_2 \in \mathrm{Mon}(D)$ by

$$m_1\, e_i > m_2\, e_j \quad :\Longleftrightarrow \quad m_1 >_1 m_2 \quad \text{or} \quad (m_1 = m_2 \quad \text{and} \quad i < j).$$

Accordingly, the *position-over-term ordering* (extending $>_1$ and $e_1 > \ldots > e_q$) is defined by

$$m_1\, e_i > m_2\, e_j \quad :\Longleftrightarrow \quad i < j \quad \text{or} \quad (i = j \quad \text{and} \quad m_1 >_1 m_2).$$

In order to apply Janet's method of partitioning multiple-closed sets of monomials into cones, we make the following assumption. It ensures that left multiplication by ∂_j has an easily predictable effect on leading monomials, namely multiplication of the leading monomial by ∂_j yields the leading monomial of the product.

Assumption 2.1.29. The term ordering $>$ on $\mathrm{Mon}(D^{1 \times q})$ has the property that for all $i = 1, \ldots, n$ and $j = 1, \ldots, l$ such that $\delta_j(z_i) \neq 0$, and all $k = 1, \ldots, q$ we have

$$z_i\, \partial_j\, e_k > \mathrm{lm}(\delta_j(z_i)\, e_k)$$

(where lm is defined with respect to $>$). We call such a term ordering *admissible*.

Example 2.1.30. If D satisfies Assumption 2.1.23, then every degree-reverse lexicographical ordering $>$ on $\mathrm{Mon}(D)$ is admissible. If, in addition, $\delta_j(z_i)$ is a polynomial in $K[z_i]$ of total degree at most one for all $i = 1, \ldots, n$, $j = 1, \ldots, l$, then every lexicographical ordering $>$ on $\mathrm{Mon}(D)$ is admissible. (For a common generalization of both types of term orderings, cf. Definition 3.1.4, p. 123.) If $>$ is an admissible term ordering on $\mathrm{Mon}(D)$, then its extensions to a term-over-position or a position-over-term ordering on $\mathrm{Mon}(D^{1 \times q})$ are admissible.

Remark 2.1.31. Let D be an Ore algebra as above, satisfying Assumption 2.1.23, and let $>$ be a term ordering on $D^{1 \times q}$. Then, for every non-zero $p \in D^{1 \times q}$, the monomials which occur with non-zero coefficient in the representation (2.8) of p form a finite sequence that is sorted with respect to $>$. Left multiplication of these monomials by any non-zero element of D produces a sequence of non-zero elements of $D^{1 \times q}$. If the term ordering $>$ satisfies Assumption 2.1.29, then the sequence that is obtained from the sequence of products by extracting the leading monomial of each element is necessarily sorted with respect to $>$. In particular, the leading monomial of every non-zero left multiple of p can easily be determined as a result of combining Assumptions 2.1.23 and 2.1.29. Moreover, in this situation the combinatorics of Janet division discussed in Subsect. 2.1.1 become applicable as follows.

Let $X := \{x_1, \ldots, x_{n+l}\}$ serve as the set of symbols used in Subsect. 2.1.1 and let

$$\Xi : \mathrm{Mon}(D) \longrightarrow \mathrm{Mon}(X)$$

be any bijection of the set $\mathrm{Mon}(D)$ onto the monoid $\mathrm{Mon}(X)$ satisfying that $\Xi(z^{\alpha_1} \partial^{\beta_1})$ divides $\Xi(z^{\alpha_2} \partial^{\beta_2})$ if and only if α_1 and β_1 are componentwise less than or equal to α_2 and β_2, respectively, where $\alpha_1, \alpha_2 \in (\mathbb{Z}_{\geq 0})^n$, $\beta_1, \beta_2 \in (\mathbb{Z}_{\geq 0})^l$. This implies that Ξ maps $\mathrm{Indet}(D)$ onto X.

Suppose that L is a subset of $D - \{0\}$. Let S be the set of leading monomials of all left multiples of elements of L by non-zero elements of D. Then $\Xi(S)$ is a multiple-closed set of monomials in X.

In order to apply Algorithms 2.1.6 and 2.1.8, which construct Janet decompositions of multiple-closed sets of monomials in X and of their complements, respectively, a total ordering on X is assumed to be chosen (independently of the choice of term ordering on $D^{1 \times q}$).

Definition 2.1.32. Let Ξ be a bijection as defined in the previous remark and let $S \subseteq \mathrm{Mon}(D^{1 \times q})$. For $k \in \{1, \ldots, q\}$ we define $S_k := \{m \in \mathrm{Mon}(D) \mid m e_k \in S\}$.

a) We call the set S *multiple-closed* if $\Xi(S_1)$, ..., $\Xi(S_q)$ are $\mathrm{Mon}(X)$-multiple-closed. A set $G \subseteq \mathrm{Mon}(D^{1 \times q})$ such that $\Xi(G_1)$, ..., $\Xi(G_q)$ are generating sets for $\Xi(S_1)$, ..., $\Xi(S_q)$, respectively, where $G_k := \{m \in \mathrm{Mon}(D) \mid m e_k \in G\}$, is called a *generating set* for S. In other words, the multiple-closed set generated by G is

$$[G] := \bigcup_{k=1}^{q} \Xi^{-1}(\mathrm{Mon}(X) \cdot \Xi(G_k)) e_k.$$

b) Let S be multiple-closed. For $k = 1, \ldots, q$, let

$$\{(m_1^{(k)}, \mu_1^{(k)}), \ldots, (m_{t_k}^{(k)}, \mu_{t_k}^{(k)})\}$$

be a Janet decomposition of $\Xi(S_k)$ (or of $\mathrm{Mon}(X) - \Xi(S_k)$) with respect to the chosen total ordering on X (cf. Def. 2.1.11). Then

$$\bigcup_{k=1}^{q} \left\{ \left(\Xi^{-1}(m_1^{(k)})e_k, \Xi^{-1}(\mu_1^{(k)}) \right), \ldots, \left(\Xi^{-1}(m_{t_k}^{(k)})e_k, \Xi^{-1}(\mu_{t_k}^{(k)}) \right) \right\}$$

is called a *Janet decomposition* of S (resp. of $\mathrm{Mon}(D^{1 \times q}) - S$). The *cones* of the Janet decomposition are given by

$$\Xi^{-1}(\mathrm{Mon}(\mu_i^{(k)})m_i^{(k)})e_k, \quad i = 1, \ldots, t_k, \quad k = 1, \ldots, q.$$

If the Janet decomposition is constructed from the generating set G for S, then we call the set of generators $\Xi^{-1}(m_i^{(k)})e_k$ of the cones the *Janet completion* of G.

For the rest of this section, let D be an Ore algebra as described in the beginning of this subsection which satisfies Assumption 2.1.23, and let $>$ be an admissible term ordering on $D^{1 \times q}$ (i.e., satisfying Assumption 2.1.29). We fix a bijection $\Xi : \mathrm{Mon}(D) \to \mathrm{Mon}(X)$ as above and a total ordering on X such that the Janet completion of any set $G \subseteq \mathrm{Mon}(D^{1 \times q})$ is uniquely defined.

Let M be a submodule of $D^{1 \times q}$. Starting with a finite generating set L of M, Janet's algorithm possibly removes elements from L and inserts new elements of M into L repeatedly in order to finally achieve that

$$[\mathrm{lm}(L)] = \mathrm{lm}(M).$$

An element $p \in L$ is removed if it is reduced to zero by subtraction of suitable left multiples of other elements of L. Before describing the process of auto-reduction we define when a coefficient in K is reducible modulo another one. This notion depends on whether K is a field or not.

Definition 2.1.33. Let $a, b \in K$, $b \neq 0$. If K is a field, then a is said to be *reducible modulo* b if $a \neq 0$. If $K = \mathbb{Z}$, then a is said to be *reducible modulo* b if $|a| \geq |b|$. In both cases, if a is not reducible modulo b, then the element a is also said to be *reduced modulo* b.

Definition 2.1.34. A subset L of $D^{1 \times q}$ is said to be *auto-reduced* if $0 \notin L$ holds, and for every $p_1, p_2 \in L$, $p_1 \neq p_2$, there exists no monomial $m \in \mathrm{Mon}(D^{1 \times q})$ such that the following two conditions are satisfied.

a) We have $\Xi(\mathrm{lm}(p_2)) \mid \Xi(m)$.
b) The coefficient c of m in the representation of p_1 as left K-linear combination of monomials is reducible modulo $\mathrm{lc}(p_2)$.

Given any finite subset L of $D^{1 \times q}$, there is an obvious way of computing an auto-reduced subset L' of $D^{1 \times q}$ which generates the same submodule of $D^{1 \times q}$ as L, namely by subtracting suitable left multiples of elements of L from other elements of L. We denote by ${}_D\langle L \rangle$ the submodule of $D^{1 \times q}$ generated by L.

Algorithm 2.1.35 (*Auto-reduce*).

Input: $L \subseteq D^{1 \times q}$ finite and an admissible term ordering $>$ on $D^{1 \times q}$

Output: $L' \subseteq D^{1 \times q} - \{0\}$ finite such that $_D \langle L' \rangle = _D \langle L \rangle$ and L' is auto-reduced

Algorithm:

1: $L' \leftarrow L - \{0\}$

2: **while** there exist $p_1, p_2 \in L'$, $p_1 \neq p_2$ and $m \in \mathrm{Mon}(D^{1 \times q})$ occurring with co-efficient c in the representation (2.8) of p_1 such that $\Xi(\mathrm{lm}(p_2)) \mid \Xi(m)$ and c is reducible modulo $\mathrm{lc}(p_2)$ **do**

3:     $L' \leftarrow L' - \{p_1\}$

4:     subtract a suitable left multiple of p_2 from p_1 such that the coefficient of m in the representation (2.8) of the result r is reduced modulo $\mathrm{lc}(p_2)$

5:     **if** $r \neq 0$ **then**

6:         $L' \leftarrow L' \cup \{r\}$

7:     **end if**

8: **end while**

9: **return** L'

Remark 2.1.36. Termination and the result of Algorithm 2.1.35 depend on the order in which reductions are performed. Our intention is to construct any auto-reduced set L' satisfying $[\mathrm{lm}(L)] \subseteq [\mathrm{lm}(L')]$ (and $_D \langle L' \rangle = _D \langle L \rangle$). By the choice of reductions, the result of Algorithm 2.1.35 is auto-reduced. Since only elements are removed from or replaced in L' whose leading monomial m satisfies $\Xi(\mathrm{lm}(p_2)) \mid \Xi(m)$ for a different element $p_2 \in L'$, the property $[\mathrm{lm}(L)] \subseteq [\mathrm{lm}(L')]$ is ensured as well (cf. also Def. 2.1.32 a)). Clearly, the assertion $_D \langle L' \rangle = _D \langle L \rangle$ also holds. Moreover, it is easy to see that, if in each round of the loop, the monomial m in step 2 is chosen as large as possible with respect to $>$, then Algorithm 2.1.35 terminates because $>$ is a well-ordering. In fact, if K is a field, then step 4 can be understood as replacing the term $c \cdot m$ of p_1 with a sum of terms whose monomials are smaller than m with respect to $>$. In case $K = \mathbb{Z}$, either the same kind of substitution takes place or this substitution also adds a term with monomial m, whose coefficient, however, is smaller in absolute value than c; this can be repeated only finitely many times.

In case $K = \mathbb{Z}$, computing the coefficient of m in r amounts to Euclidean division for integers. If m is the leading monomial of p_1, then it is more efficient in practice to apply the extended Euclidean algorithm to $\mathrm{lc}(p_1)$ and $\mathrm{lc}(p_2)$ in order to obtain a representation of the greatest common divisor g of $\mathrm{lc}(p_1)$ and $\mathrm{lc}(p_2)$ as linear combination of these. If $\mathrm{lc}(p_1)$ is not a multiple of $\mathrm{lc}(p_2)$, then the corresponding linear combination r of p_1 and p_2 is computed such that the leading coefficient of r equals g. Then both p_1 and r are inserted into L'. In this context, p_2 in step 2 should be chosen with the least possible absolute value of $\mathrm{lc}(p_2)$ among the candidates with the same leading monomial. In this way, Euclid's algorithm and polynomial division in the sense of Janet are interwoven.

Next we describe a reduction process which takes the Janet division into account. If a divisor of the leading monomial exists in a set defined as follows, then it is uniquely determined due to the disjointness of a cone decomposition.

Definition 2.1.37. Let $T = \{ (b_1, \mu_1), (b_2, \mu_2), \ldots, (b_t, \mu_t) \}$, where $b_i \in D^{1 \times q} - \{0\}$ and $\mu_i \subseteq \mathrm{Indet}(D)$, $i = 1, \ldots, t$.

a) The set T is said to be *Janet complete* if $\{ \mathrm{lm}(b_1), \mathrm{lm}(b_2), \ldots, \mathrm{lm}(b_t) \}$ equals its Janet completion[5] and, for each $i \in \{1, \ldots, t\}$, μ_i is the set of multiplicative variables of the cone with generator $\mathrm{lm}(b_i)$ in the Janet decomposition $\{ (\mathrm{lm}(b_1), \mu_1), \ldots, (\mathrm{lm}(b_t), \mu_t) \}$ of $[\mathrm{lm}(b_1), \ldots, \mathrm{lm}(b_t)]$ (cf. Def. 2.1.32).

b) An element $p \in D^{1 \times q}$ is said to be *Janet reducible modulo T* if there exist $(b, \mu) \in T$ and a monomial $m \in \mathrm{Mon}(D^{1 \times q})$ which occurs with coefficient c in the representation of p as left K-linear combination of monomials such that

$$\Xi(m) \in \mathrm{Mon}(\Xi(\mu)) \, \Xi(\mathrm{lm}(b))$$

and c is reducible modulo $\mathrm{lc}(b)$. In this case, (b, μ) is called a *Janet divisor* of p. Otherwise, p is also said to be *Janet reduced modulo T*.

The following algorithm subtracts suitable multiples of Janet divisors from a given element $p \in D^{1 \times q}$ as long as a term in p is Janet reducible modulo T.

Algorithm 2.1.38 (*Janet-reduce*).

Input: $p \in D^{1 \times q}$, $T = \{ (b_1, \mu_1), \ldots, (b_t, \mu_t) \}$, and an admissible term ordering $>$ on $D^{1 \times q}$, where T is Janet complete (with respect to $>$, cf. Def. 2.1.37)

Output: $r \in D^{1 \times q}$ such that $r + {}_D\langle b_1, \ldots, b_t \rangle = p + {}_D\langle b_1, \ldots, b_t \rangle$ and r is Janet reduced modulo T

Algorithm:

1: $p' \leftarrow p$; $r \leftarrow 0$

2: **while** $p' \neq 0$ **do**

3:     **if** there exists a Janet divisor (b, μ) of $\mathrm{lc}(p') \, \mathrm{lm}(p')$ in T **then**

4:         subtract a suitable left multiple of b from p' such that the coefficient of $\mathrm{lm}(p')$ in the result is reduced modulo $\mathrm{lc}(b)$; replace p' with this result

5:     **else**

6:         subtract the term of p' with monomial $\mathrm{lm}(p')$ from p' and add it to r

7:     **end if**

8: **end while**

9: **return** r

---

[5] More generally, a set of monomials is said to be *complete* (with respect to an involutive division), if it consists of the generators of the cones in a cone decomposition of the multiple-closed set they generate, where multiplicative variables for each cone are defined according to the involutive division (cf. Def. 2.1.5 for the case of Janet division). Here we confine ourselves to the complete sets of monomials which are constructed by Algorithm 2.1.6.

Remarks 2.1.39. a) Algorithm 2.1.38 terminates because, as long as p' is non-zero, the leading monomial of p' decreases with respect to the term ordering $>$, which is a well-ordering, or, if $K = \mathbb{Z}$, the absolute value of its coefficient decreases. Its correctness is clear. The result r of Algorithm 2.1.38 is uniquely determined for the given input because every monomial has at most one Janet divisor in T, and also the course of Algorithm 2.1.38 is uniquely determined as opposed to reduction procedures which apply multivariate polynomial division without distinguishing between multiplicative and non-multiplicative variables.

b) Let p_1, $p_2 \in D^{1 \times q}$ and T be as in the input of Algorithm 2.1.38. In general, the equality $p_1 + {}_D\langle b_1, \ldots, b_t \rangle = p_2 + {}_D\langle b_1, \ldots, b_t \rangle$ does not imply that the results of applying *Janet-reduce* to p_1 and p_2, respectively, are equal. But later on (cf. Thm. 2.1.43 d)) it is shown that, if T is a Janet basis, then the result of *Janet-reduce* constitutes a unique representative for every coset in $D^{1 \times q}/{}_D\langle b_1, \ldots, b_t \rangle$. This unique representative of $p_1 + {}_D\langle b_1, \ldots, b_t \rangle$ is called the *Janet normal form of p_1 modulo T*. For the sake of conciseness, we write $\mathrm{NF}(p, T, >)$ for *Janet-reduce$(p, T, >)$*, even if T is not a Janet basis.

Definition 2.1.40. A Janet complete set $T = \{(b_1, \mu_1), \ldots, (b_t, \mu_t)\}$ (as in Definition 2.1.37 a)) is said to be *passive* if

$$\mathrm{NF}(v \cdot b_i, T, >) = 0 \qquad \text{for all} \quad v \in \overline{\mu_i}, \quad i = 1, \ldots, t \tag{2.9}$$

(where we recall that $\mathrm{NF}(p, T, >)$ is the result of Algorithm 2.1.38 (*Janet-reduce*) applied to p, T, $>$). If T is passive, then it is called a *Janet basis for* ${}_D\langle b_1, \ldots, b_t \rangle$, and $\{b_1, \ldots, b_t\}$ is often referred to as a Janet basis for ${}_D\langle b_1, \ldots, b_t \rangle$ as well.

The term "passive" can be understood as the property of T which ensures that taking left D-linear combinations of $b_1, \ldots, b_t$ does not produce any $p \in D^{1 \times q} - \{0\}$ such that $\mathrm{lm}(p) \notin [\mathrm{lm}(b_1), \ldots, \mathrm{lm}(b_t)]$ (cf. also Remark 2.1.41 below).

More generally, an *involutive basis* is defined by replacing the reference to Janet completeness in the previous definition with a possibly different way of partitioning multiple-closed sets of monomials into cones, as determined by an involutive division (cf. the paragraphs before and after Def. 2.1.5, p. 10). For instance, *Pommaret bases*, i.e., involutive bases with respect to Pommaret division (cf. [Pom94, p. 90], [Jan29, no. 58]) are investigated, e.g., in [Sei10]. Pommaret bases are guaranteed to be finite only in coordinate systems that are sufficiently generic (so-called δ-regular coordinates). Essentially the same technique has been applied to a study of homogeneous ideals in a commutative algebra context by Mutsumi Amasaki in [Ama90] (where this form of Gröbner basis in generic coordinates is referred to as *a system of Weierstraß polynomials*).

Remark 2.1.41. Let M be a submodule of $D^{1 \times q}$. Then $\mathrm{lm}(M)$ is multiple-closed (cf. Rem. 2.1.31 and Def. 2.1.32). Janet's algorithm (cf. Alg. 2.1.42 below) constructs an ascending chain of multiple-closed subsets of $\mathrm{lm}(M)$, which terminates by Lemma 2.1.2. In each round, a cone decomposition is computed for the current multiple-closed set S generated by the leading monomials of an auto-reduced generating set G for M. Note that, if K is a field, these leading monomials form not just any generating set, but the minimal generating set for S.

The Janet decomposition is constructed by applying Algorithm 2.1.6, p. 11, directly to G, in the sense that its run is determined by the monomials $\Xi(\mathrm{lm}(g))$, $g \in G$, but left multiplications of such a monomial by y are replaced with left multiplications of g by $\Xi^{-1}(y)$. Accordingly, the result $J = \{(b_1, \mu_1), \ldots, (b_t, \mu_t)\}$ consists of pairs of a non-zero element b_i of $D^{1 \times q}$ and a subset μ_i of $\mathrm{Indet}(D)$. In the following algorithm, this adapted version of Algorithm 2.1.6 (*Decompose*) will be applied (using the given total ordering on X).

Since $\{b_1, \ldots, b_t\}$ is a generating set for M, every element of M is a left D-linear combination of $b_1, \ldots, b_t$. We assume that J is passive. Let $k_i m_i b_i$ be a summand in such a linear combination, where $k_i \in K$ and $m_i \in \mathrm{Mon}(D)$. If m_i involves some variable which is non-multiplicative for b_i, then this summand can be replaced with a left K-linear combination of elements in $\mathrm{Mon}(\mu_1)b_1, \ldots, \mathrm{Mon}(\mu_t)b_t$. Using (2.9), this can be achieved by applying successively Algorithm 2.1.38 to terms involving only one non-multiplicative variable. This substitution process should deal with the largest term with respect to $>$ first. Elimination of all non-multiplicative variables demonstrates that the leading monomial of every element of $M - \{0\}$ has a Janet divisor in J. We conclude that passivity of the Janet complete set J is equivalent to

$$[\mathrm{lm}(b_1), \ldots, \mathrm{lm}(b_t)] = \mathrm{lm}(M).$$

Now Janet's algorithm is presented, which computes a Janet basis for a submodule of $D^{1 \times q}$, given in terms of a finite generating set. Note that we ignore efficiency issues in favor of a concise formulation of the algorithm (cf. also Subsect. 2.1.6).

For any set S we denote by $\mathscr{P}(S)$ the power set of S.

Algorithm 2.1.42 (*JanetBasis*).

Input: A finite set $L \subseteq D^{1 \times q}$, an admissible term ordering $>$ on $D^{1 \times q}$, and a total ordering on $\Xi(\mathrm{Indet}(D)) = X$ (used by *Decompose*)

Output: A finite subset J of $D^{1 \times q} \times \mathscr{P}(\mathrm{Indet}(D))$ which is a Janet basis for the left D-module ${}_D\langle p \mid (p, \mu) \in J\rangle = {}_D\langle L\rangle$ (and $J = \emptyset$ if and only if ${}_D\langle L\rangle = \{0\}$)

Algorithm:

1: $G \leftarrow L$

2: **repeat**

3:      $G \leftarrow \textit{Auto-reduce}(G, >)$ // cf. Alg. 2.1.35

4:      $J \leftarrow \textit{Decompose}(G)$ // cf. Rem. 2.1.41

5:      $P \leftarrow \{\mathrm{NF}(v \cdot p, J, >) \mid (p, \mu) \in J, v \in \overline{\mu}\}$ // cf. Alg. 2.1.38

6:      $G \leftarrow \{p \mid (p, \mu) \in J\} \cup P$

7: **until** $P \subseteq \{0\}$

8: **return** J

Theorem 2.1.43. *a) Algorithm 2.1.42 terminates and is correct.*

b) A K-basis of $_D\langle L \rangle$ *is given by* $\biguplus\limits_{(p,\mu)\in J} \mathrm{Mon}(\mu)p$, *where J is the result of Algorithm 2.1.42. In particular, every* $r \in {_D\langle L \rangle}$ *has a unique representation*

$$ r = \sum_{(p,\mu)\in J} c_{(p,\mu)}\, p, $$

where each $c_{(p,\mu)} \in D$ *is a left K-linear combination of elements in* $\mathrm{Mon}(\mu)$.

c) The cosets in $D^{1\times q}/{_D\langle L \rangle}$ *with representatives in*

$$ \mathrm{Mon}(D^{1\times q}) - [\mathrm{lm}(p) \mid (p,\mu) \in J,\, 1\ \text{is reducible modulo}\ \mathrm{lc}(p)] $$

form a generating set for the left K-module $D^{1\times q}/{_D\langle L \rangle}$, *and the cosets with representatives in* $C := \mathrm{Mon}(D^{1\times q}) - [\mathrm{lm}(p) \mid (p,\mu) \in J]$ *form the unique maximal K-linearly independent subset.*

Let $C_1, \ldots, C_k$ *be the cones of a Janet decomposition of C (cf. Fig. 2.2, p. 15, for an illustration). If K is a field, then the cosets with representatives in* $C_1 \uplus \ldots \uplus C_k$ *form a basis for the left K-vector space* $D^{1\times q}/{_D\langle L \rangle}$.

d) For every $r_1, r_2 \in D^{1\times q}$ *the following equivalence holds.*

$$ r_1 + {_D\langle L \rangle} = r_2 + {_D\langle L \rangle} \iff \mathrm{NF}(r_1, J, >) = \mathrm{NF}(r_2, J, >). $$

Proof. a) First we show that *JanetBasis* terminates. For the result G of *Auto-reduce* in step 3, $[\mathrm{lm}(G)]$ contains the multiple-closed set generated by the leading monomials of the previous set G. *Decompose* only augments the generating set G by elements $p \in D^{1\times q}$ satisfying $\mathrm{lm}(p) \in [\mathrm{lm}(G)]$ if it is necessary for the chosen strategy of decomposing $[\mathrm{lm}(G)]$ into disjoint cones. In any case it ensures $[\mathrm{lm}(p) \mid (p,\mu) \in J] = [\mathrm{lm}(G)]$. If all Janet normal forms in step 5 are zero, then the algorithm terminates. If $P \not\subseteq \{0\}$, then $G' := G \cup P$ satisfies $[\mathrm{lm}(G)] \subsetneq [\mathrm{lm}(G')]$. By Lemma 2.1.2, after finitely many steps we have $[\mathrm{lm}(G)] = [\mathrm{lm}(G')]$, which is equivalent to $P \subseteq \{0\}$ (cf. Rem. 2.1.41). Therefore, *JanetBasis* terminates in any case.

In order to prove correctness of *JanetBasis*, we note that the result J of step 4 is Janet complete. Therefore $\mathrm{NF}(v \cdot p, J, >)$ in step 5 is well-defined. Once $P \subseteq \{0\}$ holds in step 7, the set J is passive, thus a Janet basis. The equality of left D-modules $_D\langle\, p \mid (p,\mu) \in J\,\rangle = {_D\langle L \rangle}$ is an invariant of the loop.

b) Set $B := \bigcup_{(p,\mu)\in J} \mathrm{Mon}(\mu)p$.

For the K-linear independence of B we note first that $0 \notin B$ holds because J is constructed as Janet completion of an auto-reduced set. Furthermore, we have $\mathrm{lm}(p_1) \neq \mathrm{lm}(p_2)$ for all $p_1, p_2 \in B$ with $p_1 \neq p_2$ because J is Janet complete, which proves that B is K-linearly independent.

We are going to show that B is a generating set for the free left K-module $_D\langle L \rangle$. Let $0 \neq r \in {_D\langle L \rangle}$. Then r is Janet reducible modulo J by Remark 2.1.41. Now, $\mathrm{NF}(r, J, >) \in {_D\langle L \rangle}$, and $\mathrm{NF}(r, J, >)$ is not Janet reducible modulo J. The previous argument implies $\mathrm{NF}(r, J, >) = 0$. Hence, we have $r \in {_K\langle B \rangle}$.

c) Let $0 \neq r + {}_D\langle L\rangle \in D^{1 \times q}/{}_D\langle L\rangle$. Then $\mathrm{NF}(r, J, >)$ is a representative of the residue class $r + {}_D\langle L\rangle$ as well, and $\mathrm{NF}(r, J, >) \neq 0$ because otherwise $r \in {}_D\langle L\rangle$. Janet reduction (Alg. 2.1.38) ensures that if a term $c \cdot m$ in $\mathrm{NF}(r, J, >)$, where $c \in K$ and $m \in \mathrm{Mon}(D^{1 \times q})$, has a Janet divisor (p, μ) in J, then c is reduced modulo $\mathrm{lc}(p)$. Therefore, the cosets represented by those monomials $\mathrm{lm}(p)$, $(p, \mu) \in J$, for which 1 is reducible modulo $\mathrm{lc}(p)$, are not needed to generate $D^{1 \times q}/{}_D\langle L\rangle$ as a left K-module.

Due to the equality $[\mathrm{lm}(p) \mid (p, \mu) \in J] = \mathrm{lm}({}_D\langle L\rangle)$ (cf. Rem. 2.1.41) we have $C = \mathrm{Mon}(D^{1 \times q}) - \mathrm{lm}({}_D\langle L\rangle)$. The cosets with representatives in C are K-linearly independent because no K-linear combination of these has a non-zero representative with leading monomial in $\mathrm{lm}({}_D\langle L\rangle)$. The rest is clear.

d) It remains to show that the normal form $\mathrm{NF}(r, J, >)$ is uniquely determined by the coset $r + {}_D\langle L\rangle \in D^{1 \times q}/{}_D\langle L\rangle$. But if n_1, $n_2 \in D^{1 \times q}$ are Janet normal forms of the same coset $r + {}_D\langle L\rangle$, then $n_1 - n_2 \in {}_D\langle L\rangle$, and $n_1 - n_2$ is Janet reduced modulo J because n_1 and n_2 are so. The same argument as in the last part of b) shows that $n_1 - n_2 = 0$. $\qquad\square$

We present a couple of examples demonstrating Janet's algorithm.

Example 2.1.44. Let $D = K[x, y]$ be the commutative polynomial algebra over a field K of arbitrary characteristic or over $K = \mathbb{Z}$. We choose the degree-reverse lexicographical ordering on $\mathrm{Mon}(D)$ satisfying $x > y$ (cf. Ex. 2.1.27). Let the ideal I of D be generated by

$$g_1 := \underline{x^2} - y, \quad g_2 := \underline{xy} - y.$$

Then the method of Subsect. 2.1.1 (using the total ordering on $\{x, y\}$ for which x is greater than y) constructs the following cone decomposition of the multiple-closed set which is generated by the (underlined) leading monomials of g_1 and g_2:

$$\{ (x^2, \{x, y\}), (xy, \{y\}) \}.$$

This result indicates that we need to check whether $f := x \cdot g_2$ can be written as

$$f = c_1 \cdot (x^2 - y) + c_2 \cdot (xy - y), \qquad c_1 \in K[x, y], \quad c_2 \in K[y]. \tag{2.10}$$

The monomials appearing in $f = x^2 y - xy \in I$ lie in the cones $(x^2, \{x, y\})$ and $(xy, \{y\})$, respectively. Reduction yields $g_3 := y^2 - y \in I$, which does not have a representation as in (2.10). So, we include g_3 in our list of generators, and for this example, we already arrive at the (minimal) Janet basis

$$\{ (g_1, \{x, y\}), (g_2, \{y\}), (g_3, \{y\}) \}$$

for I.

No division by any coefficient was necessary to arrive at a Janet basis for I. The statements above therefore hold for a field K of any characteristic and for $K = \mathbb{Z}$. In Example 2.1.47, the relevance of Janet bases with integer coefficients for constructing matrix representations of finitely presented groups is demonstrated.

Example 2.1.45. Let us consider the system

$$\frac{\partial u}{\partial x} = x \frac{\partial u}{\partial y}, \qquad u(x-1,y) = u(x,y) \tag{2.11}$$

of one linear partial differential equation and one linear delay equation for one unknown smooth function u of two independent variables x and y. According to the types of functional operators occurring in the system (2.11) we define the Ore algebra[6] $D = \mathbb{Q}[x,y][\partial_x; \mathrm{id}, \delta_1][\partial_y; \mathrm{id}, \delta_2][\delta_x; \sigma, \delta_3]$, where the derivations δ_1 and δ_2 are defined by partial differentiation with respect to x and y, respectively, where σ is the $\mathbb{Q}$-algebra automorphism of D defined by

$$a(x,y,\partial_x,\partial_y,\delta_x) \longmapsto a(x-1,y,\partial_x,\partial_y,\delta_x),$$

and where δ_3 is the zero map. By writing the equations in (2.11) as

$$(\partial_x - x\partial_y)\,u = 0, \qquad (\delta_x - 1)\,u = 0,$$

we are led to study the left ideal I of D which is generated by

$$\{\partial_x - x\partial_y,\ \delta_x - 1\}.$$

Janet's algorithm can be applied for determining all linear consequences of (2.11). We choose the degree-reverse lexicographical ordering (cf. Ex. 2.1.27) on $\mathrm{Mon}(D)$ satisfying $\partial_x > \partial_y > \delta_x > x > y$. The multiple-closed set which is generated by the leading monomials ∂_x and δ_x is partitioned into cones. Using the total ordering on $\{\partial_x, \partial_y, \delta_x, x, y\}$ which is defined by $\partial_x > \partial_y > \delta_x > x > y$, all indeterminates are assigned as multiplicative variables to the first generator g_1, whereas ∂_x is a non-multiplicative variable for the second generator g_2. Janet reduction of $\partial_x(\delta_x - 1)$ yields

$$g_3 := \partial_x(\delta_x - 1) - (\delta_x - 1)(\partial_x - x\partial_y) - (x-1)\partial_y(\delta_x - 1) = -\partial_y.$$

After adding $-g_3$ to the generating set and updating the Janet decomposition, Janet reduction replaces g_1 with $g_1 - xg_3 = \partial_x$. It can be easily checked that the resulting Janet complete set is passive. Therefore, the (minimal) Janet basis is given by

$$\frac{\partial u}{\partial x} = 0,\ \{\,\partial_x, \partial_y, \delta_x,\ x\ ,\ y\ \},$$

$$\frac{\partial u}{\partial y} = 0,\ \{\ *\ ,\partial_y, \delta_x,\ x\ ,\ y\ \},$$

$$u(x-1,y) - u(x,y) = 0,\ \{\ *\ ,\ *\ ,\delta_x,\ x\ ,\ y\ \}.$$

The system (2.11) only admits constant solutions.

---

[6] The computations performed in this example do not change if we replace $\mathbb{Q}[x,y]$ with its field of fractions $\mathbb{Q}(x,y)$ in the definition of D.

The following example applies Janet's algorithm to a system of linear partial differential equations which arises from linearization of a nonlinear PDE system. Linearization is a common simplification technique for studying differential equations. Being an approximation, the linearized system reflects only a few properties of the original system in general. It can be understood as the first order term of the Taylor expansion of the nonlinear system around a chosen solution (e.g., a critical point for ordinary differential equations) using the Fréchet derivative (cf., e.g., [Olv93, Sect. 5.2], [Rob06, Sect. 3.2]). By computing these derivatives symbolically, we do not need any particular solution of the nonlinear system, but work with a symbol which is subject to the nonlinear equations. We refer to the resulting linear system as the *general linearization*. (Alternatively, the linearization of algebraic differential equations can also be expressed in terms of Kähler differentials, cf. also Subsect. 3.3.3.)

In the given example all real analytic solutions of the nonlinear PDE system are available explicitly, which is obviously a very special case, but which allows a comparison of the solutions of the linearized system and those of the original one. (For notation that concerns notions of differential algebra, we refer to Sect. A.3. More details on the notion of general linearization can be found in [Rob06, Sect. 3.2].)

Example 2.1.46. [Rob06, Ex. 3.3.9] We consider the system of nonlinear PDEs

$$\frac{\partial u}{\partial x} - u^2 = 0, \qquad \frac{\partial^2 u}{\partial y^2} - u^3 = 0 \qquad (2.12)$$

for one unknown real analytic function u of two independent variables x and y. Note that

$$u(x,y) = \frac{2}{-2x \pm \sqrt{2}y + c}, \qquad c \in \mathbb{R}, \qquad (2.13)$$

are explicit solutions of (2.12). We are going to apply Janet's algorithm to the general linearization

$$\frac{\partial U}{\partial x} - 2uU = 0, \qquad \frac{\partial^2 U}{\partial y^2} - 3u^2 U = 0 \qquad (2.14)$$

of (2.12), which is a system of linear PDEs for an unknown real analytic function U of x and y, and where the function u is subject to (2.12). Since Janet's algorithm has to decide whether coefficients of polynomials are equal to zero or not, it is required to bring the nonlinear system (2.12) into a form that allows effective computation with u. Applying the techniques to be discussed in Sect. 2.2 to (2.12) yields a Thomas decomposition of that system (where subscripts indicate differentiation):

$\underline{u_x} - u^2 = 0 \ \{ \partial_x, \partial_y \}$	
$2u_y{}^2 - u^4 = 0 \ \{ *, \partial_y \}$	
$u \neq 0$	$u = 0 \ \{ \partial_x, \partial_y \}$

We are interested in the first case, solve the second equation for u_y, and use

$$u_x = u^2, \qquad u_y = \pm \frac{\sqrt{2}}{2} u^2 \qquad (2.15)$$

as rewriting rules for the coefficients in Janet's algorithm. For the second rule a choice of sign should be made and used consistently in what follows. Let $\mathbb{Q}(\sqrt{2})\{u\}$ be the differential polynomial ring in one differential indeterminate u with coefficients in $\mathbb{Q}(\sqrt{2})$ and commuting derivations δ_x, δ_y (cf. Sect. A.3). Moreover, let I be the differential ideal of $\mathbb{Q}(\sqrt{2})\{u\}$ which is generated by $u_x - u^2$ and $u_y \mp \frac{\sqrt{2}}{2} u^2$. Then $\mathbb{Q}(\sqrt{2})\{u\}/I$ is a domain because it is isomorphic to $\mathbb{Q}(\sqrt{2})[u]$. We denote by K the field of fractions of $\mathbb{Q}(\sqrt{2})\{u\}/I$, which is a differential field with derivations extending δ_x and δ_y (using the quotient rule). Now the left hand sides of the input (2.14) for Janet's algorithm are to be understood as elements of the skew polynomial ring $K\langle \partial_x, \partial_y \rangle = K[\partial_x; \mathrm{id}, \delta_1][\partial_y; \mathrm{id}, \delta_2]$, where the derivations δ_1 and δ_2 are defined as the extensions of δ_x and δ_y to $K\langle \partial_x, \partial_y \rangle$ satisfying $\delta_1(\partial_x) = \delta_1(\partial_y) = 0$ and $\delta_2(\partial_x) = \delta_2(\partial_y) = 0$, respectively (cf. also Def. 2.1.14).

In this example the passivity check only involves the partial derivative with respect to x of the second equation in (2.14), whose normal form is computed by subtracting the second partial derivative with respect to y of the first equation. After simplification using the rewriting rules (2.15) we obtain

$$\pm 2\sqrt{2} u^2 U_y - 4u^3 U = 0,$$

and since $u \neq 0$, a Janet basis for the linearized system is

$$U_x - 2uU = 0, \ \{ \partial_x, \partial_y \},$$
$$U_y \mp \sqrt{2} uU = 0, \ \{ *, \partial_y \}.$$

Substituting (2.13) for u in this Janet basis results in a system of linear PDEs for U whose analytic solutions are given by

$$U(x,y) = \frac{C}{(-2x \pm \sqrt{2}y + c)^2}, \qquad C \in \mathbb{R}. \qquad (2.16)$$

If we consider the map (between Banach spaces) which associates with ε in a small real interval containing 0 the explicit solution in (2.13) with constant $c + \varepsilon$, then the solution (2.16) for a certain value of C coincides, as expected, with the coefficient of ε in the Taylor expansion of this (sufficiently differentiable) map around 0. (We refer to [Rob06, Sect. 3.2] for more details.)

For applications of Janet bases with integer coefficients, e.g., for constructing matrix representations of finitely presented groups (without specifying the characteristic of the field in advance), for a constructive version of the Quillen-Suslin Theorem, and for primary decomposition, we refer to [PR06], [Fab09], [FQ07], [Jam11]. The following example is an application of the first kind.

Example 2.1.47. [Rob07, Ex. 5.1] We would like to construct matrix representations of degree 3 over various fields K of the finitely presented group

$$G_{2,3,13;4} := \langle a, b \mid a^2, b^3, (ab)^{13}, [a,b]^4 \rangle,$$

where $[a,b] := aba^{-1}b^{-1}$. To this end, we write the images of (the residue classes of) a and b under such a representation as 3×3 matrices A and B with indeterminate entries. The relators a^2, b^3, $(ab)^{13}$, $[a,b]^4$ of the above presentation are translated into relations for commutative polynomials obtained from the entries of the matrix equations

$$A^2 = I_3, \quad B^3 = I_3, \quad (AB)^{13} = I_3, \quad [A,B]^4 = I_3, \tag{2.17}$$

where I_3 is the identity matrix in $GL(3,K)$. (We refer to [PR06] for more details on this approach.) We may choose a K-basis (v_1, v_2, v_3) of $K^{3\times 1}$ with respect to which the K-linear action on $K^{3\times 1}$ of (the residue classes of) a and b in $G_{2,3,13;4}$ is represented by A and B, respectively. We let v_1 be an eigenvector of $A \cdot B$ with eigenvalue λ, possibly in an algebraic extension field of K, and let $v_2 := Bv_1$ and $v_3 := Bv_2$. We confine ourselves to finding irreducible representations, which implies that (v_1, v_2, v_3) is K-linearly independent. By using (v_1, v_2, v_3) as a basis for $K^{3\times 1}$, we may assume without loss of generality that A and B have the form

$$A := \begin{pmatrix} 0 & c_2 & c_3 \\ c_1 & 0 & c_4 \\ 0 & 0 & c_5 \end{pmatrix}, \quad B := \begin{pmatrix} 0 & 0 & 1 \\ 1 & 0 & 0 \\ 0 & 1 & 0 \end{pmatrix},$$

where $c_1 = \lambda^{-1}$, $c_2 = \lambda$, because $ABv_1 = \lambda v_1$ implies $v_2 = Bv_1 = A^2Bv_1 = \lambda Av_1$ and we have $B^3v_1 = v_1$ due to the given relations (2.17). Moreover, we derive from (2.17) a system of algebraic equations for $c_1, \ldots, c_5$. We compute

$$A^2 = \begin{pmatrix} c_1c_2 & 0 & c_2c_4 + c_3c_5 \\ 0 & c_1c_2 & c_1c_3 + c_4c_5 \\ 0 & 0 & c_5^2 \end{pmatrix}.$$

The determinant of A equals $-c_1c_2c_5$. Now, by (2.17) we have $\det(A^2) = 1$, $\det(B) = 1$, and $\det((AB)^{13}) = 1$, which implies $\det(A) = 1$. Due to $c_1c_2 = 1$ we have $c_5 = -1$. Hence, we substitute -1 for c_5 in A and we are left with four unknowns. We define L to be the set of all entries of the matrices $A^2 - I_3$, $(AB)^{13} - I_3$, $(ABAB^2)^2 - (BAB^2A)^2$. Thus L consists of polynomials in c_1, c_2, c_3, c_4 with integer coefficients. We compute a Janet basis for the ideal of $\mathbb{Q}[c_1, c_2, c_3, c_4]$ which is generated by L. The result consists of the polynomial 1 only. This shows that the above system of algebraic equations for c_1, c_2, c_3, c_4 has no solution in $\mathbb{C}^4$, hence there exists no irreducible matrix representation $G_{2,3,13;4} \to GL(3,\mathbb{C})$.

Next we check whether there are such matrix representations of $G_{2,3,13;4}$ in positive characteristic. To this end, we compute a Janet basis J with respect to the degree-reverse lexicographical ordering extending $c_1 > c_2 > c_3 > c_4$ (and using the same total ordering of variables for determining Janet decompositions) for the ideal of $\mathbb{Z}[c_1, c_2, c_3, c_4]$ which is generated by L:

$15,$	$\{ \, * \, , \, * \, , \, * \, , \, * \, \},$
$15 c_4,$	$\{ \, * \, , \, * \, , \, * \, , \, * \, \},$
$15 c_3,$	$\{ \, * \, , \, * \, , \, * \, , \, * \, \},$
$15 c_2,$	$\{ \, * \, , \, * \, , \, * \, , \, * \, \},$
$c_1 + 4 c_2 + c_3 + 4 c_4,$	$\{ \, c_1 \, , \, c_2 \, , \, c_3 \, , \, c_4 \, \},$
$15 c_4^2,$	$\{ \, * \, , \, * \, , \, * \, , \, * \, \},$
$15 c_3 c_4,$	$\{ \, * \, , \, * \, , \, * \, , \, * \, \},$
$c_2 c_4 - c_3,$	$\{ \, * \, , \, * \, , \, * \, , \, c_4 \, \},$
$c_3^2 + 4 c_3 c_4 + c_4^2 + c_3 + c_4 + 4,$	$\{ \, * \, , \, * \, , \, c_3 \, , \, c_4 \, \},$
$c_2 c_3 - 4 c_4^2 - 4 c_3 - 1,$	$\{ \, * \, , \, * \, , \, c_3 \, , \, c_4 \, \},$
$c_2^2 + c_4^2 + 2 c_3 - 7,$	$\{ \, * \, , \, c_2 \, , \, c_3 \, , \, c_4 \, \},$
$c_4^3 + 2 c_3 c_4 + 4 c_4^2 + 4 c_3 - 7 c_4 + 1,$	$\{ \, * \, , \, * \, , \, * \, , \, c_4 \, \},$
$c_3 c_4^2 - 4 c_3 c_4 - c_4^2 + c_2 + 7 c_3 - 2 c_4 - 4,$	$\{ \, * \, , \, * \, , \, * \, , \, c_4 \, \}.$

We find that solutions of the system of algebraic equations exist only if $15 = 0$, i.e., possibly in characteristic 3 or 5. We are going to check both possibilities. It turns out that replacing each coefficient of the above polynomials with its remainder modulo 3 (resp. 5) yields (after removing zero polynomials) the minimal Janet basis for the algebraic system over $\mathbb{Z}/3\mathbb{Z}$ (resp. $\mathbb{Z}/5\mathbb{Z}$) with the same multiplicative variables. A Janet decomposition of the complement in $\mathrm{Mon}(\{c_1, c_2, c_3, c_4\})$ of the multiple-closed set generated by the leading monomials is given by

$$\{ (1, \emptyset), (c_4, \emptyset), (c_3, \emptyset), (c_2, \emptyset), (c_4^2, \emptyset), (c_3 c_4, \emptyset) \}.$$

Denoting by F either $\mathbb{Z}/3\mathbb{Z}$ or $\mathbb{Z}/5\mathbb{Z}$, and by I the ideal of $F[c_1, c_2, c_3, c_4]$ which is generated by L (modulo 3 resp. 5), we conclude that $R := F[c_1, c_2, c_3, c_4]/I$ is 6-dimensional as an F-vector space. By the Chinese Remainder Theorem, the residue class ring of R modulo its radical is isomorphic to a direct sum of at most six fields, which define at most six solutions to the above algebraic system over an algebraic closure of F, the bound being attained precisely if R has no nilpotent elements. For the present example we obtain quickly the Janet basis

$c_4^6 + c_4^4 + c_4 + 1,$	$\{ \, * \, , \, * \, , \, * \, , \, c_4 \, \},$
$c_3 + 2 c_4^5 + 2 c_4^4 + c_4^3 + 2 c_4 + 2,$	$\{ \, * \, , \, * \, , \, c_3 \, , \, c_4 \, \},$
$c_2 + c_4^5 + 2 c_4^4 + c_4^2,$	$\{ \, * \, , \, c_2 \, , \, c_3 \, , \, c_4 \, \},$
$c_1 + 2 c_4^4 + 2 c_4^3 + 2 c_4^2 + 2 c_4 + 1,$	$\{ \, c_1 \, , \, c_2 \, , \, c_3 \, , \, c_4 \, \}$

for I with respect to the lexicographical ordering extending $c_1 > c_2 > c_3 > c_4$, which allows to solve for c_1, c_2, c_3 in terms of c_4. In $(\mathbb{Z}/3\mathbb{Z})[c_4]$ we have

$$c_4^6 + c_4^4 + c_4 + 1 = (c_4^3 + c_4^2 + 2)(c_4^3 + 2 c_4^2 + 2 c_4 + 2),$$

the factors on the right hand side being irreducible. Hence, we have found matrix representations of $G_{2,3,13;4}$ of degree 3 over the fields $(\mathbb{Z}/3\mathbb{Z})[\xi]/(\xi^3 + \xi^2 + 2)$ and

$(\mathbb{Z}/3\mathbb{Z})[\xi]/(\xi^3 + 2\xi^2 + 2\xi + 2)$. For instance, in the first case we obtain

$$A = \begin{pmatrix} 0 & 2\xi + 1 & 2\xi^2 + \xi \\ \xi^2 + 2\xi + 2 & 0 & \xi \\ 0 & 0 & 2 \end{pmatrix}.$$

For $F = \mathbb{Z}/5\mathbb{Z}$ an analogous computation yields irreducible matrix representations over the fields $(\mathbb{Z}/5\mathbb{Z})[\zeta]/(\zeta^2 + 2\zeta + 4)$ and $(\mathbb{Z}/5\mathbb{Z})[\zeta]/(\zeta^4 + 3\zeta^3 + \zeta^2 + 2\zeta + 4)$; e.g., we may choose A as

$$A = \begin{pmatrix} 0 & 4\zeta^3 + \zeta^2 + 3 & 4\zeta^3 + \zeta^2 + 4 \\ \zeta + 4 & 0 & \zeta \\ 0 & 0 & 4 \end{pmatrix}$$

in the second case.

2.1.4 Comparison and Complexity

In this subsection we comment on the relationship between Janet bases and Gröbner bases and on complexity results. For surveys on the latter topic, cf., e.g., [May97], [vzGG03, Sect. 21.7].

We use the same notation as in the previous subsection.

Definition 2.1.48. Let M be a submodule of the free left D-module $D^{1 \times q}$. A finite subset $G \subseteq M - \{0\}$ is said to be a *Gröbner basis* for M (with respect to the term ordering $>$ on $D^{1 \times q}$) if the leading monomial of every non-zero element of M is the leading monomial of a left multiple of some element of G.

Remark 2.1.49. If $J = \{(p_1, \mu_1), \ldots, (p_t, \mu_t)\}$ is a Janet basis for the submodule $M = {}_D\langle p_1, \ldots, p_t \rangle$ of $D^{1 \times q}$, then the multiple-closed set $[\mathrm{lm}(p_1), \ldots, \mathrm{lm}(p_t)]$ equals $\mathrm{lm}(M)$ (cf. also Rem. 2.1.41). More generally, this equality is used as a criterion for the termination of algorithms constructing involutive bases, cf., e.g., [Ger05], and is also well-known from Buchberger's algorithm computing Gröbner bases (cf., e.g., his PhD thesis of 1965, [Buc06]). In fact, for this reason, every involutive basis is also a Gröbner basis, whenever both notions exist in the same context, but the former reflects a lot more combinatorial information about the ideal or module (cf. Subsect. 2.1.5). More precisely, in general the reduced Gröbner basis of a module (cf., e.g., [CLO07, § 2.7]) is a proper subset of a Janet basis for the same module (and with respect to the same term ordering). For another comparison of Janet and Gröbner bases, cf. also [CJMF03].

Janet's algorithm can be understood as a refinement of the original version of Buchberger's algorithm (cf., e.g., [CLO07, § 2.7], [Eis95, Sect. 15.4], [vzGG03, Sect. 21.5]). For simplicity we assume that the module is an ideal of $\mathbb{Q}[x_1, \ldots, x_n]$. Given a finite generating set L, Buchberger's algorithm forms the *S-polynomial*

$$S(p_1,p_2) := \frac{\mathrm{lcm}(\mathrm{lm}(p_1),\mathrm{lm}(p_2))}{\mathrm{lc}(p_1)\,\mathrm{lm}(p_1)}\, p_1 - \frac{\mathrm{lcm}(\mathrm{lm}(p_1),\mathrm{lm}(p_2))}{\mathrm{lc}(p_2)\,\mathrm{lm}(p_2)}\, p_2$$

for each unordered pair of (non-zero) generators p_1, p_2 in L and reduces it modulo L using multivariate polynomial division. Non-zero remainders are added to L, and this process is repeated until every S-polynomial reduces to zero. Janet's algorithm decomposes the multiple-closed set generated by the leading monomials of the generators in L into disjoint cones as described in Subsect. 2.1.1 and considers the S-polynomials which are determined by the non-multiplicative variables v of generators p and the (uniquely determined) Janet divisor of $v \cdot p$ in the current generating set (if any). This strategy avoids many S-polynomials which are examined by Buchberger's original algorithm (cf. also [Ger05, Sect. 5]). However, Buchberger's algorithm was enhanced as well by incorporating criteria which allow to neglect certain S-polynomials (cf., e.g., [Buc79], [CLO07, § 2.9]).

Algorithm 2.1.42 constructs a Janet basis which is minimal with respect to inclusion for the fixed total ordering on $\Xi(\mathrm{Indet}(D)) = X$. This Janet basis J is uniquely determined under the assumption that no term in any of its elements is Janet reducible modulo J and that, if K is a field, the coefficient of each leading monomial equals one, say. In case $K = \mathbb{Z}$, a choice for the systems of residues modulo integers, e.g., the symmetric one, should be fixed to ensure uniqueness.

Redundancy of a Janet basis (compared with the reduced Gröbner basis) is diminished by the concept of *Janet-like Gröbner basis* [GB05a, GB05b]. For each generator the partition of the set of indeterminates into sets of multiplicative and non-multiplicative variables is replaced with a map of this set into $\mathbb{Z}_{\geq 0} \cup \{\infty\}$ indicating the multiplicative degree for each indeterminate. If the image of each of these maps is required to be a subset of $\{0, \infty\}$, then the Janet division is recovered as a special case. The number of left multiples of generators by non-multiplicative variables to be included for completion is often reduced when applying Janet-like division.

Let $D = \mathbb{Q}[x_1,\ldots,x_n]$ be a commutative polynomial algebra in n variables.

The complexity of the problem to decide whether a given polynomial is an element of an ideal of D (the latter being given in terms of a finite generating set) was studied by G. Hermann [Her26]. Her result states the following.

Theorem 2.1.50. *Let $G \subset D - \{0\}$ be a finite generating set of cardinality m for an ideal I of D, and let $p \in D$. Let d be the maximum total degree of the elements of G. If p is an element of I, then p is a linear combination of the generators in G with coefficients that are either zero or polynomials of total degree at most*

$$\deg(p) + (md)^{2^n}.$$

The following upper bound on the degrees of the elements of a reduced Gröbner basis is proved, e.g., in [Dub90], where techniques of partitioning sets of monomials similar to Subsect. 2.1.1 are used.

Theorem 2.1.51. *Let $G \subset D - \{0\}$ be a finite generating set for an ideal I of D. Let d be the maximum total degree of the elements of G. Then the total degree of the elements of the reduced Gröbner basis for I with respect to any term ordering on* $\mathrm{Mon}(D)$ *is bounded by*

$$2 \left(\frac{d^2}{2} + d \right)^{2^{n-1}}.$$

Better bounds for special cases are also known. For instance, if $n = 3$, then the total degree of the polynomials constructed by Buchberger's algorithm computing a Gröbner basis for I (including the elements of G) is bounded by $(8d + 1) \cdot 2^\delta$, where δ is the minimum total degree of the elements of G [Win84].

A corresponding doubly exponential degree bound for Janet bases over the Weyl algebras was obtained in [GC08] by reducing the problem to solving linear systems over a variant of the Weyl algebra whose commutation rules have been made homogeneous by introducing an additional commuting variable (cf. also [Gri91], [Gri96]).

E. W. Mayr and A. R. Meyer constructed a family of ideals (generated by binomials) for which the doubly exponential upper bound is attained [MM82]. Further work by E. W. Mayr showed that the computation of a Gröbner basis for a general polynomial ideal is an EXPSPACE-complete problem.

Remark 2.1.52. In practice a behavior much better than the worst case has been observed for algorithms computing Gröbner or Janet bases when applied to, e.g., problems arising in algebraic geometry or systems of linear partial equations with origin in physics or the engineering sciences. In the algebraic geometry context a growth measure was introduced for the degrees of (iterated) syzygies (cf. Subsect. 3.1.5) for a given ideal I of a commutative polynomial algebra, which reflects the difficulty of computing Gröbner or Janet bases for I. This measure, called Castelnuovo-Mumford regularity, denoted by $\mathrm{reg}(I)$, is significant not only from the computational, but also from the geometric and algebraic point of view, cf., e.g., [Eis95, Eis05].

The regularity of the ideal generated by $\mathrm{lm}(I)$ is an upper bound for the regularity of I. In generic coordinates, the maximum total degree of the elements of the reduced Gröbner basis for I equals $\mathrm{reg}(\langle \mathrm{lm}(I) \rangle)$. If leading monomials are determined with respect to the degree-reverse lexicographical ordering, then we have, in generic coordinates, $\mathrm{reg}(\langle \mathrm{lm}(I) \rangle) = \mathrm{reg}(I)$, cf. [BS87]. Therefore, this term ordering is preferably used.

For lack of space, we do not discuss here recent approaches by J.-C. Faugère (cf., e.g., [Fau99]) to compute Gröbner bases using linear algebra techniques, which result in very efficient programs and applications to cryptography.

2.1.5 The Generalized Hilbert Series

In this subsection we extend the notion of generalized Hilbert series (cf. [PR05], [Rob06]) to Ore algebras and present applications of this combinatorial invariant.

Throughout this subsection, let K be a field.

Let D be an Ore algebra as described in the beginning of Subsect. 2.1.3 which satisfies Assumption 2.1.23 (p. 22). Moreover, let $q \in \mathbb{N}$ and $>$ be an admissible term ordering on $\mathrm{Mon}(D^{1 \times q})$ (cf. Assumption 2.1.29, p. 23). For combinatorial purposes we introduce a totally ordered set

$$X := \{x_1, \ldots, x_{n+l}\}$$

of indeterminates and we choose a bijection $\Xi \colon \mathrm{Mon}(D) \to \mathrm{Mon}(X)$ as in Remark 2.1.31 (p. 24), where it was used to apply the combinatorics of Janet division to a set of monomials of the Ore algebra D.

Definition 2.1.53. For any subset S of $\mathrm{Mon}(D^{1 \times q})$, the *generalized Hilbert series* of S is defined by

$$H_S(x_1, \ldots, x_{n+l}) := \sum_{se_k \in S} \Xi(s) f_k \in \bigoplus_{k=1}^{q} \mathbb{Z}[[x_1, \ldots, x_{n+l}]] f_k,$$

where the symbols $f_1, \ldots, f_q$ form a basis of a free left $\mathbb{Z}[[x_1, \ldots, x_{n+l}]]$-module of rank q. For $k = 1, \ldots, q$, we define $H_{S,k}(x_1, \ldots, x_{n+l})$ by

$$H_S(x_1, \ldots, x_{n+l}) = \sum_{k=1}^{q} H_{S,k}(x_1, \ldots, x_{n+l}) f_k,$$

and we identify $H_S(x_1, \ldots, x_{n+l})$ with $H_{S,1}(x_1, \ldots, x_{n+l})$ in case $q = 1$.

Remark 2.1.54. Let M be a submodule of $D^{1 \times q}$ and let J be a Janet basis for M with respect to some term ordering on $D^{1 \times q}$. We denote by S the multiple-closed set generated by $\{ \mathrm{lm}(p) \mid (p, \mu) \in J \}$ (cf. Def. 2.1.32).

a) According to Theorem 2.1.43 b), the (disjoint) union of the cones $\mathrm{Mon}(\mu) p$, $(p, \mu) \in J$, is a K-basis of M. Thus, the generalized Hilbert series $H_S(x_1, \ldots, x_{n+l})$ enumerates a K-basis of M, in the sense that its terms enumerate the leading monomials of the above K-basis via Ξ^{-1}.

b) Similarly, by Theorem 2.1.43 c), a K-basis of the factor module $D^{1 \times q}/M$ is given by the cosets in $D^{1 \times q}/M$ which are represented by the monomials in $C_1 \uplus \ldots \uplus C_k$, where $C_1, \ldots, C_k$ are the cones of a Janet decomposition of $\overline{S} := \mathrm{Mon}(D^{1 \times q}) - S$. Therefore, the generalized Hilbert series $H_{\overline{S}}(x_1, \ldots, x_{n+l})$ enumerates a K-basis of $D^{1 \times q}/M$ via Ξ^{-1}.

When the Janet basis J for M is clear from the context, we also call H_S and $H_{\overline{S}}$ the *generalized Hilbert series* of M and $D^{1 \times q}/M$, respectively.

The next remark shows that the generalized Hilbert series of a set S of monomials has a succinct representation in finite terms if a cone decomposition of S is available.

Remark 2.1.55. Let (C, μ) be a monomial cone, i.e., $C \subseteq \mathrm{Mon}(X)$, $\mu \subseteq X$, and

$$S := C - \mathrm{Mon}(\mu) \cdot v$$

for some $v \in C$. We use the (formal) geometric series

$$\frac{1}{1-x} = \sum_{i \in \mathbb{Z}_{\geq 0}} x^i$$

to write down the generalized Hilbert series $H_S(x_1, \ldots, x_{n+l})$ as follows:

$$H_S(x_1, \ldots, x_{n+l}) = \frac{v}{\prod_{x \in \mu}(1-x)}.$$

More generally, every cone decomposition of a multiple-closed set S allows to compute the generalized Hilbert series of S by adding the generalized Hilbert series of the cones. In an analogous way this remark applies to the complements of multiple-closed sets.

Example 2.1.56. Let R be the commutative polynomial algebra $K[x_1, x_2, x_3]$ over any field K and S the multiple-closed set generated by $\{x_1 x_2, x_1^3 x_3\}$. The Janet decomposition of S which is computed in Example 2.1.7 yields the generalized Hilbert series

$$H_S(x_1, x_2, x_3) = \frac{x_1^3 x_2}{(1-x_1)(1-x_2)(1-x_3)} + \frac{x_1^3 x_3}{(1-x_1)(1-x_3)}$$

$$+ \frac{x_1^2 x_2}{(1-x_2)(1-x_3)} + \frac{x_1 x_2}{(1-x_2)(1-x_3)}.$$

The Janet decomposition of the complement $\overline{S} = \mathrm{Mon}(\{x_1, x_2, x_3\}) - S$ of S obtained in Example 2.1.10 yields

$$H_{\overline{S}}(x_1, x_2, x_3) = \frac{1}{(1-x_2)(1-x_3)} + \frac{x_1}{1-x_3} + \frac{x_1^2}{1-x_3} + \frac{x_1^3}{1-x_1}.$$

Note that the sum of the two Hilbert series equals $1/((1-x_1)(1-x_2)(1-x_3))$, i.e.,

$$H_S + H_{\overline{S}} = H_{\mathrm{Mon}(\{x_1, x_2, x_3\})}.$$

Next we describe the relationship between the generalized Hilbert series and the Hilbert series of filtered and graded modules.

Definition 2.1.57. Let A be a (not necessarily commutative) K-algebra and assume that $F = (F_i)_{i \in \mathbb{Z}_{\geq 0}}$ is an (exhaustive) increasing filtration of A (cf., e.g., [Bou98b]), i.e., each F_i is a (left) K-subspace of A and we have $1 \in F_0$,

$$\bigcup_{i \in \mathbb{Z}_{\geq 0}} F_i = A, \qquad \text{and} \qquad F_i \subseteq F_{i+1}, \qquad F_i \cdot F_j \subseteq F_{i+j} \qquad \text{for all} \quad i, j \in \mathbb{Z}_{\geq 0}.$$

Moreover, let M be a left A-module endowed with an (exhaustive) increasing F-filtration $\Phi = (\Phi_i)_{i \in \mathbb{Z}}$, i.e., each Φ_i is a (left) K-subspace of M such that

$$\bigcup_{i \in \mathbb{Z}} \Phi_i = M \quad \text{and} \quad \Phi_i \subseteq \Phi_{i+1}, \quad F_i \cdot \Phi_j \subseteq \Phi_{i+j} \quad \text{for all} \quad i \in \mathbb{Z}_{\geq 0}, \quad j \in \mathbb{Z}.$$

We assume that M is finitely generated and that each Φ_i has finite K-dimension. Then the *Hilbert series of M with respect to Φ* is defined by the (formal) Laurent series

$$H_{M,\Phi}(\lambda) := \sum_{i \in \mathbb{Z}} (\dim_K \Phi_i) \, \lambda^i \in \mathbb{Z}((\lambda)).$$

The map

$$\mathbb{Z} \longrightarrow \mathbb{Z}_{\geq 0} : i \longmapsto \dim_K \Phi_i$$

is called the *Hilbert function of M with respect to Φ*.

Definition 2.1.58. Let A be a (not necessarily commutative) K-algebra and assume that A is positively graded, i.e., it is endowed with a family $G = (G_i)_{i \in \mathbb{Z}_{\geq 0}}$ of (left) K-subspaces of A such that

$$A = \bigoplus_{i \in \mathbb{Z}_{\geq 0}} G_i \qquad \text{and} \qquad G_i \cdot G_j \subseteq G_{i+j} \qquad \text{for all} \quad i, j \in \mathbb{Z}_{\geq 0}.$$

Moreover, let M be a left A-module with G-grading $\Gamma = (\Gamma_i)_{i \in \mathbb{Z}}$, i.e., a family of (left) K-subspaces of M such that

$$M = \bigoplus_{i \in \mathbb{Z}} \Gamma_i \qquad \text{and} \qquad G_i \cdot \Gamma_j \subseteq \Gamma_{i+j} \qquad \text{for all} \quad i \in \mathbb{Z}_{\geq 0}, \quad j \in \mathbb{Z}.$$

We assume that M is finitely generated and that each Γ_i has finite K-dimension. Then the *Hilbert series of M with respect to Γ* is defined by the (formal) Laurent series

$$H_{M,\Gamma}(\lambda) := \sum_{i \in \mathbb{Z}} (\dim_K \Gamma_i) \, \lambda^i \in \mathbb{Z}((\lambda)).$$

The map

$$\mathbb{Z} \longrightarrow \mathbb{Z}_{\geq 0} : i \longmapsto \dim_K \Gamma_i$$

is called the *Hilbert function of M with respect to Γ*.

We recall that every grading defines an increasing filtration on the same module and that every increasing filtration defines an associated graded module over the associated graded ring.

Remark 2.1.59. Given a K-algebra A with grading $G = (G_i)_{i \in \mathbb{Z}_{\geq 0}}$ and a left A-module M with G-grading $\Gamma = (\Gamma_i)_{i \in \mathbb{Z}}$ as in the previous definition, an (exhaustive) increasing filtration $F = (F_i)_{i \in \mathbb{Z}_{\geq 0}}$ of A is defined by

$$F_i := \bigoplus_{j \leq i} G_j, \qquad i \in \mathbb{Z}_{\geq 0},$$

and an (exhaustive) increasing F-filtration $\Phi = (\Phi_i)_{i \in \mathbb{Z}}$ of M is defined by

$$\Phi_i := \bigoplus_{j \leq i} \Gamma_j, \qquad i \in \mathbb{Z}.$$

If M is finitely generated and each Γ_i has finite K-dimension, then each Φ_i has finite K-dimension, and we have

$$H_{M,\Phi}(\lambda) = H_{M,\Gamma}(\lambda) \cdot \frac{1}{1 - \lambda}.$$

Conversely, in the situation of Definition 2.1.57, the *associated graded ring* is defined to be the K-algebra

$$\mathrm{gr}(A) := \bigoplus_{i \in \mathbb{Z}_{\geq 0}} (F_i / F_{i-1}) \qquad \text{(where } F_{-1} := \{0\}\text{)}$$

with multiplication

$$(p_1 + F_{i-1}) \cdot (p_2 + F_{j-1}) := p_1 \cdot p_2 + F_{i+j-1}, \qquad p_1 \in F_i, \quad p_2 \in F_j,$$

and the *associated graded module* is defined by

$$\mathrm{gr}(M) := \bigoplus_{i \in \mathbb{Z}} (\Phi_i / \Phi_{i-1})$$

with left $\mathrm{gr}(A)$-action

$$(p + F_{i-1})(m + \Phi_{j-1}) = p \cdot m + \Phi_{i+j-1}, \qquad p \in F_i, \quad m \in \Phi_j.$$

The grading of $\mathrm{gr}(M)$ defines again an increasing filtration of $\mathrm{gr}(M)$, but since A and $\mathrm{gr}(A)$ are non-isomorphic rings in general, the resulting filtration reflects only partial information about M. (Note also that, even if M is assumed to be finitely generated and each Φ_i has finite K-dimension, $\mathrm{gr}(M)$ is not a finitely generated $\mathrm{gr}(A)$-module in general; cf., e.g., [Bjö79, Sect. 1.2] or [Cou95, Sect. 8.3].)

The following two remarks establish a link between the Hilbert series of certain graded modules and filtered modules, respectively, and the generalized Hilbert series, which is computable via Janet bases. (The first remark will be applied in Remark 2.1.64, the second one in Remark 3.2.17.)

Remark 2.1.60. Let $D = K[x_1, \ldots, x_n]$ be a commutative polynomial algebra over the field K and assume D is positively graded. We denote by $\deg(x_i)$ the degree of x_i, $i = 1, \ldots, n$, with respect to this grading G. Let $q \in \mathbb{N}$ and let $\Gamma = (\Gamma_i)_{i \in \mathbb{Z}}$ be a G-grading of $D^{1 \times q}$ such that each Γ_i is a finite-dimensional K-vector space. For any submodule M of $D^{1 \times q}$ such that $\Gamma_i' := \Gamma_i \cap M$, $i \in \mathbb{Z}$, defines a G-grading Γ' of M, a Janet basis J for M (with respect to any term ordering) provides via the generalized Hilbert series a K-basis of M (cf. Rem. 2.1.54 a)). Then the Hilbert series of M with respect to Γ' is obtained from the generalized Hilbert series by substitution:

$$H_{M,\Gamma'}(\lambda) = \sum_{k=1}^{q} H_{S,k}(\lambda^{\deg(x_1)}, \ldots, \lambda^{\deg(x_n)}),$$

where S is the multiple-closed set generated by the leading monomials of elements of J.

In this case, $D^{1 \times q}/M$ has the G-grading $\Gamma'' = (\Gamma_i'')_{i \in \mathbb{Z}}$, where Γ_i'' is defined as the image of Γ_i under the canonical projection $D^{1 \times q} \to D^{1 \times q}/M$. The generalized Hilbert series of the complement $\overline{S}$ of S in $\mathrm{Mon}(D^{1 \times q})$ yields the Hilbert series of $D^{1 \times q}/M$ with respect to Γ'':

$$H_{D^{1 \times q}/M,\Gamma''}(\lambda) = \sum_{k=1}^{q} H_{\overline{S},k}\left(\lambda^{\deg(x_1)}, \ldots, \lambda^{\deg(x_n)}\right)$$

(cf. Rem. 2.1.54 b)).

The maximum number of multiplicative variables of cones in a Janet decomposition of $D^{1 \times q}/M$, and therefore the order of 1 as a pole of the corresponding generalized Hilbert series, equals the Krull dimension of $D^{1 \times q}/M$ (cf., e.g., [Sta96, I.5] or [SW91]).

Remark 2.1.61. Let D be an Ore algebra as before, $q \in \mathbb{N}$, and M a submodule of $D^{1 \times q}$. We define an (exhaustive) increasing filtration $F = (F_i)_{i \in \mathbb{Z}_{\geq 0}}$ on D by

$$F_i := \{ p \in D \mid p = 0 \text{ or } \deg(p) \leq i \}, \qquad i \in \mathbb{Z}_{\geq 0},$$

where deg denotes the total degree, and an (exhaustive) increasing F-filtration on $D^{1 \times q}$ by

$$\Phi_i := \{ t \in D^{1 \times q} \mid t = 0 \text{ or } \deg(t) \leq i \}, \qquad i \in \mathbb{Z},$$

where $\deg(t)$ is defined as the maximum of the total degrees of the non-zero components of the tuple t. (In case of the Weyl algebras $A_n(K)$ this filtration is known as the Bernstein filtration; cf., e.g., [Bjö79] or [Cou95].) Intersecting with M defines an F-filtration $\Phi' := (\Phi_i \cap M)_{i \in \mathbb{Z}}$ of M.

Assumption 2.1.23 implies that the associated graded ring $\mathrm{gr}(D)$ is isomorphic to the commutative polynomial algebra $K[\xi_1, \ldots, \xi_n, \eta_1, \ldots, \eta_l]$ with standard grading G (since the degrees of the indeterminates of D are all equal to one), and the $\mathrm{gr}(D)$-module $\mathrm{gr}(M)$ is isomorphic to a graded $K[\xi_1, \ldots, \xi_n, \eta_1, \ldots, \eta_l]$-module. Let Γ be the G-grading of $\mathrm{gr}(M)$.

Let J be a Janet basis for M with respect to an admissible term ordering which is compatible with the total degree. Then the corresponding generalized Hilbert series $H_S(x_1, \ldots, x_{n+l})$ enumerates a K-basis of M (cf. Rem. 2.1.54 a)), where S is the multiple-closed set generated by the leading monomials of elements of J. Since the term ordering is compatible with the total degree,

$$\biguplus_{(p,\mu) \in J} \mathrm{Mon}(\mu)\, (\mathrm{lm}(p) + \Phi'_{\deg(p)-1})$$

is a K-basis of $\mathrm{gr}(M)$, so that the coefficient of λ^i in the formal power series $H_S(\lambda, \ldots, \lambda)$ equals $\dim_K \Gamma'_i$ for all $i \in \mathbb{Z}_{\geq 0}$. (Of course, we have $\dim_K \Gamma'_i = 0$ for all $i \in \mathbb{Z}_{<0}$.) Therefore, we have

$$H_{\mathrm{gr}(M),\Gamma'}(\lambda) = \sum_{k=1}^{q} H_{S,k}(\lambda, \ldots, \lambda).$$

More generally, if degrees (not necessarily equal to one) are assigned to the indeterminates $z_1, \ldots, z_n, \partial_1, \ldots, \partial_l$ of D, the corresponding Hilbert series of $\mathrm{gr}(M)$ with respect to Γ is obtained from the generalized Hilbert series of S by substituting $\lambda^{\deg(z_i)}$ for x_i, $i = 1, \ldots, n$, and $\lambda^{\deg(\partial_j)}$ for x_{n+j}, $j = 1, \ldots, l$.

An (exhaustive) increasing F-filtration of the factor module $D^{1 \times q}/M$ is given by $\Phi'' := (\Phi''_i)_{i \in \mathbb{Z}}$, where Φ''_i is the image of Φ_i under the canonical projection $D^{1 \times q} \to D^{1 \times q}/M$. Note that $\mathrm{gr}(D^{1 \times q}/M)$ is a finitely generated $\mathrm{gr}(D)$-module. Let Γ'' be the grading of $\mathrm{gr}(D^{1 \times q}/M)$. If C is a cone decomposition of the complement $\overline{S}$ of S in $\mathrm{Mon}(D^{1 \times q})$ (cf. Rem. 2.1.54 b)), then

$$\biguplus_{(t,v) \in C} \mathrm{Mon}(v)\,((t + M) + \Gamma''_{\deg(t)-1})$$

is a K-basis of $\mathrm{gr}(D^{1 \times q}/M)$ and we obtain the Hilbert series of $\mathrm{gr}(D^{1 \times q}/M)$ with respect to Γ'' as follows:

$$H_{\mathrm{gr}(D^{1 \times q}/M),\Gamma''}(\lambda) = \sum_{k=1}^{q} H_{\overline{S},k}(\lambda, \ldots, \lambda).$$

Remark 2.1.62. Let K be a field, $D = K[x_1, \ldots, x_n]$ a commutative polynomial algebra over K which is positively graded, and $q \in \mathbb{N}$. We denote by G the grading of D and let $\Gamma = (\Gamma_i)_{i \in \mathbb{Z}}$ be a G-grading of the free left D-module $D^{1 \times q}$ such that each Γ_i is a finite-dimensional K-vector space. Let M be a submodule of $D^{1 \times q}$ such that

$$\Gamma'_i := \Gamma_i \cap M, \qquad i \in \mathbb{Z},$$

defines a G-grading Γ' of M. Moreover, let

$$J = \{(p_1, \mu_1), \ldots, (p_t, \mu_t)\}$$

be a Janet basis for M with respect to any admissible term ordering on $\mathrm{Mon}(D^{1 \times q})$. Then the Hilbert series of M with respect to Γ' is given by

$$
\begin{aligned}
H_{M,\Gamma'}(\lambda) &= \sum_{i \in \mathbb{Z}} (\dim_K \Gamma_i') \lambda^i \\
&= \sum_{k=1}^{t} \frac{\lambda^{\deg(p_k)}}{(1 - \lambda)^{|\mu_k|}} \qquad\qquad (2.18) \\
&= \sum_{k=1}^{t} \lambda^{\deg(p_k)} \sum_{j \geq 0} \binom{|\mu_k| + j - 1}{j} \lambda^j.
\end{aligned}
$$

For $i \geq \max\{\deg(p_k) \mid k = 1, \dots, t\}$, we have

$$
\dim_K \Gamma_i' = \sum_{k=1}^{t} \binom{|\mu_k| + i - \deg(p_k) - 1}{i - \deg(p_k)} = \sum_{k=1}^{t} \binom{|\mu_k| + i - \deg(p_k) - 1}{|\mu_k| - 1},
$$

which is a polynomial in i of degree less than $n + l$. In other words, the Hilbert function of M with respect to Γ' is a polynomial function when restricted to integers greater than or equal to $\max\{\deg(p_k) \mid k = 1, \dots, t\}$. This polynomial is called the *Hilbert polynomial of M with respect to Γ'*. In a similar way the notion of Hilbert polynomial is defined for a residue class module of $D^{1 \times q}$ with respect to some G-grading and for submodules and residue class modules of $D^{1 \times q}$ with respect to (exhaustive) increasing filtrations.

Note that if non-standard degrees are assigned to the indeterminates of D, these have to be taken into account in the right hand side of (2.18) in terms of the corresponding powers of λ. Thus, in general, the Hilbert function is asymptotically polynomial on residue classes (also called quasipolynomial, cf. [Sta96, Sect. 0.1]).

Remark 2.1.63. In the situation of the previous remark let I be a homogeneous ideal of the commutative polynomial algebra D with standard grading, i.e., $q = 1$. Then the degree d of the Hilbert polynomial of D/I equals the dimension of the corresponding projective variety in projective $(n - 1)$-space defined over an algebraic closure $\overline{K}$ of K (cf., e.g., [Eis95]). The product of the leading coefficient of the Hilbert polynomial and $d!$ is called the degree of the corresponding projective variety and coincides with the number of points in which the variety intersects a generic projective subspace of dimension $n - 1 - d$.

For the case of an algebra D of differential operators with rational function coefficients (cf. Ex. 2.1.18 b), p. 19) and general q an upper bound for this product in terms of the numbers of independent and dependent variables, the number of equations, the maximum differential order, and the degree of the Hilbert polynomial of the given system was derived in [Gri05].

In the following remark we outline one application of the generalized Hilbert series to commutative algebra and algebraic geometry. This application provides a constructive and deterministic approach to the Noether normalization lemma.

Remark 2.1.64. For a given finitely generated commutative algebra over a field, the Noether normalization lemma (cf., e.g., [Eis95], [Vas98]) ensures the existence of a maximal subset of algebraically independent elements such that the given algebra is an integral extension of the polynomial ring generated by this system of parameters. An affine variety whose coordinate ring is isomorphic to the given algebra is therefore shown to be a branched covering of some affine space.

The normalization lemma can be proved in a constructive manner, but most of the computational approaches today perform a random change of coordinates producing very large polynomials, which are difficult to handle afterwards.

Given an ideal I of a commutative polynomial algebra $D = K[x_1, \ldots, x_n]$ over a field K, the generalized Hilbert series of D/I can be used effectively to construct sparse coordinate changes which achieve Noether normal position for the given ideal [Rob09].

The maximum number of multiplicative variables of cones in a Janet decomposition of D/I equals the Krull dimension d of D/I (cf. Rem. 2.1.60). Let v be the union of all sets of multiplicative variables of the cones of such a decomposition. The crucial observation is that a variable is not an element of v if and only if a power of that variable is a leading monomial of an element of the Janet basis J for I. If $|v| = d$, then the set Y of residue classes of the elements of v in D/I is a maximal subset of algebraically independent elements of D/I such that D/I is an integral extension of $K[Y]$. Otherwise, we have $|v| > d$ and coordinates should be changed in such a way that the Janet basis for the transformed ideal has more elements whose leading monomial is a power of a variable. It turns out that a good strategy is to investigate an element $p \in J$ such that $\mathrm{lm}(p) \in \mathrm{Mon}(v)$ and $\mathrm{lm}(p)$ involves the least number of variables and is maximal with respect to the chosen term ordering $>$ among these candidates. Then the coordinate transformation is chosen in such a way that each variable dividing $\mathrm{lm}(p)$ is mapped to a linear combination of variables in which the $>$-greatest variable in v has non-zero coefficient[7]. We demonstrate this procedure in the following example and refer to [Rob09] for more details.

Example 2.1.65. Let $D = \mathbb{Q}[w, x, y, z]$ be the commutative polynomial algebra and choose the degree-reverse lexicographical ordering $>$ on D which extends the ordering $w > x > y > z$. Let I be the ideal of D which is generated by

$$L := \{wxy^2 - y^2z,\, xyz - wz^2,\, y^2z - wx^2yz\}.$$

It is not radical and has five minimal associated primes of dimensions 1, 2, 2, 2, 2, respectively, and one embedded associated prime of dimension 1. All the following computations were done in the computer algebra system Maple in a couple of seconds using the package `Involutive` [BCG$^+$03a] (cf. also Subsect. 2.1.6).

Let J_1 be a Janet basis for I with respect to the term ordering $>$ (using the total ordering $w > x > y > z$ for determining the multiplicative variables). The Janet decomposition of the complement of $\mathrm{lm}(I)$ in $\mathrm{Mon}(\{w, x, y, z\})$ (determined by Alg. 2.1.8,

---

[7] If the ground field K is finite and not large enough, it may be necessary to use polynomials of higher degree to define the coordinate transformation.

p. 14) yields the generalized Hilbert series[8] of D/I:

$$\frac{1}{1-z} + \frac{y}{1-z} + \frac{x}{(1-x)(1-z)} + \frac{w}{1-z} + \frac{y^2}{1-y} + \frac{xy}{(1-x)(1-y)} + wy + \frac{wx}{(1-x)(1-z)} +$$

$$w^2 + \frac{y^2z}{1-y} + wyz + w^2z + wy^2 + \frac{wxy}{1-x} + w^2y + \frac{w^2x}{1-x} + \frac{w^3}{1-w} + \frac{y^2z^2}{1-y} + wyz^2 + w^2z^2 +$$

$$wy^2z + w^2yz + \frac{w^2xz}{1-x} + \frac{w^3z}{1-w} + \frac{wy^3}{1-y} + w^2y^2 + \frac{w^2xy}{1-x} + \frac{w^3y}{1-w} + \frac{w^3x}{(1-w)(1-x)} +$$

$$\frac{y^2z^3}{1-y} + w^2z^3 + wy^2z^2 + \frac{w^3z^2}{1-w} + w^2y^2z + \frac{w^3yz}{1-w} + \frac{w^3xz}{(1-w)(1-x)} + \frac{w^2y^3}{1-y} + \frac{w^3y^2}{1-w} +$$

$$\frac{w^3xy}{(1-w)(1-x)} + \frac{y^2z^4}{1-y} + \frac{w^3y^2z}{1-w} + \frac{w^3y^3}{(1-w)(1-y)}.$$

Hence, the sets of multiplicative variables μ_i of the Janet decomposition are among the following ones:

$$\emptyset, \quad \{w\}, \quad \{x\}, \quad \{y\}, \quad \{z\}, \quad \{w,x\}, \quad \{w,y\}, \quad \{x,y\}, \quad \{x,z\}.$$

The Krull dimension d of D/I equals 2. We have $v_1 := \bigcup \mu_i = \{w,x,y,z\}$, and so $|v_1| = 4 > d$. In order to keep the coordinate transformation sparse, it is advisable to choose

$$p_1 = w^2z^4 - wy^2z^2 \in J_1,$$

whose leading monomial $\mathrm{lm}(p_1) = w^2z^4$ involves only two variables. We choose the automorphism $\psi_1 : D \to D$ (restricting to the identity on K) which maps z to $z - w$ and fixes all other variables.

Let J_2 be a Janet basis for $\psi_1(I)$ (with the same specifications as above). The generalized Hilbert series of $D/\psi_1(I)$ is given by

$$\frac{1}{(1-y)(1-z)} + \frac{x}{1-z} + \frac{w}{1-z} + \frac{xy}{1-z} + \frac{wy}{1-z} + \frac{x^2}{1-z} + \frac{wx}{(1-x)(1-z)} + \frac{w^2}{1-z} + xy^2 +$$

$$\frac{wy^2}{1-z} + \frac{x^2y}{1-z} + \frac{wxy}{(1-x)(1-z)} + \frac{w^2y}{1-z} + \frac{x^3}{(1-x)(1-z)} + \frac{w^2x}{1-z} + xy^2z + xy^3 + \frac{wy^3}{1-y} +$$

$$x^2y^2 + w^2y^2 + \frac{x^3y}{(1-x)(1-z)} + \frac{w^2xy}{1-z} + \frac{w^2x^2}{(1-x)(1-z)} + xy^3z + \frac{wy^3z}{1-y} + x^2y^2z + w^2y^2z +$$

$$\frac{xy^4}{1-y} + \frac{x^2y^3}{1-y} + \frac{x^3y^2}{(1-x)(1-y)} + \frac{wy^3z^2}{1-y} + w^2y^2z^2 + \frac{wy^3z^3}{1-y}.$$

In particular, the Janet decomposition of $\mathrm{Mon}(\{w,x,y,z\}) - \mathrm{lm}(\psi_1(I))$ has sets of multiplicative variables among the following ones:

$$\emptyset, \quad \{y\}, \quad \{z\}, \quad \{x,y\}, \quad \{x,z\}, \quad \{y,z\}.$$

We have $v_2 := \{x,y,z\}$, and therefore $|v_2| = 3 > d$. Now we choose the polynomial

---

[8] Using the package `Involutive` (cf. Subsect. 2.1.6), the generalized Hilbert series can be obtained with the command `FactorModuleBasis`, after applying `InvolutiveBasis` to L.

$$p_2 = xy^2z^2 - w^2y^2 + wy^3 + 3wy^2z - y^3z - 2y^2z^2 \in J_2$$

with $\mathrm{lm}(p_2) = xy^2z^2$, and the automorphism ψ_2 of D mapping y to $y - x$, z to $z - x$, and fixing w and x.

Finally, we compute a Janet basis J_3 for $(\psi_2 \circ \psi_1)(I)$. The generalized Hilbert series of $D/(\psi_2 \circ \psi_1)(I)$ is given by

$$\frac{1}{(1-y)(1-z)} + \frac{x}{(1-y)(1-z)} + \frac{w}{(1-y)(1-z)} + \frac{x^2}{(1-y)(1-z)} + \frac{wx}{(1-y)(1-z)} +$$

$$\frac{w^2}{(1-y)(1-z)} + \frac{x^3}{(1-y)(1-z)} + \frac{wx^2}{1-z} + \frac{w^2x}{1-z} + \frac{wx^2y}{1-y} + \frac{w^2xy}{1-z} + x^4 + w^2x^2 +$$

$$\frac{wx^2yz}{1-y} + x^4z + w^2x^2z + w^2xy^2 + \frac{wx^2yz^2}{1-y} + w^2x^2z^2 + \frac{wx^2yz^3}{1-y}.$$

The Janet decomposition of the complement of $\mathrm{lm}((\psi_2 \circ \psi_1)(I))$ in $\mathrm{Mon}(\{w,x,y,z\})$ consists of cones having sets of multiplicative variables among the following ones:

$$\emptyset, \quad \{y\}, \quad \{z\}, \quad \{y,z\}.$$

Thus, $v_3 := \{y,z\}$ and $|v_3| = d$, and we are done[9]. The coordinate change $\psi_2 \circ \psi_1$ is defined by

$$w \mapsto w, \quad x \mapsto x, \quad y \mapsto y - x, \quad z \mapsto z - x - w.$$

The maximum number of summands of a polynomial in J_3 is 102. The coefficient in J_3 of largest absolute value equals 40.

A typical coordinate transformation to Noether normal position returned by the (randomized) command `noetherNormal` of the computer algebra system Singular (version 3-1-6) [DGPS12] is defined by

$$w \mapsto w, \quad x \mapsto 10w + x, \quad y \mapsto 6w + 10x + y, \quad z \mapsto 8w + 4x + 3y + z,$$

which in this case results in a Gröbner basis of the transformed ideal with coefficient of largest absolute value of more than 30 decimal digits and maximum number of summands 123.

For more details about this approach to Noether normalization and a more systematic comparison of some existing implementations, we refer to [Rob09].

In the rest of this subsection we discuss the relevance of the generalized Hilbert series for systems of linear partial differential equations. Computation of a Janet basis for such a system produces an equivalent system which is ensured to be *formally integrable*, i.e., it admits a straightforward method of determining all formal power series solutions from the equations of the system (which is in some sense similar to back substitution applied to the result of Gaussian elimination). In general, two distinct equations may yield a non-trivial consequence of lower differentiation order

---

[9] Note that neither of the Janet bases J_1, J_2, J_3 is a Pommaret basis, i.e., Noether normalization is achieved using a sparse transformation which does not define δ-regular coordinates.

when the highest terms in a suitable linear combination of certain of their derivatives cancel. If the system is not formally integrable, the computation of a power series solution from the given equations may miss the conditions implied by such consequences. Since Janet's algorithm determines the multiple-closed set of monomials which occur as leading monomials of consequences of the system, a Janet basis reveals all conditions on Taylor coefficients of a solution.

We recall that the vector space which is dual to a polynomial algebra over a field is given by the algebra of formal power series in the same number of indeterminates. This relationship will be generalized to Ore algebras in the following remark.

Remark 2.1.66. Let $D := A[\partial_1; \sigma_1, \delta_1] \ldots [\partial_l; \sigma_l, \delta_l]$ be an Ore algebra, where the domain A is an algebra over the field K, and define

$$\mathscr{F} := \hom_K(D, K).$$

Since multiplication in K is commutative, the set $\mathscr{F}$ of all homomorphisms from the left K-vector space D to K is a left K-vector space. Moreover, $\mathscr{F}$ is a left D-module in virtue of

$$D \times \mathscr{F} \longrightarrow \mathscr{F} : (d, f) \longmapsto (a \mapsto f(a \cdot d)),$$

and this left action of D restricts to the left action of K because every element of K commutes with every element of D. We have a pairing of D and $\mathscr{F}$, i.e., a K-bilinear form

$$(\ , \ ) : D \times \mathscr{F} \longrightarrow K : (d, f) \longmapsto f(d) \tag{2.19}$$

which is non-degenerate in both arguments. With respect to this pairing, D and $\mathscr{F}$ can be considered as dual to each other. Moreover, the linear map $D \to D$ defined by right multiplication by a fixed element $d \in D$ and the linear map $\mathscr{F} \to \mathscr{F}$ given by left multiplication by the same element d are adjoint to each other:

$$(a \cdot d, f) = f(a \cdot d) = (d \cdot f)(a) = (a, d \cdot f), \qquad a \in D, \quad f \in \mathscr{F}. \tag{2.20}$$

Since every homomorphism $f \in \mathscr{F} = \hom_K(D, K)$ is uniquely determined by its values for the elements of the K-basis $\mathrm{Mon}(D)$ of D, we can write f in a unique way as a (not necessarily finite) formal sum

$$\sum_{m \in \mathrm{Mon}(D)} (m, f) \, m. \tag{2.21}$$

Due to (2.20), for every $d \in D$ the representation of $d \cdot f$ can be obtained from

$$\sum_{m \in \mathrm{Mon}(D)} (m, d \cdot f) \, m = \sum_{m \in \mathrm{Mon}(D)} (m \cdot d, f) \, m. \tag{2.22}$$

It is reasonable to write the monomials in the sum (2.21) using new indeterminates, which will be done in the following remark dealing with the case of commutative polynomial algebras.

Remark 2.1.67. Let D be the commutative polynomial algebra $K[\partial_1, \ldots, \partial_n]$ over a field K of characteristic zero and $>$ a term ordering on D. Then $(\partial^\beta \mid \beta \in (\mathbb{Z}_{\geq 0})^n)$ is a K-basis of D. We define $\mathscr{F} := \hom_K(D, K)$ with (left) D-module structure and the pairing in (2.19) as in the previous remark. The discussion leading to (2.21) shows that $\mathscr{F}$ can be considered as the K-algebra $K[[z_1, \ldots, z_n]]$ of formal power series in the same number n of indeterminates. Moreover, it follows from (2.22) that the (left) action on $\mathscr{F}$ of any monomial in D effects a shift of the coefficients of the power series according to the exponent vector of the monomial, which is the same action as the one defined by partial differentiation. Therefore, we establish the identification of $\mathscr{F}$ with $K[[z_1, \ldots, z_n]]$ in such a way that

$$(z^\alpha / \alpha! \mid \alpha \in (\mathbb{Z}_{\geq 0})^n) \quad \text{and} \quad (\partial^\beta \mid \beta \in (\mathbb{Z}_{\geq 0})^n)$$

are dual to each other with respect to the pairing (2.19), i.e.,

$$\left(\partial^\beta, \sum_{\alpha \in (\mathbb{Z}_{\geq 0})^n} c_\alpha \frac{z^\alpha}{\alpha!} \right) = c_\beta, \qquad \beta \in (\mathbb{Z}_{\geq 0})^n, \qquad \alpha! := \alpha_1! \cdot \ldots \cdot \alpha_n!.$$

Suppose that a system of (homogeneous) linear PDEs with constant coefficients for one unknown function of n arguments is given. We compute a Janet basis J for the ideal of D which is generated by the left hand sides p of these equations with respect to the term ordering $>$. The differential equations are considered as linear equations for (∂^β, f), $\beta \in (\mathbb{Z}_{\geq 0})^n$, where $f \in \mathscr{F}$ is a formal power series solution, and using the term ordering $>$, we may solve each of these equations for $(\mathrm{lm}(p), f)$. Then Janet's algorithm partitions $\mathrm{Mon}(D)$ into a set of monomials m for which $(m, f) \in K$ can be chosen arbitrarily and a set S of monomials for which $(\mathrm{lm}(p), f) \in K$ is uniquely determined by these choices. The latter set is the multiple-closed subset

$$S := [\mathrm{lm}(p) \mid (p, \mu) \in J]$$

of $\mathrm{Mon}(D)$. In particular, the K-dimension of the space of formal power series solutions, if finite, can be computed as the number of monomials in the complement C of S in $\mathrm{Mon}(D)$. In fact, the generalized Hilbert series $H_C(\partial_1, \ldots, \partial_n)$ of C enumerates a basis for the Taylor coefficients (∂^β, f) of f whose values can be assigned freely.

M. Janet called the monomials ∂^β in $\mathrm{Mon}(D) - S$ *parametric derivatives* because the corresponding Taylor coefficients (∂^β, f) of a formal power series solution f can be chosen arbitrarily. The monomials in S are called *principal derivatives* [Jan29, e.g., no. 22, no. 38]. The Taylor coefficients (∂^β, f) which correspond to principal derivatives ∂^β are uniquely determined by K-linear equations in terms of the Taylor coefficients of parametric derivatives. Of course, the extension of this method of determining the formal power series solutions of a system of linear partial differential equations is extended to the case of more than one unknown function in a straightforward way by using submodules of $D^{1 \times q}$ instead of ideals of D.

Note that convergence of series solutions is to be investigated separately.

For a similar treatment of partial difference equations, we refer to [OP01].

Example 2.1.68. [Jan29, no. 23] The left hand side of the heat equation

$$\frac{\partial u}{\partial t} - \frac{\partial^2 u}{\partial x^2} = 0 \tag{2.23}$$

for an unknown real analytic function u of t and x is represented by the polynomial

$$p := \partial_t - \partial_x^2 \in D := K[\partial_t, \partial_x],$$

where $K = \mathbb{Q}$ or $\mathbb{R}$. Choosing a degree-reverse lexicographical term ordering on the polynomial algebra D, the leading monomial of p is ∂_x^2. The polynomial p forms a Janet basis for the ideal of D it generates, and the parametric derivatives are given by ∂_t^i, $\partial_t^j \partial_x$, $i, j \in \mathbb{Z}_{\geq 0}$. Hence, any choice of formal power series in t for $u(t,0)$ and $\frac{\partial u}{\partial x}(t,0)$ uniquely determines a formal power series solution u to (2.23). In this case, every choice of convergent power series yields a convergent series solution u. On the other hand, using the lexicographical term ordering extending $t > x$, the parametric derivatives are given by ∂_x^i, $i \in \mathbb{Z}_{\geq 0}$. Now, the choice

$$u(0,x) = \sum_{i \geq 0} x^i$$

determines a divergent series solution u.

The following example demonstrates how a Janet decomposition (and the resulting generalized Hilbert series) of the complement of the set of principal derivatives in $\mathrm{Mon}(D)$ allows to collect the parametric derivatives in such a way as to express the solutions in terms of arbitrary functions and constants.

Example 2.1.69. For illustrative reasons, we consider the system of linear partial differential equations for one unknown analytic function u of x, y, z which corresponds to the set of monomials dealt with in Examples 2.1.7, p. 12, and 2.1.10:

$$\frac{\partial^2 u}{\partial x \partial y} = 0, \quad \frac{\partial^4 u}{\partial x^3 \partial z} = 0. \tag{2.24}$$

The Janet completion already yields the (minimal) Janet basis

$$\frac{\partial^2 u}{\partial x \partial y} = 0, \ \{ *, \partial_y, \partial_z \},$$

$$\frac{\partial^3 u}{\partial x^2 \partial y} = 0, \ \{ *, \partial_y, \partial_z \},$$

$$\frac{\partial^4 u}{\partial x^3 \partial z} = 0, \ \{ \partial_x, *, \partial_z \},$$

$$\frac{\partial^4 u}{\partial x^3 \partial y} = 0, \ \{ \partial_x, \partial_y, \partial_z \}$$

and we obtain the following Janet decomposition of the set of parametric derivatives (cf. also Ex. 2.1.10 and Fig. 2.2, p. 15):

$$1, \{ *, \partial_y, \partial_z \},$$
$$\partial_x, \{ *, *, \partial_z \},$$
$$\partial_x^2, \{ *, *, \partial_z \},$$
$$\partial_x^3, \{ \partial_x, *, * \}.$$

The corresponding generalized Hilbert series is

$$\frac{1}{(1 - \partial_y)(1 - \partial_z)} + \frac{\partial_x}{1 - \partial_z} + \frac{\partial_x^2}{1 - \partial_z} + \frac{\partial_x^3}{1 - \partial_x}.$$

Accordingly, a formal power series solution u of (2.24) is uniquely determined as

$$u(x,y,z) = f_0(y,z) + x f_1(z) + x^2 f_2(z) + x^3 f_3(x)$$

by any choice of formal power series f_0, f_1, f_2, f_3 of the indicated variables.

In general, the expression of the solutions in terms of arbitrary functions and constants depends on the choices of the coordinate system, the term ordering, and the total ordering which is used for determining the Janet decomposition. However, the maximum number of arguments of functions which occur in such an expression is invariant because it is the Krull dimension of the corresponding (graded) module (over the associated graded ring defined in Rem. 2.1.61), cf. also Rem. 2.1.60. The number of cones in a Janet decomposition having a fixed number of multiplicative variables and generator of a certain degree is also referred to as *Cartan character*.

Remark 2.1.70. The statements of Remark 2.1.67 also apply to systems of linear PDEs whose coefficients are rational functions in $z_1, \ldots, z_n$, i.e., $D = K[\partial_1, \ldots, \partial_n]$ is replaced with $B_n(K) = K(z_1, \ldots, z_n)\langle \partial_1, \ldots, \partial_n \rangle$ (introduced also in Ex. 2.1.18 b) using Ore algebra notation), where K is the subfield of constants of $K(z_1, \ldots, z_n)$.

Let M be the submodule of $B_n(K)^{1 \times p}$ which is generated by the left hand sides of the equations (for p unknown functions) and let J be a Janet basis for M. Now, a formal power series solution is determined by any choice of Taylor coefficients for the parametric derivatives, if the left hand sides of the given PDE system are also defined over $A\langle \partial_1, \ldots, \partial_n \rangle^{1 \times p}$, where A is a K-subalgebra of $K(z_1, \ldots, z_n)$ whose elements do not have a pole in $0 \in K^n$, if J is computed within $A\langle \partial_1, \ldots, \partial_n \rangle^{1 \times p}$, and if 0 is not a zero of the leading coefficient of any element of J. In other words, if 0 is not a zero of any denominator arising in the course of Janet's algorithm applied to the PDE system and is not a zero of any leading coefficient, then all power series solutions are obtained in this way. Accordingly, having computed a Janet basis for M, the center $c \in K^n$ for the Taylor series expansion of an analytic solution has to be chosen in such a way that the previous conditions are met with 0 replaced with c.

Similar remarks hold for the case of coefficients in a field of meromorphic functions on a connected open subset of $\mathbb{C}^n$.

2.1.6 Implementations

Work by the author of this monograph on implementations of techniques related to Janet's algorithm is summarized in this subsection. We also give references to software serving similar purposes. However, because of the large number of implementations of Buchberger's algorithm, a complete review is not aimed for.

The formulation of Algorithm 2.1.42, p. 29, computing Janet bases (and of the algorithms on which it depends) ignores the matter of realizing these techniques as efficient computer programs. For instance, the alternating use of *Auto-reduce* and *Decompose* in Algorithm 2.1.42 clearly removes left multiples of generators by non-multiplicative variables which may be added again if required by the Janet decomposition. Moreover, the computation of the Janet normal form of left multiples of generators by non-multiplicative variables should not be performed a second time when it is clear that the reduction steps will not differ from the previous computation.

The *involutive basis algorithm* (cf., e.g., [Ger05]), developed in work of V. P. Gerdt, Y. A. Blinkov, and A. Y. Zharkov, provides an efficient method to compute Janet bases. It builds on the more general concept of involutive division, which allows for other ways of defining multiplicative variables for generators than the pattern named after Janet. (We refer to [GB11] and ongoing work for a recent development of an even more efficient involutive division.) Using a very small part of the history of an involutive basis computation, the reduced Gröbner basis for the same ideal or module (and with respect to the same term ordering) can be extracted as a subset of the involutive basis without further computation. Moreover, analogues of Buchberger's criteria [Buc79] in the context of involutive division avoid unnecessary passivity checks. Heuristic strategies determining in which order the left multiples of generators by non-multiplicative variables should be considered for reduction are incorporated into the involutive approach.

A software package ALLTYPES realizing Riquier's and Janet's theory in the computer algebra system REDUCE [Hea99] has been developed by F. Schwarz (cf. [Sch08b, Sch84]). Another implementation in the programming language REFAL was described in [Top89]. The REDUCE package INVSYS, which computes Janet bases for ideals of commutative polynomial algebras, was developed by A. Y. Zharkov and Y. A. Blinkov [ZB96]. A program computing involutive bases for monomial ideals in Mathematica [Wol99] was reported on in [GBC98]. Moreover, involutive basis techniques have been implemented in MuPAD [CGO04] by M. Hausdorf and W. M. Seiler [HS02].

Symmetry analysis of systems of partial differential equations (cf., e.g., [Olv93], [Pom78], [Vin84], [BCA10], [Sch08a]) is an important area of application of Riquier's and Janet's theory. G. Reid and collaborators have been developing the *rif* algorithm and have been applying it in the symmetry analysis context, cf. [RWB96] and the references therein, and also [MRC98], where it had been combined with the differential Gröbner basis package of E. Mansfield [Man91]. By repeated prolongation and elimination steps as described in geometric approaches to differential systems (cf., e.g., [Pom78] and the references therein), the *rif* algorithm transforms a system of nonlinear PDEs into *reduced involutive form*, which is formally integrable. An implementation by A. Wittkopf is available as a Maple package. For a review of further symbolic software for symmetry analysis of differential equations, cf. also [Her97].

Implementations of Buchberger's algorithm for computing Gröbner bases are available in many computer algebra systems and more specialized software, e.g., in AXIOM [JS92], Maple [MAP], Mathematica [Wol99], Magma [BCP97], REDUCE [Hea99], Singular [DGPS12], Macaulay2 [GS], CoCoA [CoC] (cf. also [CLO07, Appendix C] for a further discussion of such implementations). A variant of Buchberger's algorithm for systems of linear differential or difference equations and algorithms for computing Hilbert polynomials along with an implementation in the programming language REFAL were described in [Pan89].

An implementation of the involutive basis technique for commutative polynomial algebras over fields and for $K\langle\partial_1,\ldots,\partial_n\rangle$ as packages `Involutive` and `Janet`, respectively, for the computer algebra system Maple has been started by C. F. Cid at Lehrstuhl B für Mathematik, RWTH Aachen, in 2000. Here, $K\langle\partial_1,\ldots,\partial_n\rangle$ is the skew polynomial ring of differential operators with coefficients in a differential field K (of characteristic zero), whose arithmetic is implemented in Maple.

Since the year 2001 the author of this monograph has been adapting the packages `Involutive` and `Janet` to more recent versions of the involutive basis algorithm (with the help of V. P. Gerdt and Y. A. Blinkov) and has been extending these packages with new features.

Starting in 2003, the author of this monograph has been developing a Maple package `JanetOre`, which implements the involutive basis technique for certain iterated Ore extensions of a commutative polynomial algebra (as in Subsect. 2.1.3).

In collaboration with V. P. Gerdt a Maple package `LDA` (for "linear difference algebra") has been developed since 2005, which computes involutive bases for left ideals of (and left modules over) rings of difference operators with coefficients in a difference field (of characteristic zero), whose arithmetics are supported by Maple. For applications of the package `LDA` to formal computational consistency checks of finite difference approximation of linear PDE systems, cf. [GR10].

We refer to [BCG+03a, BCG+03b, Rob07, GR06, GR12], the Maple help pages accompanying these packages, and the related web pages for more information.

The package `Involutive` computes Janet bases and Janet-like Gröbner bases (cf. Rem. 2.1.49, p. 37) for submodules of finitely generated free modules over commutative polynomial algebras with coefficients in $\mathbb{Z}$ or finitely generated extension fields of $\mathbb{Q}$ or finite fields that are supported by Maple. The implementation of the involutive basis algorithm has the additional feature that its computations may be performed in a parallel way on auxiliary data, which yields a means to record the history of a Janet basis computation. Syzygies and free resolutions, cf. Subsect. 3.1.5, can be computed by `Involutive`. Further procedures implementing module-theoretic constructions build on this possibility. (We also refer to the package `homalg` [BR08], which implements methods of homological algebra in an abstract way, and to which `Involutive` can be connected. Delegating ring arithmetics to separate software, the package `homalg` provides an additional layer of abstraction. Meanwhile, `homalg` has been redesigned by M. Barakat as a package in GAP4 [GAP] and has been widely extended, e.g., being capable now of computing certain spectral sequences.) The package `Involutive` has been extended with functionality improving computation with rational function coefficients by M. Schröer and with procedures dealing with localizations at maximal ideals by M. Lange-Hegermann.

The Maple package `Janet` computes Janet bases and Janet-like Gröbner bases for submodules of finitely generated free left modules over the skew polynomial ring $K\langle \partial_1, \ldots, \partial_n \rangle$ of partial differential operators. Apart from implementing the counterpart of the module-theoretic methods of `Involutive` for the ring $K\langle \partial_1, \ldots, \partial_n \rangle$, it provides, e.g., procedures which compute (truncated) formal power series solutions and polynomial solutions up to a given degree of systems of linear PDEs. The package `Janet` uses some data structures and procedures of the Maple package `jets` developed by M. Barakat [Bar01] and can be combined with `jets`, in order to compute symmetries of differential equations (cf., e.g., [Olv93]).

The functionality of the packages `JanetOre` and `LDA`, although handling different types of algebras, is analogous to that of `Involutive` and `Janet`, respectively.

Each of the above mentioned Maple packages provides combinatorial tools like the generalized Hilbert series (cf. Subsect. 2.1.5), Hilbert polynomials, Cartan characters, etc.

A very useful feature of these packages is the possibility to collect all expressions (typically arising as coefficients of polynomials) by which a Janet basis computation divided. Hence, the applicability of the performed computation for special values of parameters can be checked and singular configurations can be determined afterwards (cf. also Rem. 2.1.70).

The open source software package `ginv` implements the involutive basis technique in C++, using Python as an interpreter in addition [BG08]. Its development was initiated by V. P. Gerdt and Y. A. Blinkov. Contributions have been made at Lehrstuhl B für Mathematik, RWTH Aachen, during the last seven years, in particular by S. Jambor and the author of this monograph.

Interfaces between the Maple package `Involutive` and `ginv` are available including the possibility to delegate involutive basis computations during the current session of `Involutive` to the considerably faster C++ routines.

The author of this monograph implemented some parts of the involutive basis technique as a package `InvolutiveBases` [Rob] in Macaulay2 [GS].

Another Maple implementation of the involutive basis technique for linear PDEs is described in [ZL04].

2.2 Thomas Decomposition of Differential Systems

A *system of polynomial partial differential equations and inequations*

$$p_1 = 0, \quad \ldots, \quad p_s = 0, \quad q_1 \neq 0, \quad \ldots, \quad q_t \neq 0 \qquad (s, t \in \mathbb{Z}_{\geq 0}) \qquad (2.25)$$

for m unknown smooth functions of independent variables $z_1, \ldots, z_n$ is given by differential polynomials $p_1, \ldots, p_s, q_1, \ldots, q_t$ in $u_1, \ldots, u_m$, i.e., elements of the differential polynomial ring $K\{u_1, \ldots, u_m\}$ with commuting derivations $\partial_1, \ldots, \partial_n$, where K is a differential field of characteristic zero. (For definitions of these notions of differential algebra, cf. Sect. A.3.) Similarly to Sect. 2.1 we will concentrate on analytic solutions.

Every solution of (2.25) satisfies all consequences of (2.25); the consequences we consider here are given by linear combinations of arbitrary partial derivatives of system equations with coefficients in $K\{u_1, \ldots, u_m\}$, polynomial factors of (left hand sides of) such equations and of inequations, and quotients of (the respective left hand sides of) equations by inequations. Leaving aside for a moment the inequations, we then deal with the radical differential ideal of $K\{u_1, \ldots, u_m\}$ which is generated by $p_1, \ldots, p_s$. Taking the inequations into account, the present section is concerned with an effective procedure which constructs a finite set of differential systems as in (2.25), whose sets of solutions form a partition of the solution set of (2.25), and such that all consequences of each resulting system can easily be described.

Let us assume for simplicity that (2.25) already has the same quality as each of these resulting systems. Then all consequences of (2.25) in terms of equations are

$$\{ p = 0 \mid p \in \sqrt{E : q^\infty} \}, \qquad (2.26)$$

where E is the differential ideal of $R := K\{u_1, \ldots, u_m\}$ which is generated by the polynomials $p_1, \ldots, p_s$, the differential polynomial q is the product of $q_1, \ldots, q_t$, and

$$E : q^\infty := \{ p \in R \mid q^r \cdot p \in E \text{ for some } r \in \mathbb{Z}_{\geq 0} \}$$

is the *saturation* of E with respect to q. The solutions of (2.25) form an open subset of the set of solutions of (2.26) with respect to a certain topology. (It is, in fact, a dense subset, cf. Lemma 2.2.62.)

Let $\mathscr{F}$ be a differential algebra over K, whose elements we think of as candidates for solutions of (2.26). Every homomorphism $\varphi \colon K\{u_1, \ldots, u_m\} \to \mathscr{F}$ of differential algebras over K is uniquely determined by its values $f_1, \ldots, f_m$ for $u_1, \ldots, u_m$, and every choice of these values defines such a homomorphism. Now, $(f_1, \ldots, f_m)$ solves (2.26) if and only if the corresponding homomorphism φ of differential algebras factors over $K\{u_1, \ldots, u_m\}/\sqrt{E : q^\infty}$. Thus, the set of homomorphisms

$$K\{u_1, \ldots, u_m\}/\sqrt{E : q^\infty} \longrightarrow \mathscr{F}$$

of differential algebras over K is in one-to-one correspondence with the set of solutions $(f_1, \ldots, f_m) \in \mathscr{F}^m$ of (2.26).

This structural description of the solutions of (2.26) is analogous to the linear case (cf. the introduction to Sect. 2.1). However, only to some extent does it incorporate the conditions on solutions of (2.25) that are imposed by the given inequations. Moreover, even if no inequations are present in the given system, inequations emerge naturally. As it turns out, an equivalent form of (2.25) which allows to keep track of all of its consequences effectively, requires splittings into complementary systems. The approach we pursue here introduces inequations, which results in a partition of the solution set.

In this section we describe a method introduced by the American mathematician Joseph Miller Thomas (1898–1979) to deal in an effective way with systems of polynomial differential equations and inequations [Tho37, Tho62]. It belongs to the class of triangular decomposition methods (cf., e.g., the survey papers [Hub03a, Hub03b] by Evelyne Hubert) and can be used to compute characteristic sets (cf. also Subsect. A.3.2). Each system in the resulting decomposition admits an effective membership test for the corresponding differential ideal. A first implementation of this decomposition method was realized in the computer algebra system Maple by Dongming Wang [Wan98, LW99, Wan01, Wan04].

While the development of differential algebra following Joseph Fels Ritt in the twentieth century, in particular the work by Ellis R. Kolchin, did not seem to adapt the ideas of Thomas, they have been revived in recent years by Vladimir P. Gerdt [Ger08]. In the context of algebraic equations, Wilhelm Plesken introduced a univariate polynomial which is a counting invariant of a quasi-affine or quasi-projective variety (in given coordinates) in the sense that it counts the (closed) points using the indeterminate ∞ for the cardinality of the affine line [Ple09a]. Markus Lange-Hegermann defined a differential counting polynomial and generalized the differential dimension polynomial, which had been introduced by E. R. Kolchin [Kol64] for prime differential ideals and which had been elaborated by Joseph Johnson [Joh69a], to differential systems which result from Thomas' method [LH14]. For further applications of the algebraic Thomas decomposition to algebraic varieties, to algebraic groups, and to linear codes and hyperplane arrangements we refer to [Ple09b], [PB14], [Bäc14]. An application of the differential Thomas decomposition to nonlinear control systems is developed in [LHR13].

In joint work of T. Bächler, V. P. Gerdt, M. Lange-Hegermann, and the author of this monograph the algorithmic details of J. M. Thomas' method of decomposing algebraic and differential systems into simple systems in combination with the notion of passive differential system following M. Janet have been worked out [BGL+10, BGL+12]. Implementations in Maple have been developed by T. Bächler and M. Lange-Hegermann (cf. also Subsect. 2.2.6).

The characteristic set method developed by J. F. Ritt and Wen-tsün Wu provides another decomposition algorithm, which, however, depends on the possibility to factor polynomials (cf., e.g., [Rit50] and also Subsect. A.3.2 for the rudiments of this theory, and [Wu89] for another variant). For algebraic systems, Wu's method competes with Janet and Gröbner basis techniques and has been applied to automated proving of theorems in geometry (cf., e.g., [Wan04]).

For applications of the characteristic set method to systems theory, we refer to work by Sette Diop, cf., e.g., [Dio92].

Abraham Seidenberg developed an elimination method for differential algebra [Sei56] by using the same splitting technique as J. M. Thomas. As a result, a constructive analog of Hilbert's Nullstellensatz for differential algebra was obtained. For the case of ordinary differential equations an algorithm with improved complexity was given by Dmitry Grigoryev in [Gri89].

Combining Seidenberg's theory and Buchberger's algorithm, the Rosenfeld-Gröbner algorithm, described in [BLOP95, BLOP09], computes a representation of a radical differential ideal as finite intersection of certain differential ideals, each of which also allows an effective membership test. The interactions of the relevant differential and algebraic constructions were investigated in [Hub00]. This approach is based on Rosenfeld's Lemma in differential algebra [Ros59], which is also applicable in the context of Thomas' theory. However, the assumption of coherence of an auto-reduced set of differential polynomials is replaced here with a passivity condition in the sense of Janet (cf. Sect. 2.1). The Rosenfeld-Gröbner algorithm is implemented in the Maple package `DifferentialAlgebra` (formerly `diffalg`). A description of its foundation based on Kolchin's book [Kol73] was given in [Sad00]. Another approach to characteristic sets using Gröbner bases was presented in [BKRM01].

Yet another direction of research tries to adapt the notion of Gröbner basis to the case of a differential polynomial ring, cf., e.g., [CF07]. In general a differential ideal may admit only infinite differential Gröbner bases as defined by Giuseppa Carrà Ferro or infinite standard bases as defined by François Ollivier in this context [Oll91]. Elizabeth L. Mansfield developed an algorithm for the computation of a different kind of (finite) differential Gröbner basis (cf. [Man91]), which applies pseudo-reductions, but does not analyze the initials of divisors, and which therefore may result in a basis which cannot be used to decide membership to the given differential ideal.

Subsection 2.2.1 is devoted to the Thomas decomposition of systems of algebraic equations and inequations, its geometric properties, and its construction. Subsection 2.2.2 builds on the algebraic techniques of the previous subsection and develops Thomas' algorithm for systems of differential equations and inequations. The combinatorics of Janet's algorithm (cf. Subsect. 2.1.1) are used here to ensure formal integrability for each simple system in the resulting Thomas decomposition. After defining and discussing the notion of the generic simple system of a Thomas decomposition of a prime (algebraic or differential) ideal in Subsect. 2.2.3, which will be an essential ingredient for the elimination methods in Sect. 3.3, the following subsection comments on the relationship of simple systems and other types of triangular sets and on the complexity of differential elimination. Subsection 2.2.5 introduces the generalized Hilbert series for simple differential systems. In the last subsection implementations of J. M. Thomas' ideas are discussed and references to related packages are given.

2.2.1 Simple Algebraic Systems

Let K be a computable field of characteristic zero and $R := K[x_1,\ldots,x_n]$ a commutative polynomial algebra with standard grading. We assume that the set $\{x_1,\ldots,x_n\}$ is totally ordered, without loss of generality

$$x_1 > x_2 > \ldots > x_n,$$

and we denote by $\overline{K}$ an algebraic closure of K.

This subsection presents the approach of J. M. Thomas [Tho37] transforming a given system of polynomial equations and inequations in $x_1, \ldots, x_n$, defined over K, into a finite collection of so-called simple systems, each of which can in principle be solved recursively by determining roots of univariate polynomials according to the recursive structure of the solution set as finite-sheeted covering. In other words, the set V of solutions in $\overline{K}^n$ of the given system is partitioned into finitely many subsets $V_1, \ldots, V_m$ in such a way that, for each i, the projection of the last $k+1$ coordinates of V_i onto the last k coordinates has fibers of the same finite or co-finite cardinality (where the cardinality may depend on i and where k ranges from $n-1$ down to 1).

The corresponding decomposition of differential systems (cf. Subsect. 2.2.2) is based on the decomposition of algebraic systems discussed here, but the algebraic part is interesting and of high value in itself.

In the present context we adopt a recursive representation of the elements of $R = K[x_1,\ldots,x_n]$ as follows.

Definition 2.2.1. For $p \in R - K$ we denote by $\mathrm{ld}(p)$ the $>$-greatest variable such that p is a non-constant polynomial in that variable. According to standard terminology in differential algebra we call it the *leader* of p (although often *main variable* is also used when dealing with algebraic systems). The coefficient of the highest power of $\mathrm{ld}(p)$ occurring in p is called the *initial* of p and denoted by $\mathrm{init}(p)$. Finally, the *discriminant* of p is defined in terms of the resultant of p and its partial derivative with respect to its leader as

$$\mathrm{disc}(p) := (-1)^{d(d-1)/2} \cdot \mathrm{res}\left(p, \frac{\partial p}{\partial \mathrm{ld}(p)}, \mathrm{ld}(p)\right) / \mathrm{init}(p),$$

where d is the degree of p in $\mathrm{ld}(p)$ and $\mathrm{res}(p_1, p_2, x)$ denotes the resultant of the polynomials p_1 and p_2 with respect to the indeterminate x. Recall that the above resultant is divisible by $\mathrm{init}(p)$ because every entry of the first column of the Sylvester matrix of p and $\partial p/\partial \mathrm{ld}(p)$ is so. The discriminant of p is used to determine those values for the indeterminates smaller than $\mathrm{ld}(p)$ with respect to $>$ for which p as a polynomial in $\mathrm{ld}(p)$ has zeros of multiplicity greater than one.

Every non-constant polynomial $p \in R$ is now considered as univariate polynomial in $\mathrm{ld}(p)$, whose coefficients are univariate polynomials in their leaders (if not constant) and so on, i.e., we regard R as $K[x_n][x_{n-1}]\ldots[x_1]$.

Definition 2.2.2. Let

$$S = \{ p_i = 0, q_j \neq 0 \mid i \in I, j \in J \}, \qquad p_i, q_j \in R,$$

be a system of algebraic equations and inequations, where I and J are index sets. We define the *set of solutions* or *variety of S in $\overline{K}^n$* by

$$\mathrm{Sol}_{\overline{K}}(S) := \{ a \in \overline{K}^n \mid p_i(a) = 0, q_j(a) \neq 0 \text{ for all } i \in I, j \in J \}$$

($a_1, \ldots, a_n$ are substituted for $x_1, \ldots, x_n$, respectively). For $k \in \{0, 1, \ldots, n-1\}$ let

$$\pi_k : \overline{K}^n \longrightarrow \overline{K}^{n-k} : (a_1, a_2, \ldots, a_n) \longmapsto (a_{k+1}, a_{k+2}, \ldots, a_n)$$

be the projection onto the last $n - k$ components (i.e., the first k components are dropped).

Remark 2.2.3. By Hilbert's Basis Theorem (cf., e.g., [Eis95]), the index set I may be assumed to be finite without loss of generality. In general, the set of inequations cannot be replaced with an equivalent finite set of inequations. Since we aim at effective methods for dealing with algebraic systems, both index sets I and J will be assumed to be finite. The subsets of affine space $\overline{K}^n$ which are of the form $\mathrm{Sol}_{\overline{K}}(S)$ for systems S of algebraic equations defined over $\overline{K}$, i.e., $J = \emptyset$, are the closed sets of the *Zariski topology* on $\overline{K}^n$.

The notion of simple system, central for constructing partitions of varieties as proposed by J. M. Thomas, can now be defined using the projections π_k as follows.

Definition 2.2.4. A system S of algebraic equations and inequations

$$p_1 = 0, \quad \ldots, \quad p_s = 0, \quad q_1 \neq 0, \quad \ldots, \quad q_t \neq 0,$$

where $p_1, \ldots, p_s, q_1, \ldots, q_t \in R - K$, $s, t \in \mathbb{Z}_{\geq 0}$, is said to be *simple* if the following three conditions are satisfied.

a) The leaders of $p_1, \ldots, p_s, q_1, \ldots, q_t$ are pairwise distinct.
b) For every $r \in \{p_1, \ldots, p_s, q_1, \ldots, q_t\}$, if $\mathrm{ld}(r) = x_k$, then the equation $\mathrm{init}(r) = 0$ has no solution in $\pi_k(\mathrm{Sol}_{\overline{K}}(S))$.
c) For every $r \in \{p_1, \ldots, p_s, q_1, \ldots, q_t\}$, if $\mathrm{ld}(r) = x_k$, then the equation $\mathrm{disc}(r) = 0$ has no solution in $\pi_k(\mathrm{Sol}_{\overline{K}}(S))$.

(In b) and c), we have $\mathrm{init}(r), \mathrm{disc}(r) \in K[x_{k+1}, \ldots, x_n]$.)

Remark 2.2.5. A set of polynomials satisfying condition a) is called *triangular set* (cf., e.g., [Hub03a]). This condition implies that $s + t \leq n$.

Furthermore, a simple system S admits the following recursive solution procedure. We introduce the notations $S_{<x_k}$ and $S_{\leq x_k}$ for the subsets of S consisting of the equations and inequations with leader smaller than x_k and with leader smaller than or equal to x_k, respectively. For every $k \in \{1, 2, \ldots, n-1\}$, every tuple

$$(a_{k+1}, a_{k+2}, \ldots, a_n) \in \overline{K}^{n-k}$$

which is a solution of $S_{<x_k}$ can be extended to a solution

$$(a_k, a_{k+1}, \ldots, a_n) \in \overline{K}^{n-(k-1)}$$

of $S_{\leq x_k}$, and every solution $a \in \overline{K}^n$ of S with projection $\pi_k(a) = (a_{k+1}, a_{k+2}, \ldots, a_n)$ is obtained through this process. The possible values of a_k are determined exactly by the equation or inequation in S with leader x_k if it exists, and a_k may take an arbitrary value in $\overline{K}$ otherwise. Condition b) of Definition 2.2.4 implies that the degree of the equation or inequation in S in its leader x_k, if it exists, does not depend on the choice of the values $a_{k+1}, a_{k+2}, \ldots, a_n$ of $x_{k+1}, x_{k+2}, \ldots, x_n$. The result of substituting $x_{k+1} = a_{k+1}, \ldots, x_n = a_n$ into the left hand side of the equation or inequation is a square-free polynomial by condition c). Therefore, the fibers of the projection of $\pi_{k-1}(\mathrm{Sol}_{\overline{K}}(S))$ onto $\pi_k(\mathrm{Sol}_{\overline{K}}(S))$ have the same finite or co-finite cardinality, which is given by the degree in x_k of the equation or inequation, respectively.

Geometrically speaking, the solution set of S is identified recursively as a branched covering. If the variety of interest has a non-trivial ramification locus as a branched covering, then the Thomas decomposition represents it as a partition into solution sets of several simple systems.

Before giving a precise definition and describing the algorithmic construction of a Thomas decomposition, we draw an algebraic consequence that will also be relevant for the differential case. First we recall the notion of vanishing ideal.

Definition 2.2.6. For any $X \subseteq \overline{K}^n$ we define the *vanishing ideal of X in R* by

$$\mathscr{I}_R(X) := \{ p \in R \mid p(x) = 0 \text{ for all } x \in X \}.$$

It is a radical ideal of $R = K[x_1, \ldots, x_n]$. By Hilbert's Nullstellensatz (cf., e.g., [Eis95]), the closed sets of the Zariski topology on $\overline{K}^n$ are in one-to-one and inclusion-reversing correspondence with the radical ideals of $\overline{K}[x_1, \ldots, x_n]$. Therefore, $\mathrm{Sol}_{\overline{K}}(\mathscr{I}_R(X))$ is the closure of X in $\overline{K}^n$ with respect to the Zariski topology.

Proposition 2.2.7. *Let a simple algebraic system S over R be given by*

$$p_1 = 0, \quad \ldots, \quad p_s = 0, \quad q_1 \neq 0, \quad \ldots, \quad q_t \neq 0.$$

Let E be the ideal of R which is generated by $p_1, \ldots, p_s$ and define q to be the product of all $\mathrm{init}(p_i)$, $i = 1, \ldots, s$. *Then we have the equality*

$$E : q^\infty := \{ p \in R \mid q^r \cdot p \in E \text{ for some } r \in \mathbb{Z}_{\geq 0} \} = \mathscr{I}_R(\mathrm{Sol}_{\overline{K}}(S)).$$

In particular, $E : q^\infty$ is a radical ideal. A polynomial $p \in R$ is an element of $E : q^\infty$ if and only if the remainder of pseudo-reduction of p modulo $p_1, \ldots, p_s$ is zero.

Remark 2.2.8. Since $\mathrm{Sol}_{\overline{K}}(E : q^\infty) = \mathrm{Sol}_{\overline{K}}(\mathscr{I}_R(\mathrm{Sol}_{\overline{K}}(S)))$ is the closure of $\mathrm{Sol}_{\overline{K}}(S)$ in $\overline{K}^n$ with respect to the Zariski topology, the inequations $q_1 \neq 0, \ldots, q_t \neq 0$ do not figure on the left hand side of the equality asserted in Proposition 2.2.7.

Proof (of Proposition 2.2.7). From the discussion in Remark 2.2.5 it follows that $\mathrm{Sol}_{\overline{K}}(S)$ is not empty. Hence, the vanishing ideal $\mathscr{I} := \mathscr{I}_R(\mathrm{Sol}_{\overline{K}}(S))$ is contained in a maximal ideal of R, and, in particular, we have $\mathscr{I} \cap K = \{0\}$. The inclusion "$\subseteq$" in the assertion of the proposition is clear. Moreover, if $\mathscr{I} = \{0\}$, then the reverse inclusion is also clear. Otherwise, let $p \in \mathscr{I} - \{0\}$ and $x_k := \mathrm{ld}(p)$. Let

$$(a_{k+1}, \ldots, a_n) \in \overline{K}^{n-k}$$

be a solution of $S_{<x_k}$ (possibly the empty tuple). As in Remark 2.2.5, this tuple can be extended to a solution

$$(a_k, a_{k+1}, \ldots, a_n) \in \overline{K}^{n-(k-1)}$$

of $S_{\leq x_k}$. If S contains no equation with leader x_k or contains an inequation with that leader, then the set of possible a_k is infinite, which is a contradiction to the fact that the equation $p = 0$ allows only $\deg_{x_k}(p)$ values for a_k. Hence, S contains an equation $p_i = 0$ with $\mathrm{ld}(p_i) = x_k$ and $\deg_{x_k}(p_i) \leq \deg_{x_k}(p)$. Now, pseudo-division of p modulo p_i (i.e., Euclidean division of $c \cdot p$ modulo p_i for a suitable power c of $\mathrm{init}(p_i)$) yields a polynomial p' which is either zero or has smaller degree in x_k than p and which is an element of $\mathscr{I}$. Iteration of this argument shows that pseudo-reduction of p modulo equations in S yields the zero polynomial. Hence, $p \in E : q^{\infty}$, which proves the inclusion "$\supseteq$". $\square$

Remark 2.2.9. The same argument as in the proof of Proposition 2.2.7 shows that the residue classes in $R/(E : q^{\infty})$ of the variables in $\{x_1, \ldots, x_n\}$ that are not leaders of an equation in a simple system S form a maximal subset of $R/(E : q^{\infty})$ that is algebraically independent over K. In other words, these residue classes form a system of parameters for the coordinate ring $R/(E : q^{\infty})$ of the Zariski closure V of $\mathrm{Sol}_{\overline{K}}(S)$, in the sense that their number equals the dimension of the affine variety V and any choice of values for these "coordinates" defines a point on (one branch of) the variety.

Definition 2.2.10. Let

$$S = \{ p_i = 0, q_j \neq 0 \mid i \in I, j \in J \}, \qquad p_i, q_j \in R,$$

be a system of algebraic equations and inequations, where I and J are index sets and J is finite. A *Thomas decomposition* of S or of $\mathrm{Sol}_{\overline{K}}(S)$ is a finite collection of simple systems $S_1, \ldots, S_k$ such that

$$\mathrm{Sol}_{\overline{K}}(S) = \mathrm{Sol}_{\overline{K}}(S_1) \uplus \ldots \uplus \mathrm{Sol}_{\overline{K}}(S_k)$$

is a partition of $\mathrm{Sol}_{\overline{K}}(S)$.

Remark 2.2.11. We outline *Thomas' algorithm* (for algebraic systems), which computes a Thomas decomposition for any given system of finitely many algebraic equations and inequations (defined over the computable field K) in finitely many steps. A more precise description will be given on pages 67–87.

First of all, systems containing an equation whose left hand side is a non-zero constant or an inequation with zero left hand side are inconsistent and will be discarded. On the other hand, an equation with zero left hand side and an inequation whose left hand side is a non-zero constant are supposed to be removed from each system. In what follows, we therefore assume that the left hand side of every equation and inequation is a non-constant polynomial.

According to the recursive representation of polynomials, Euclidean pseudo-division is applied to (the left hand sides of) pairs of distinct equations with the same leader, i.e., if $p_1 = 0$, $p_2 = 0$ are distinct equations of the system satisfying $\mathrm{ld}(p_1) = \mathrm{ld}(p_2) =: x$ and $\deg_x(p_1) \geq \deg_x(p_2)$, then usual Euclidean division is performed on $c \cdot p_1$ modulo p_2, where the polynomial c is chosen as (a suitable power of) the initial of p_2 such that division without fractions is made possible.

Let the result of the pseudo-division be p_3. When $p_1 = 0$ is replaced with $p_3 = 0$ in the system S under consideration, a sufficient condition for the set of solutions of S to be unaltered is that c does not vanish for any solution of S. In order to guarantee that the solution set is not changed, the algorithm actually replaces S with two systems S' and S'' and continues to work with S' and S'' separately in the same way as it did with S. The systems S' and S'' are obtained from S by replacing $p_1 = 0$ with $p_3 = 0$ and inserting the inequation $c \neq 0$ in case of S' and the equation $c = 0$ in case of S''.

For each pair $p = 0$, $q \neq 0$ in S with $\mathrm{ld}(p) = \mathrm{ld}(q)$, the greatest common divisor[10] r of p and q is computed. To this end, pseudo-divisions are performed, assuming that the initials of the divisors do not vanish[11], which possibly generates new case distinctions. If q is a multiple of p, then S is inconsistent and will be discarded. If r is a non-zero constant, then $q \neq 0$ is removed from S. Otherwise, $p = 0$ is replaced with $p/r = 0$.

If $q_1 \neq 0$, $q_2 \neq 0$ are two inequations in S with $\mathrm{ld}(q_1) = \mathrm{ld}(q_2)$, then these are replaced with $q_3 \neq 0$, where q_3 is the least common multiple of q_1 and q_2. The computation of the least common multiple involves pseudo-divisions and case distinctions according to vanishing of initials as above.

In the same way as Euclid's algorithm terminates with a single polynomial (being the greatest common divisor of the input polynomials), after finitely many steps the systems produced by Thomas' algorithm will be triangular sets (i.e., condition a) in Def. 2.2.4 will be satisfied), and initials of equations and inequations of each system will not vanish for any solution of the respective system (condition b)). Condition c) in Def. 2.2.4 is accomplished as follows. Since the field of definition K is of characteristic zero, the square-free part of a non-constant polynomial r can be determined as quotient of r by the greatest common divisor of r and the partial derivative of r

---

[10] The terms *greatest common divisor* and *least common multiple* should actually be used with care here because the coefficients of the polynomials in question will be considered subject to equations and inequations with smaller leader so that these notions may not be uniquely defined. For a more precise description, we refer to pages 67–87.

[11] Using subresultant polynomial remainder sequences (cf., e.g., [Mis93]) to compute greatest common divisors often reduces the growth of initials and therefore the number of case distinctions. For more details, cf. [BGL+12, Sect. 2].

with respect to its leader. Again, coefficients of r must be handled with care, and computation of this greatest common divisor usually involves case distinctions. By equating some element of the polynomial remainder sequence with zero, the possible cases for the greatest common divisor are dealt with separately, which in general produces new systems to be treated again in the same way as above.

There are a number of possible ways how to combine these steps. One strategy is to deal in each system with the polynomials of least leader first. For each variable x at most one equation or inequation with leader x is registered which is guaranteed to have non-vanishing initial in the above sense, where equations are preferred to inequations. The next equation or inequation in the current algebraic system to be processed is reduced modulo the registered equations. If the resulting left hand side is not a constant and if an equation or inequation with the same leader is registered, then this pair is treated as discussed above. Splittings of systems regarding initials and square-free parts result in new equations and inequations with smaller leader. Since a registered equation is only replaced with an equation of smaller degree (in the same leader) and since inequations are replaced with equations if possible or with the least common multiple of inequations with the same leader, this strategy terminates after finitely many steps.

The result of the algorithm is a Thomas decomposition of the given algebraic system. It depends on the chosen ordering of the variables $x_1, \ldots, x_n$ and on the order in which the steps of Thomas' algorithm are carried out. Moreover, polynomial factorization of left hand sides of equations is often favorable because proper factors lead to a splitting of the system into smaller systems, each of which is obtained by replacing the original equation with one of its factors of smaller degree.

Thomas' algorithm returns an empty result if and only if no solution (defined over $\overline{K}$) exists for the input system. The result being $\{\emptyset\}$ (i.e., a set consisting of one empty system) is equivalent to the solution set being $\overline{K}^n$.

Example 2.2.12. [BGL$^+$12, Ex. 2.5] Let us examine

$$ax^2 + bx + c = 0, \tag{2.27}$$

a quadratic equation in x with parameters a, b, c. In order to discuss the well-known types of solution sets (in an algebraic closure of $\mathbb{Q}$ or in $\mathbb{C}$) such an equation can have, we consider the left hand side p of (2.27) as element of $\mathbb{Q}[x, a, b, c]$, where $x > c > b > a$, and apply Thomas' algorithm to this algebraic system.

The initial of p equals a. The given system is therefore replaced with

$$S_1 := \{p = 0, a \neq 0\}, \quad S_2 := \{p = 0, a - 0\}.$$

Conditions a) and b) in Definition 2.2.4 are already satisfied for S_1. Euclid's algorithm applied to p and $\frac{\partial p}{\partial \mathrm{ld}(p)}$ (as polynomials in $\mathrm{ld}(p) = x$) computes the polynomial remainder sequence

$$p, \quad \frac{\partial p}{\partial \mathrm{ld}(p)}, \quad 4ac - b^2.$$

Multiplication by a for pseudo-division is harmless because a is assumed not to vanish. Note that the last polynomial equals the discriminant of p (up to sign). Therefore, we replace S_1 with

$$S_{1,1} := \{p = 0, 4ac - b^2 \neq 0, a \neq 0\}, \quad S_{1,2} := \{2ax + b = 0, 4ac - b^2 = 0, a \neq 0\},$$

where $2ax + b$ is the square-free part of p in case $4ac - b^2 = 0$. These two systems are simple.

On the other hand, S_2 is not a triangular set. Euclidean division simplifies $p = 0$ to $bx + c = 0$, whose initial equals b. Thus S_2 is split into two systems

$$S_{2,1} := \{bx + c = 0, b \neq 0, a = 0\}, \quad S_{2,2} := \{bx + c = 0, b = 0, a = 0\},$$

which are easily dealt with. The final result is given by the following four simple systems, where leaders of polynomials are underlined, where not obvious:

$a\underline{x}^2 + b\underline{x} + c = 0$	$2a\underline{x} + b = 0$	$b\underline{x} + c = 0$	
$4a\underline{c} - b^2 \neq 0$	$4a\underline{c} - b^2 = 0$		$c = 0$
		$b \neq 0$	$b = 0$
$a \neq 0$	$a \neq 0$	$a = 0$	$a = 0$

We give another example, which shows that an algebraic system may be simple, although it contains no inequations.

Example 2.2.13. Let $R = \mathbb{Q}[x, y]$ and $x > y$. Then

$$(y + 1)x = 0, \quad y(y - 1) = 0$$

is a simple algebraic system S over R. Using the factorization of the second equation, a splitting of this system into

$$\{(y + 1)x = 0, y = 0\}, \quad \{(y + 1)x = 0, y - 1 = 0\}$$

makes further reductions possible, which results in another Thomas decomposition

$$\{x = 0, y = 0\}, \quad \{x = 0, y = 1\}$$

of the same system S. Using the factorization of the first equation yields the same answer after removing inconsistent systems.

Remark 2.2.14. Let us assume that a system S of algebraic equations and inequations, defined over $\mathbb{Z}$, is given. A variant of Thomas' algorithm (neglecting square-freeness) allows to compute a finite collection of systems from which partitions of the solution sets $\mathrm{Sol}_{\overline{\mathbb{Q}}}(S)$ and $\mathrm{Sol}_{\overline{\mathbb{F}_p}}(S)$ can be extracted for algebraic closures $\overline{\mathbb{Q}}$ of $\mathbb{Q}$ and $\overline{\mathbb{F}_p}$ of $\mathbb{F}_p = \mathbb{Z}/p\mathbb{Z}$, where p is a prime number. To this end, vanishing of initials has to be checked also if these are integers, and division by non-invertible integers must be prevented. This leads to new splittings, in particular when the greatest common divisor of two polynomials is an integer of absolute value at least 2. For instance, a system could be split into two systems which include a new equation $6 = 0$ and a new inequation $6 \neq 0$, respectively. Integer factorization can be used to split the first system again.

Example 2.2.15. We consider the system of algebraic equations

$$y\underline{x}^2 - \underline{x} + 1 = 0, \qquad y^2\underline{x} - y^3 + 2 = 0, \qquad \underline{x} + y = 0,$$

which is defined over $\mathbb{Z}$ and where $x > y$. Euclidean division of the first and the second modulo the third polynomial yields

$$-2\underline{y}^3 + 2 = 0, \qquad \underline{y}^3 + \underline{y} + 1 = 0, \qquad \underline{x} + y = 0. \qquad (2.28)$$

The result of applying Euclidean division to the first polynomial modulo the second one is $2y + 4$. In order to be able to replace the second polynomial by its pseudo-remainder modulo the new polynomial without changing the solution set of the system, we assume that $2 \neq 0$ holds. Then the pseudo-division yields $18 = 0$, which is equivalent to $9 = 0$. Since we consider the solutions in an algebraic closure of a field $\mathbb{F}_p$, the final result in this case is

$$\underline{x} + 1 = 0, \qquad \underline{y} + 2 = 0, \qquad 3 = 0.$$

If $2 = 0$, only the second and third equation in (2.28) remain, and the final result in this case is

$$\underline{x} + y = 0, \qquad \underline{y}^3 + \underline{y} + 1 = 0, \qquad 2 = 0.$$

For a different approach to decomposing algebraic systems into simple systems in positive characteristic, cf. [LMW10, MLW13].

We finish this subsection by giving a more precise description of the algebraic part of Thomas' algorithm, ignoring, however, efficiency issues. The total ordering $>$ on the set of indeterminates $\{x_1, \ldots, x_n\}$ of R is part of the input. It determines the leader of each non-constant polynomial in $x_1, \ldots, x_n$.

Definition 2.2.16. Let $p \in R$, $q \in R - K$, and $G \subseteq R - K$.

a) The polynomial p is said to be *reduced with respect to q* if $p \in K$ or if $p \in R - K$ and we have $\deg_v(p) < \deg_v(q)$ for $v := \operatorname{ld}(q)$.
b) The polynomial p is said to be *reduced with respect to G* if $p \in K$ or if $p \in R - K$ and p is reduced with respect to each element of G, and if each coefficient of p (as a polynomial in its leader) is reduced with respect to each element of G.
c) An equation or inequation (with zero right hand side) is said to be *reduced with respect to q* or *reduced with respect to G* if its left hand side is so.

Given a polynomial r in R and a finite set G of non-constant polynomials in R, the following algorithm subtracts from r suitable multiples of polynomials in G with the same leader as r until the result is reduced with respect to each polynomial in G with that leader. It treats the coefficients of the result, which are polynomials with smaller leader, if not constant, in the same way. This recursive reduction is essential for the description of Thomas' algorithm below because we suppose that the highest term of the left hand side of $p = 0$ (or of $p \neq 0$) will be canceled if $\mathrm{init}(p) = 0$ is an equation of the same algebraic system. However, the reduction of coefficients of terms of lower degree could be omitted (which would require an adaptation of Definition 2.2.16).

Algorithm 2.2.17 *(Reduce).*

Input: $r \in R$, $G = \{ p_1, p_2, \ldots, p_s \} \subseteq R - K$, and a total ordering $>$ on $\{x_1, \ldots, x_n\}$

Output: $r' \in R$ and an element b of the multiplicatively closed set generated by $\bigcup_{i=1}^{s} \{ \mathrm{init}(p_i) \} \cup \{1\}$ such that r' is reduced with respect to G, and such that $r' = r$, $b = 1$ if $G = \emptyset$, and $r' + \langle p_1, \ldots, p_s \rangle = b \cdot r + \langle p_1, \ldots, p_s \rangle$ otherwise

Algorithm:

1: $r' \leftarrow r$

2: $b \leftarrow 1$

3: **if** $r' \notin K$ **then**

4:      $v \leftarrow \mathrm{ld}(r')$

5:      **while** $r' \notin K$ and there exists $p \in G$ with $\mathrm{ld}(p) = v$, $\deg_v(r') \geq \deg_v(p)$ **do**

6:          $r' \leftarrow \mathrm{init}(p) \cdot r' - \mathrm{init}(r') \cdot v^{d-d'} \cdot p$, where $d := \deg_v(r')$ and $d' := \deg_v(p)$

7:          $b \leftarrow \mathrm{init}(p) \cdot b$

8:      **end while**

9:      **while** there exists a coefficient c of r' (as a polynomial in v) which is not reduced with respect to G **do**

10:          $(r'', b') \leftarrow Reduce(c, G, >)$

11:          replace the coefficient $b' \cdot c$ in $b' \cdot r'$ with r'' and replace r' with this result

12:          $b \leftarrow b' \cdot b$

13:      **end while**

14: **end if**

15: **return** (r', b)

Remarks 2.2.18. a) The loop in steps 5–8 ensures that r' is reduced with respect to each $p \in G$ with $\mathrm{ld}(p) = v$. Termination of Algorithm 2.2.17 follows from the facts that the coefficients c which are dealt with recursively in step 10 are either constant or have leaders which are smaller than v with respect to $>$ and that the property of r' which is achieved by the loop in steps 5–8 is retained by the recursion. The asserted equation follows recursively from the updates of b. Note that in general, if $b \neq 1$, then r and r' are not in the same residue class of $R/(\langle p_1, \ldots, p_s \rangle : q^{\infty})$ (cf. also the following example).

b) Let r_1, $r_2 \in R$ and $G = \{p_1, p_2, \ldots, p_s\}$ be as in the input of Algorithm 2.2.17, and define q to be the product of all $\text{init}(p_i)$, $i = 1, \ldots, s$. In general, the equality

$$r_1 + \langle p_1, \ldots, p_s \rangle : q^{\infty} = r_2 + \langle p_1, \ldots, p_s \rangle : q^{\infty}$$

does not imply that the results of applying *Reduce* to r_1 and r_2, respectively, are equal. However, Proposition 2.2.7 shows that, if $p_1 = 0$, $p_2 = 0$, $\ldots$, $p_s = 0$ are the equations of a simple algebraic system, then the result r' of applying *Reduce* to r_1 is zero if and only if we have $r_1 \in \langle p_1, \ldots, p_s \rangle : q^{\infty}$.

Example 2.2.19. Let $R = \mathbb{Q}[x, y]$ and $x > y$. Then

$$yx - 1 = 0, \quad y \neq 0$$

is a simple algebraic system over R. Algorithm 2.2.17 (*Reduce*) applied to $r := x$, $G := \{yx - 1\}$, and $>$ computes

$$r' := yr - (yx - 1) = 1,$$

and the output is $(r', b) = (1, y)$. Note that r and r' are not in the same residue class of $R/(\langle yx - 1 \rangle : q^{\infty})$, where $q := y$, but $b \cdot r$ and r' are. Moreover, the result of applying *Reduce* to yx, which is in the same residue class as 1, is (y, y). Hence, for different representatives of the same residue class, the first component of the output of *Reduce* may be different in general.

The following description of the algebraic part of Thomas' algorithm deals with triples (L, M, N) of finite algebraic systems over R which are gathered in a set Q. Initially this set contains only the triple $(S, \emptyset, \emptyset)$, where S is the input system, more triples will usually be inserted into Q as such triples are processed, and after finitely many steps the set Q will be empty. Another set T collects the simple algebraic systems of the Thomas decomposition to be constructed.

The second and third component of every triple have the following properties throughout the algorithm. The left hand side p of every equation and inequation in M is non-constant and $\text{init}(p) \neq 0$ holds if a solution of the algebraic system $L \cup M \cup N$ is substituted for x_1, $\ldots$, x_n. Similarly, the left hand side p of every equation and inequation in N is non-constant and both $\text{init}(p) \neq 0$ and $\text{disc}(p) \neq 0$ hold if a solution of $L \cup M \cup N$ is substituted for x_1, $\ldots$, x_n. Moreover, for every $v \in \{x_1, \ldots, x_n\}$, $M \cup N$ contains at most one equation or inequation with leader v.

For any algebraic system

$$S = \{p_i = 0, q_j \neq 0 \mid i \in I, j \in J\}, \qquad p_i, q_j \in R,$$

where I and J are index sets, we denote by

$$S^{=} := \{p_i \mid i \in I\}$$

the set of left hand sides of equations in S.

Algorithm 2.2.20 (*AlgebraicThomasDecomposition*).

Input: A finite algebraic system S over R and a total ordering $>$ on $\{x_1, \ldots, x_n\}$

Output: A Thomas decomposition of S

Algorithm:

1: $Q \leftarrow \{(S, \emptyset, \emptyset)\}$
2: $T \leftarrow \emptyset$
3: **repeat**
4:     choose $(L, M, N) \in Q$ and remove (L, M, N) from Q
5:     replace the left hand side p of each equation and inequation in L with the first
        entry of the result of $Reduce(p, M^= \cup N^=, >)$ // cf. Alg. 2.2.17
6:     remove $0 = 0$ and $p \neq 0$ from L for any $p \in K - \{0\}$
7:     **if** L does neither contain $p = 0$ with $p \in K - \{0\}$ nor $0 \neq 0$ **then**
8:         **if** $L = \emptyset$ **then**
9:             **if** $M = \emptyset$ **then**
10:                insert N into T
11:            **else**
12:                $Q \leftarrow ProcessDiscriminant((L, M, N), Q, >)$ // cf. Alg. 2.2.23
13:            **end if**
14:        **else**
15:            $Q \leftarrow ProcessInitial((L, M, N), Q, >)$ // cf. Alg. 2.2.21
16:        **end if**
17:    **end if**
18: **until** $Q = \emptyset$
19: **return** T

The proof that Algorithm 2.2.20 terminates and is correct will be given after the description of the algorithms on which it depends (cf. Thm. 2.2.32, p. 79).

The following terminology will be useful for the rest of this subsection. For a triple (L, M, N) of algebraic systems over R we refer to $\mathrm{Sol}_{\overline{K}}(L \cup M \cup N)$ as the *solution set* of the triple (L, M, N), and for a set Q of such triples we denote by

$$\mathrm{Sol}_{\overline{K}}(Q) := \bigcup_{(L, M, N) \in Q} \mathrm{Sol}_{\overline{K}}(L \cup M \cup N)$$

the union of the solution sets of all triples in Q.

Moreover, let

$$S = \{ p_i = 0, q_j \neq 0 \mid i \in I, j \in J \}, \qquad p_i, q_j \in R,$$

be an algebraic system, where no p_i and no q_j is constant, and let $v \in \{x_1, \ldots, x_n\}$. Then $S_{\geq v}$ (resp. $S_{<v}$) is a notation for the subset of S which consists of the equations and inequations with leader greater than or equal to (resp. smaller than) v with respect to $>$. We are also going to write $S_{<v}^=$ instead of $(S_{<v})^=$ (in Remarks 2.2.26).

Algorithm 2.2.21 (*ProcessInitial*).

Input: A triple (L,M,N) of finite algebraic systems over R, a finite set P of such triples, and a total ordering $>$ on $\{x_1,\ldots,x_n\}$, where $L \neq \emptyset$, the left hand sides of elements of $L \cup M \cup N$ are non-constant, those of $M \cup N$ having pairwise distinct leaders, those of L being reduced with respect to $M^= \cup N^=$ (cf. Def. 2.2.16 b)), where $\text{Sol}_{\overline{K}}(L \cup M \cup N)$ and the solution sets of triples in P are pairwise disjoint

Output: A finite set $Q \supseteq P$ of triples as in P whose solution sets form a partition of $\text{Sol}_{\overline{K}}(L \cup M \cup N) \uplus \text{Sol}_{\overline{K}}(P)$ such that either

a) each triple in $Q - P$ has the property that all of its solutions satisfy $\text{init}(p) \neq 0$ or all of its solutions satisfy $\text{init}(p) = 0$, where p is the left hand side of the equation or inequation in L chosen in step 2, or

b) the triples in $Q - P$ have been inserted by Algorithm 2.2.27 (*LCMSplit*)

Algorithm:

1: $Q \leftarrow P$
2: among the elements of L with least possible leader v with respect to $>$ choose one with left hand side p of least possible degree in v, preferably an equation
3: **if** the equation $p = 0$ is chosen **then**
4:     insert $((L - \{p = 0\}) \cup M_{\geq v} \cup N_{\geq v} \cup \{\text{init}(p) \neq 0\},$
    $(M - M_{\geq v}) \cup \{p = 0\}, N - N_{\geq v})$ into Q
5:     insert $(L \cup \{\text{init}(p) = 0\}, M, N)$ into Q
6: **else**   // *the inequation $p \neq 0$ is chosen*
7:     **if** $M \cup N$ contains an equation $q = 0$ with $\text{ld}(q) = v$ **then**
8:         $Q \leftarrow \text{GCDSplit}(q,\, p,\, (L,M,N),\, Q,\, >)$ // *cf. Alg. 2.2.25*
9:     **else if** $M \cup N$ contains an inequation $q \neq 0$ with $\text{ld}(q) = v$ **then**
10:         **if** $\deg_v(p) \geq \deg_v(q)$ **then**
11:             $Q \leftarrow \text{LCMSplit}(p,\, q,\, (L,M,N),\, Q,\, >)$ // *cf. Alg. 2.2.27*
12:         **else**
13:             insert $((L - \{p \neq 0\}) \cup \{q \neq 0, \text{init}(p) \neq 0\},$
            $(M - \{q \neq 0\}) \cup \{p \neq 0\}, N - \{q \neq 0\})$ into Q
14:             insert $(L \cup \{\text{init}(p) = 0\}, M, N)$ into Q
15:         **end if**
16:     **else**
17:         insert $((L - \{p \neq 0\}) \cup \{\text{init}(p) \neq 0\}, M \cup \{p \neq 0\}, N)$ into Q
18:         insert $(L \cup \{\text{init}(p) = 0\}, M, N)$ into Q
19:     **end if**
20: **end if**
21: **return** Q

Remark 2.2.22. Termination of Algorithm 2.2.21 follows from the fact that Algorithm 2.2.25 and Algorithm 2.2.27 terminate (cf. Lemma 2.2.28 and Lemma 2.2.29). An inspection of steps 4, 5, 13, 14, 17, and 18 and of the specifications of Algorithms 2.2.25 and 2.2.27 shows that the solution sets of triples in Q form a partition of $\mathrm{Sol}_{\overline{K}}(L \cup M \cup N) \uplus \mathrm{Sol}_{\overline{K}}(P)$. In order to show the last assertion stated in the description of the output, we observe that in each of these steps as well as in steps 6 and 7 in Algorithm 2.2.25, where r_{i+2} is equal to p_2 in the first round of the loop, either the inequation $\mathrm{init}(p) \neq 0$ or the equation $\mathrm{init}(p) = 0$ is imposed.

Algorithm 2.2.23 (*ProcessDiscriminant*).

Input: (L, M, N), P, and $>$ with the same specification as in Algorithm 2.2.21 and satisfying $L = \emptyset$ and $M \neq \emptyset$

Output: A finite set $Q \supseteq P$ of triples as in P whose solution sets form a partition of $\mathrm{Sol}_{\overline{K}}(M \cup N) \uplus \mathrm{Sol}_{\overline{K}}(P)$ such that either

a) each triple in $Q - P$ has the property that all solutions satisfy $\mathrm{disc}(p) \neq 0$, where p is the left hand side of the equation or inequation in M with least leader with respect to $>$, or

b) the triples in $Q - P$ have been inserted by Algorithm 2.2.30 (*SquarefreeSplit*)

Algorithm:

1: $Q \leftarrow P$

2: let $p = 0$ or $p \neq 0$ be the equation or inequation in M with least leader with respect to $>$ and let v be its leader

3: **if** $\deg_v(p) = 1$ **then**

4:     **if** M contains $p = 0$ **then**

5:         insert $(\emptyset, M - \{p = 0\}, N \cup \{p = 0\})$ into Q

6:     **else**   // *M contains* $p \neq 0$

7:         insert $(\emptyset, M - \{p \neq 0\}, N \cup \{p \neq 0\})$ into Q

8:     **end if**

9: **else**

10:     $Q \leftarrow SquarefreeSplit(p, (\emptyset, M, N), Q, >)$ // *cf. Alg. 2.2.30*

11: **end if**

12: **return** Q

Remark 2.2.24. Termination of Algorithm 2.2.23 follows from the fact that Algorithm 2.2.30 terminates (cf. Lemma 2.2.31). It is easily checked by considering steps 5 and 7 and the specification of Algorithm 2.2.30 that the solution sets of triples in Q form a partition of $\mathrm{Sol}_{\overline{K}}(M \cup N) \uplus \mathrm{Sol}_{\overline{K}}(P)$. The last assertion which is stated in the description of the output is shown as follows. A solution of a triple in $Q - P$ satisfies $\mathrm{disc}(p) = 0$ if and only if the univariate polynomial $\overline{p}$ which is obtained by substituting this solution for $x_1, \ldots, x_n$ except $\mathrm{ld}(p)$ in p has multiple roots. But in steps 5 and 7 the polynomial $\overline{p}$ has degree one.

Algorithm 2.2.25 (*GCDSplit*).

Input: $p_1, p_2 \in R - K$ with the same leader v and (L, M, N), P, $>$ with the same specification as in Algorithm 2.2.21, where $p_1 = 0$ is in $M \cup N$, $p_2 \neq 0$ is in L, $\deg_v(p_1) \geq \deg_v(p_2)$, and p_2 is reduced with respect to $M^= \cup N^=$

Output: A finite set $Q \supseteq P$ of triples as in P whose solution sets form a partition of $\text{Sol}_{\overline{K}}(L \cup M \cup N) \uplus \text{Sol}_{\overline{K}}(P)$ such that for each triple in $Q - P$ we have either

a) the polynomials $\overline{p}_1$ and $\overline{p}_2$ which are obtained from p_1 and p_2 by substituting a solution of the triple for $x_1, \ldots, x_n$ except v have a greatest common divisor whose degree does not depend on the choice of the solution of the triple, or

b) the triple has been inserted in step 6

Algorithm:

1: $Q \leftarrow P$; $U \leftarrow \emptyset$
2: $v \leftarrow \text{ld}(p_1)$; $i \leftarrow 0$
3: $r_1 \leftarrow p_1$; $c_1 \leftarrow 0$
4: $r_2 \leftarrow p_2$; $c_2 \leftarrow 1$
5: **repeat**
6:     insert $(L \cup \{\text{init}(r_{i+2}) = 0\} \cup U, M, N)$ into Q
7:     $U \leftarrow U \cup \{\text{init}(r_{i+2}) \neq 0\}$
8:     $i \leftarrow i + 1$
9:     $r_{i+2} \leftarrow a_i \cdot r_i - q_i \cdot r_{i+1}$, where a_i is a power of $\text{init}(r_{i+1})$ and $q_i \in R$ such that
       $r_{i+2} = 0$ or $\deg_v(r_{i+2}) < \deg_v(r_{i+1})$
10:     $(r_{i+2}, b_{i+2}) \leftarrow Reduce(r_{i+2}, M^= \cup N^=, >)$ // cf. Alg. 2.2.17
11:     $c_{i+2} \leftarrow b_{i+2} \cdot (a_i \cdot c_i + q_i \cdot c_{i+1})$
12: **until** $r_{i+2} = 0$ **or** $\deg_v(r_{i+2}) = 0$
13: insert $(L \cup \{c_{i+2} = 0, r_{i+2} = 0\} \cup U, M - \{p_1 = 0\}, N - \{p_1 = 0\})$ into Q
14: insert $((L - \{p_2 \neq 0\}) \cup \{r_{i+2} \neq 0\} \cup U, M, N)$ into Q
15: **return** Q

The proof that Algorithm 2.2.25 terminates and is correct (cf. Lemma 2.2.28) is based on the following remarks.

Remarks 2.2.26. a) The triples which are inserted into Q in steps 6, 13, and 14 in Algorithm 2.2.25 define inconsistent algebraic systems if $\text{init}(r_{i+2})$ in step 6 or r_{i+2} in step 13 is a non-zero constant or if r_{i+2} is the zero polynomial in step 14. These triples should be discarded right away. For the sake of conciseness these case distinctions are omitted here.

b) Since we have $r_1 = p_1$ and $\deg_v(r_{i+2}) < \deg_v(r_{i+1})$ after step 9 and since $p_1 = 0$ is the unique equation with leader v in $M \cup N$, the reduction in step 10 considers only left hand sides of equations with leader smaller than v as pseudo-divisors.

c) Algorithm 2.2.25 is a variant of Euclid's Algorithm with bookkeeping, where (coefficients of) intermediate results are also reduced with respect to $M_{<v}^{=} \cup N_{<v}^{=}$. Steps 9–11 ensure that the following congruence holds for all $i \in \mathbb{Z}_{\geq 0}$:

$$c_{i+2} \cdot r_{i+1} \equiv \left(\prod_{j=1}^{i} a_j\right) \cdot \left(\prod_{k=3}^{i+2} b_k\right) \cdot p_1 - c_{i+1} \cdot r_{i+2} \quad \mathrm{mod} \ \langle M_{<v}^{=} \cup N_{<v}^{=}\rangle. \quad (2.29)$$

Its significance derives from the following special case. If, for all $i \in \mathbb{Z}_{\geq 0}$, both sides are not merely congruent modulo $\langle M_{<v}^{=} \cup N_{<v}^{=}\rangle$, but equal, and if i is minimal with the property that r_{i+2} is the zero polynomial, then r_{i+1} is the greatest common divisor of p_1 and p_2 in $\mathrm{Quot}(K[x \mid v > x])[v]$, where we denote by $\mathrm{Quot}(K[x \mid v > x])$ the field of fractions of the polynomial ring $K[x \mid v > x]$. Then c_{i+2} is the quotient of $a_1 \cdot a_2 \cdot \ldots \cdot a_i \cdot b_3 \cdot b_4 \cdot \ldots \cdot b_{i+2} \cdot p_1$ divided by r_{i+1}.
We prove (2.29) by induction on i. Indeed, for $i = 0$ we have $c_2 \cdot r_1 = p_1$ by steps 3 and 4 (where an empty product is equal to 1 by convention). Let $i > 0$. After step 11 we have

$$\left. \begin{aligned} r_{i+2} &\equiv b_{i+2} \cdot (a_i \cdot r_i - q_i \cdot r_{i+1}) \quad \mathrm{mod} \ \langle M_{<v}^{=} \cup N_{<v}^{=}\rangle, \\ c_{i+2} &= b_{i+2} \cdot (a_i \cdot c_i + q_i \cdot c_{i+1}). \end{aligned} \right\} \quad (2.30)$$

The induction hypothesis states that we have

$$c_{i+1} \cdot r_i \equiv \left(\prod_{j=1}^{i-1} a_j\right) \cdot \left(\prod_{k=3}^{i+1} b_k\right) \cdot p_1 - c_i \cdot r_{i+1} \quad \mathrm{mod} \ \langle M_{<v}^{=} \cup N_{<v}^{=}\rangle. \quad (2.31)$$

Using (2.30) and (2.31), we deduce

$$c_{i+2}\, r_{i+1} \equiv b_{i+2}\, (a_i\, c_i + q_i\, c_{i+1})\, r_{i+1}$$

$$\equiv b_{i+2}\, a_i\, c_i\, r_{i+1} + c_{i+1}\, b_{i+2}\, q_i\, r_{i+1}$$

$$\equiv b_{i+2}\, a_i\, c_i\, r_{i+1} + c_{i+1}\, (b_{i+2}\, a_i\, r_i - r_{i+2})$$

$$\equiv b_{i+2}\, a_i\, c_i\, r_{i+1} - c_{i+1}\, r_{i+2} + b_{i+2}\, a_i \left(\left(\prod_{j=1}^{i-1} a_j\right) \left(\prod_{k=3}^{i+1} b_k\right) p_1 - c_i\, r_{i+1} \right)$$

$$\equiv \left(\prod_{j=1}^{i} a_j\right) \left(\prod_{k=3}^{i+2} b_k\right) p_1 - c_{i+1}\, r_{i+2}$$

modulo $\langle M_{<v}^{=} \cup N_{<v}^{=}\rangle$, which proves (2.29).
Similarly, if we set $d_2 := 0$ and $d_3 := 1$ and update, if $i > 1$,

$$d_{i+2} \leftarrow b_{i+2} \cdot (a_i \cdot d_i + q_i \cdot d_{i+1})$$

after step 11, then the following congruence holds for all $i \in \mathbb{Z}_{\geq 1}$:

$$d_{i+2} \cdot r_{i+1} \equiv \left(\prod_{j=2}^{i} a_j \right) \cdot \left(\prod_{k=4}^{i+2} b_k \right) \cdot p_2 - d_{i+1} \cdot r_{i+2} \quad \mathrm{mod} \; \langle M_{<v}^= \cup N_{<v}^= \rangle. \quad (2.32)$$

This is proved in the same way as (2.29).

The next algorithm applies a reduction, analogous to the one used in the previous algorithm, to a pair of inequations $p_1 \neq 0$, $p_2 \neq 0$ instead of $p_1 = 0$ and $p_2 \neq 0$.

Algorithm 2.2.27 (*LCMSplit*).

Input: p_1, $p_2 \in R - K$ with the same leader v and (L, M, N), P, $>$ with the same specification as in Algorithm 2.2.21, where $p_1 \neq 0$ is in L, $p_2 \neq 0$ is in $M \cup N$, and $\deg_v(p_1) \geq \deg_v(p_2)$

Output: A finite set $Q \supseteq P$ of triples as in P whose solution sets form a partition of $\mathrm{Sol}_{\overline{K}}(L \cup M \cup N) \uplus \mathrm{Sol}_{\overline{K}}(P)$ such that for each triple in $Q - P$ we have either

a) the polynomials $\overline{p}_1$ and $\overline{p}_2$ which are obtained from p_1 and p_2 by substituting a solution of the triple for $x_1, \ldots, x_n$ except v have a least common multiple whose degree does not depend on the choice of the solution of the triple, or

b) the triple has been inserted in step 11

Algorithm:

1: $Q \leftarrow P$; $\; U \leftarrow \emptyset$
2: $v \leftarrow \mathrm{ld}(p_1)$; $\; i \leftarrow 0$
3: $r_1 \leftarrow p_1$; $\; c_1 \leftarrow 0$
4: $r_2 \leftarrow p_2$; $\; c_2 \leftarrow 1$
5: **repeat**
6:     $i \leftarrow i + 1$
7:     $r_{i+2} \leftarrow a_i \cdot r_i - q_i \cdot r_{i+1}$, where a_i is a power of $\mathrm{init}(r_{i+1})$ and $q_i \in R$ such that $r_{i+2} = 0$ or $\deg_v(r_{i+2}) < \deg_v(r_{i+1})$
8:     $(r_{i+2}, b_{i+2}) \leftarrow Reduce(r_{i+2}, M^= \cup N^=, >)$ // cf. Alg. 2.2.17
9:     $c_{i+2} \leftarrow b_{i+2} \cdot (a_i \cdot c_i + q_i \cdot c_{i+1})$
10:     **if** $r_{i+2} \neq 0$ **and** $\deg_v(r_{i+2}) > 0$ **then**
11:         insert $(L \cup \{\mathrm{init}(r_{i+2}) = 0\} \cup U, M, N)$ into Q
12:         $U \leftarrow U \cup \{\mathrm{init}(r_{i+2}) \neq 0\}$
13:     **end if**
14: **until** $r_{i+2} = 0$ **or** $\deg_v(r_{i+2}) = 0$
15: insert $((L - \{p_1 \neq 0\}) \cup \{c_{i+2} \cdot p_2 \neq 0, r_{i+2} = 0\} \cup U, M - \{p_2 \neq 0\},$ $N - \{p_2 \neq 0\})$ into Q
16: insert $((L - \{p_1 \neq 0\}) \cup \{p_1 \cdot p_2 \neq 0, r_{i+2} \neq 0\} \cup U, M - \{p_2 \neq 0\},$ $N - \{p_2 \neq 0\})$ into Q
17: **return** Q

Lemma 2.2.28. *Algorithm 2.2.25 (on page 73) terminates and is correct.*

Proof. Termination of Algorithm 2.2.25 follows from the fact that the degree in v of the elements of the sequence $r_2, r_3, r_4, \ldots$ is decreasing.

The solution set of (L, M, N) is partitioned into solution sets of several triples in the result Q due to steps 6, 13, and 14. In the beginning of each round of the loop the splitting of the current triple into the one defined in step 6 and complementary ones incorporating the update of U in step 7 ensures that in step 9 the inequation $\text{init}(r_{i+1}) \neq 0$ holds if a solution of the current triple is substituted for $x_1, \ldots, x_n$. This also implies $a_i \neq 0$. Since the initials of left hand sides of elements of $M \cup N$ do not vanish on solutions of the current triple, the inequation $b_{i+2} \neq 0$ holds as well.

In step 13 the condition $r_{i+2} = 0$ is imposed, which is complemented by $r_{i+2} \neq 0$ in the triple defined in step 14. Note that the first component of the triple in step 13 contains the inequation $p_2 \neq 0$, so that the inequation $r_{i+1} \neq 0$ holds for all solutions of this triple because of (2.32), $r_{i+2} = 0$, and $a_2 \cdot a_3 \cdot \ldots \cdot a_i \cdot b_4 \cdot b_5 \cdot \ldots \cdot b_{i+2} \neq 0$. Then, by (2.29), the equation $c_{i+2} = 0$ holds for all solutions of the triple. Conversely, the equations $c_{i+2} = 0$ and $r_{i+2} = 0$ and the inequation $a_1 \cdot a_2 \cdot \ldots \cdot a_i \cdot b_3 \cdot b_4 \cdot \ldots \cdot b_{i+2} \neq 0$ imply $p_1 = 0$. Therefore, the solution set of $L \cup U \cup M \cup N$ is not changed if $p_1 = 0$ is replaced with $c_{i+2} = 0$.

Finally, in step 14 the condition $r_{i+2} \neq 0$ is imposed. Since r_{i+2} is an R-linear combination of p_2 and the left hand sides of the equations in $M \cup N$, this condition implies $p_2 \neq 0$, so that the latter inequation is dispensable for the updated triple.

Let $\overline{p}_1$ and $\overline{p}_2$ be obtained from p_1 and p_2, respectively, by substituting a solution of the triple in step 13 or 14 for $x_1, \ldots, x_n$ except v. The same substitution specializes the sequence of polynomials $r_1, r_2, r_3, \ldots$ to the one (up to non-zero constant factors) which is computed by Euclid's algorithm for the univariate polynomials $\overline{p}_1$ and $\overline{p}_2$, because $\text{init}(r_i)$ does not vanish for any polynomial r_i preceding the final one. This shows the last assertion stated in the description of the output. $\square$

Lemma 2.2.29. *Algorithm 2.2.27 (on page 75) terminates and is correct.*

Proof. Termination is shown exactly as in the proof of Lemma 2.2.28.

The solution set of (L, M, N) is partitioned into solution sets of several triples in the result Q due to steps 11, 15, and 16. As opposed to the input of Algorithm 2.2.25, the inequation $p_2 \neq 0$ is an element of $M \cup N$ rather than L. This ensures that $\text{init}(r_{i+1}) \neq 0$ holds if a solution of the current triple in step 7 in the first round of the loop is substituted for $x_1, \ldots, x_n$. The splitting of algebraic systems in step 11 arranges for the corresponding property in the next round.

Similarly to step 13 in Algorithm 2.2.25, in step 15 the condition $r_{i+2} = 0$ is imposed, which is complemented by $r_{i+2} \neq 0$ in the triple defined in step 16. Again, the inequation $r_{i+1} \neq 0$ holds for all solutions of the triple in step 15 because of (2.32), $r_{i+2} = 0$, $a_1 \cdot a_2 \cdot \ldots \cdot a_i \cdot b_3 \cdot b_4 \cdot \ldots \cdot b_{i+2} \neq 0$, and $p_2 \neq 0$ (imposed by the first entry of the triple). Given these conditions, the inequations $c_{i+2} \neq 0$ and $p_1 \neq 0$ are equivalent by (2.29). Hence, replacing $p_1 \neq 0$ and $p_2 \neq 0$ with $c_{i+2} \cdot p_2 \neq 0$ in step 15 does not change the solution set of $L \cup U \cup M \cup N$. Replacing $p_1 \neq 0$ and $p_2 \neq 0$ with $p_1 \cdot p_2 \neq 0$ does not change the solution set in step 16 either.

The last assertion stated in the description of the output is proved in the same way as the corresponding one for Algorithm 2.2.25.   □

Finally, the same reduction technique is applied to determine square-free parts.

Algorithm 2.2.30 (*SquarefreeSplit*).

Input: $p \in R - K$ with degree at least 2 in its leader v and (L, M, N), P, $>$ with the same specification as in Algorithm 2.2.21, where $L = \emptyset$ and p is the left hand side of an equation or inequation in M

Output: A finite set $Q \supseteq P$ of triples as in P whose solution sets form a partition of $\mathrm{Sol}_{\overline{K}}(M \cup N) \uplus \mathrm{Sol}_{\overline{K}}(P)$ such that for each triple in $Q - P$ we have either

a) the two polynomials which are obtained from p and $\frac{\partial p}{\partial v}$ by substituting a solution of the triple for $x_1, \ldots, x_n$ except v have a greatest common divisor whose degree does not depend on the choice of the solution of the triple, or

b) the triple has been inserted in step 10

Algorithm:

1: $Q \leftarrow P$; $U \leftarrow \emptyset$; $v \leftarrow \mathrm{ld}(p)$; $i \leftarrow 0$
2: $r_1 \leftarrow p$; $c_1 \leftarrow 0$
3: $r_2 \leftarrow \frac{\partial p}{\partial v}$; $c_2 \leftarrow 1$
4: **repeat**
5:     $i \leftarrow i + 1$
6:     $r_{i+2} \leftarrow a_i \cdot r_i - q_i \cdot r_{i+1}$, where a_i is a power of $\mathrm{init}(r_{i+1})$ and $q_i \in R$ such that $r_{i+2} = 0$ or $\deg_v(r_{i+2}) < \deg_v(r_{i+1})$
7:     $(r_{i+2}, b_{i+2}) \leftarrow Reduce(r_{i+2}, M^= \cup N^=, >)$ // cf. Alg. 2.2.17
8:     $c_{i+2} \leftarrow b_{i+2} \cdot (a_i \cdot c_i + q_i \cdot c_{i+1})$
9:     **if** $r_{i+2} \neq 0$ **and** $\deg_v(r_{i+2}) > 0$ **then**
10:        insert $(\{\mathrm{init}(r_{i+2}) = 0\} \cup U, M, N)$ into Q
11:        $U \leftarrow U \cup \{\mathrm{init}(r_{i+2}) \neq 0\}$
12:    **end if**
13: **until** $r_{i+2} = 0$ **or** $\deg_v(r_{i+2}) = 0$
14: **if** M contains $p = 0$ **then**
15:    insert $(\{c_{i \mid 2} = 0, r_{i+2} = 0\} \cup U, M - \{p = 0\}, N)$ into Q
16:    insert $(\{r_{i+2} \neq 0\} \cup U, M - \{p = 0\}, N \cup \{p = 0\})$ into Q
17: **else**   // M contains $p \neq 0$
18:    insert $(\{c_{i+2} \neq 0, r_{i+2} = 0\} \cup U, M - \{p \neq 0\}, N)$ into Q
19:    insert $(\{r_{i+2} \neq 0\} \cup U, M - \{p \neq 0\}, N \cup \{p \neq 0\})$ into Q
20: **end if**
21: **return** Q

Lemma 2.2.31. *Algorithm 2.2.30 terminates and is correct.*

Proof. Again, the same argument as in the proof of Lemma 2.2.28 shows that Algorithm 2.2.30 terminates.

The solution set of (L, M, N) is partitioned into solution sets of several triples in the result Q due to steps 10 and 15, 16 or 18, 19. Since p has degree at least two in v, the initial of $\frac{\partial p}{\partial v}$ is a constant multiple of $\mathrm{init}(p)$, and since the inequation $p \neq 0$ is an element of M, the inequation $\mathrm{init}(r_{i+1}) \neq 0$ holds if a solution of the current triple in step 6 in the first round of the loop is substituted for $x_1, \ldots, x_n$. For further rounds the inequation $\mathrm{init}(r_{i+1}) \neq 0$ has been added to U in the previous round.

After step 8 the congruence (2.29) holds with p_1 replaced with p, and if the sequence d_2, d_3, d_4, ... defined in Remark 2.2.26 c) is also computed, then the congruence (2.32) holds with p_2 replaced with $\frac{\partial p}{\partial v}$.

In steps 15 and 18 the condition $r_{i+2} = 0$ is imposed, which is complemented by $r_{i+2} \neq 0$ in the triple defined in step 16 or 19, respectively. We claim that replacing the equation $p = 0$ with $c_{i+2} = 0$ does not change the solution set of the triple in step 15. First of all, by (2.29), the equations $c_{i+2} = 0$ and $r_{i+2} = 0$ and the inequation $a_1 \cdot a_2 \cdot \ldots \cdot a_i \cdot b_3 \cdot b_4 \cdot \ldots \cdot b_{i+2} \neq 0$ imply $p = 0$. Conversely, we show that every solution of $(L \cup U \cup \{r_{i+2} = 0\}, M, N)$ is a solution of $c_{i+2} = 0$. Let $\overline{p}, \overline{c}_{i+2}, \overline{r}_{i+1}, \overline{a}_j$, and $\overline{b}_k$ be obtained from p, c_{i+2}, r_{i+1}, a_j, and b_k, respectively, by substituting such a solution for $x_1, \ldots, x_n$ except v. Then (2.29) specializes to

$$\overline{c}_{i+2} \cdot \overline{r}_{i+1} = \left(\prod_{j=1}^{i} \overline{a}_j \right) \cdot \left(\prod_{k=3}^{i+2} \overline{b}_k \right) \cdot \overline{p}, \tag{2.33}$$

where the degree in v of each factor is the same as the degree in v of the corresponding factor in (2.29) because $\mathrm{init}(p)$ and $\mathrm{init}(r_{i+1})$ do not vanish. Let $\eta \in \overline{K}$ be the component of the solution which corresponds to v. If $\overline{r}_{i+1}(\eta) = 0$, then (2.33) implies $\overline{c}_{i+2}(\eta) = 0$, which proves the claim in this case. Otherwise, the corresponding specialization of (2.32) shows that η is a common root of $\overline{p}$ and its derivative. Then η is a root of $\overline{p}$ of multiplicity greater than one. Since $\overline{r}_{i+1}$ divides both $\overline{p}$ and its derivative, we conclude that η is a root of $\overline{p}/\overline{r}_{i+1}$ and hence of $\overline{c}_{i+2}$.

Next we show that replacing the inequation $p \neq 0$ with $c_{i+2} \neq 0$ does not change the solution set of the triple in step 18. Clearly, by (2.29), the equation $r_{i+2} = 0$ and the inequations $p \neq 0$ and $a_1 \cdot a_2 \cdot \ldots \cdot a_i \cdot b_3 \cdot b_4 \cdot \ldots \cdot b_{i+2} \neq 0$ imply $c_{i+2} \neq 0$. Conversely, we show that the inequation $p \neq 0$ holds for all solutions of the triple in step 18. Using the same notation as above, we have $\overline{r}_{i+1}(\eta) = 0$ or $\overline{r}_{i+1}(\eta) \neq 0$. In the former case we conclude in the same way as above that η is a common root of $\overline{p}$ and its derivative and therefore a root of $\overline{c}_{i+2}$, which is a contradiction. Hence, we have $\overline{r}_{i+1}(\eta) \neq 0$ and therefore, $p \neq 0$ holds.

The last assertion stated in the description of the output follows by the same argument as in the proof of Lemma 2.2.28. Finally, in order to justify the transfer of $p = 0$ or $p \neq 0$ from the second to the third component of the triple in step 16 or 19, we note that either r_{i+2} is the zero polynomial and the triple has no solution, or the greatest common divisor of $\overline{p}$ and its derivative is the non-zero constant which is obtained from r_{i+2} by substituting the solution that defines $\overline{p}$, which shows that $\overline{p}$ and its derivative have no common root. ☐

Theorem 2.2.32. *Algorithm 2.2.20, p. 70, terminates and is correct.*

Proof. In order to prove correctness, we note first that step 5 in Algorithm 2.2.20 (*AlgebraicThomasDecomposition*) ensures that the left hand sides of elements of L in step 15 are reduced with respect to $M^= \cup N^=$, and steps 6 and 7 guarantee that they are not constant. The property that $M \cup N$ contains at most one equation or inequation with a given leader is retained throughout.

An equation or inequation with left hand side p is only inserted into the second component M of a triple (L,M,N) if all solutions of the updated triple satisfy $\text{init}(p) \neq 0$, namely in steps 4, 13, and 17 in Algorithm 2.2.21 (*ProcessInitial*). Similarly, an equation or inequation with left hand side p is only inserted into the third component N of such a triple if it is moved there from the second component M and if all solutions of the updated triple satisfy $\text{disc}(p) \neq 0$, namely in steps 5 and 7 in Algorithm 2.2.23 (*ProcessDiscriminant*) and in steps 16 and 19 in Algorithm 2.2.30 (*SquarefreeSplit*) (cf. the end of Remark 2.2.24 and the end of the proof of Lemma 2.2.31 for justifications).

As a result of the above discussion, if an algebraic system N is inserted into T in step 10, this system is simple. The output T is a Thomas decomposition of the input system S because the solution sets of triples in Q are pairwise disjoint throughout the algorithm, the solution sets of algebraic systems in T are pairwise disjoint, and the union of $\text{Sol}_{\overline{K}}(Q)$ and the solution sets of algebraic systems in T equals $\text{Sol}_{\overline{K}}(S)$ (cf. Remarks 2.2.22 and 2.2.24, Lemmas 2.2.28, 2.2.29, and 2.2.31).

Termination of Algorithm 2.2.20 follows if we show that after finitely many steps the set Q is empty. Since every triple in Q arises from splittings of algebraic systems, whose common origin is the triple $(S,\emptyset,\emptyset)$, it is sufficient to prove that every triple is removed after finitely many steps and that no triple has infinitely many descendants. In fact, we are going to argue for each splitting that the further treatment of a new triple (L',M',N') leads to a modification of M' or N' and that only finitely many consecutive modifications are possible for each triple and its descendants.

Each triple (L,M,N) in Q is either discarded or is dealt with by Algorithm 2.2.21 (*ProcessInitial*) or Algorithm 2.2.23 (*ProcessDiscriminant*). The first case occurs if an equation or inequation with constant left hand side reveals that the algebraic system is inconsistent, or if L and M are empty, in which case N is inserted into the set T. Algorithms 2.2.21 and 2.2.23, using also Algorithms 2.2.25 (*GCDSplit*), 2.2.27 (*LCMSplit*), and 2.2.30 (*SquarefreeSplit*), insert further triples into Q whose solution sets form a partition of $\text{Sol}_{\overline{K}}(L \cup M \cup N)$.

A modification of M or N is possible precisely in the following ways:

a) An equation $p = 0$ with leader v is transferred from L to M after equations and inequations with leader greater than or equal to v have been transferred from $M \cup N$ to L (Alg. 2.2.21, step 4). If $M \cup N$ contained an equation with leader v before, then p has smaller degree in v than the left hand side of the old equation because p was reduced with respect to $M^= \cup N^=$ before the insertion.

b) An inequation $p \neq 0$ with leader v is transferred from L to M only if $M \cup N$ does not contain an equation with leader v (Alg. 2.2.21, steps 13 and 17). If $M \cup N$ contained an inequation with leader v before, then p has smaller degree in v than the left hand side of the old inequation, and the old inequation is transferred to L.

c) An equation $p_1 = 0$ with leader v is removed from M or from N and an equation $c_{i+2} = 0$ is inserted into L, where c_{i+2} is constant, but non-zero, or the leader of c_{i+2} is v, and $\deg_v(c_{i+2})$ is less than $\deg_v(p_1)$, and $\mathrm{init}(c_{i+2})$ is not in the ideal $\langle M_{<v}^{=} \cup N_{<v}^{=} \rangle$ (Alg. 2.2.25, step 13). Finitely many inequations whose left hand sides are constant, but non-zero, or have leaders which are smaller than v may be inserted into L as well.

   In order to confirm these properties, we note that c_3 in Algorithm 2.2.25 is constant if and only if we have $\deg_v(p_1) = \deg_v(p_2)$, that the degree in v of the entries of the sequence c_3, c_4, c_5, ... is increasing and the degree in v of those in r_2, r_3, r_4, ... is decreasing. If c_3 is constant, then it is non-zero because c_1 is zero, but b_3, q_1, and c_2 are not, so that the new triple defines an inconsistent algebraic system. Otherwise, c_{i+2} has leader v and degree in v less than $\deg_v(p_1)$ due to (2.29), p. 74, and because of the properties of the above sequences. The initial of c_{i+2} is not in the ideal $\langle M_{<v}^{=} \cup N_{<v}^{=} \rangle$ because the initial of p_1 is not.

   The set L in the input of Algorithm 2.2.25 contains neither equations with leader v nor equations or inequations with smaller leader. When the new triple defined in step 13 with $\deg_v(c_{i+2}) > 0$ will be further processed, inequations with leader smaller than v, contributed by the set U in step 13, if any, and equations and inequations with leader smaller than v produced by this process will be dealt with. Further splittings may occur. Since $\mathrm{init}(c_{i+2})$ is not in $\langle M_{<v}^{=} \cup N_{<v}^{=} \rangle$, a reduction may decrease the degree of c_{i+2} in v only if an equation with leader smaller than v has been inserted into the second component of the triple in question. Otherwise, (a reduced form of) the new equation $c_{i+2} = 0$ will be inserted into the second component. In all cases a modification of type a) along with the generation of a new triple as in g) below will occur.

d) Inequations $p_1 \neq 0$ and $p_2 \neq 0$ with the same leader v are removed from L and M or N, respectively, and an inequation with leader v is inserted into L (Alg. 2.2.27, steps 15 and 16). Finitely many equations and inequations whose left hand sides are constant or have leaders smaller than v may be inserted into L as well.

e) An equation or inequation with left hand side p is removed from M or N and, correspondingly, an equation or inequation with left hand side c_{i+2} is inserted into L, where c_{i+2} is constant, but non-zero, or the leader of c_{i+2} is v, and $\deg_v(c_{i+2})$ is less than $\deg_v(p)$, and $\mathrm{init}(c_{i+2})$ is not in the ideal $\langle M_{<v}^{=} \cup N_{<v}^{=} \rangle$ (Alg. 2.2.30, steps 15 and 18). An equation and finitely many inequations whose left hand sides are constant or have leaders smaller than v may be inserted into L as well. The fact that c_{i+2} has the above property follows in the same way as in c).

f) An equation or inequation with left hand side p is transferred from M to N (Alg. 2.2.23, steps 5 and 7, Alg. 2.2.30, steps 16 and 19). Finitely many inequations whose left hand sides are constant or have leaders which are smaller than v may also be inserted into L.

New triples with unmodified second and third component arise as follows:

g) An equation is inserted into L whose left hand side is the initial of a polynomial whose coefficients are reduced with respect to $M^= \cup N^=$ (Alg. 2.2.21, steps 5, 14, and 18, Alg. 2.2.25, step 6, Alg. 2.2.27, step 11, or Alg. 2.2.30, step 10). A similar argument about the insertion of equations into M as given in c) applies in this case (if the left hand side is not constant).

h) An inequation $p_2 \neq 0$ in L is replaced with finitely many inequations whose left hand sides are constant, but non-zero, or have leaders which are smaller than v (Alg. 2.2.25, step 14).

Every new triple which is inserted into Q arises in exactly one of the cases a)–h). Modifications of type c) and g) entail, after finitely many steps, modifications of type a) for each resulting triple and the creation of a new triple as in g). In this way only finitely many triples are generated because the vector $(d_1, \ldots, d_n)$ defined by

$$d_i := \begin{cases} \deg_v(p), & \text{if } M \cup N \text{ contains the equation } p = 0 \text{ with leader } v, \\ \infty, & \text{if } M \cup N \text{ contains no equation with leader } v, \end{cases}$$

where v is the i-th smallest variable with respect to $>$, decreases with respect to the lexicographical ordering as a result of a) and also as an indirect result of c) or g). Moreover, the leader of left hand sides of equations dealt with in g) decreases with respect to $>$. We claim that modifications of type b), d), e), f), and h) can be repeated (in any order) only finitely many times before a modification of type a) is applied or the algorithm stops. If an inequation in $M \cup N$ is replaced with an inequation with the same leader v, then the new inequation has smaller degree in v (cf. b)). This shows that a sequence of modifications as in the assertion contains types b) and f) only finitely many times. Modifications of the remaining types either replace two inequations with the same leader v with one inequation with leader v in $L \cup M \cup N$ (cf. d)) or remove one inequation from $L \cup M \cup N$ (cf. e) and h)), besides possibly inserting finitely many inequations into L whose left hand sides are constant or have smaller leader than the left hand side(s) of the removed inequation(s). Since no infinite sequence of non-constant polynomials exists in which each polynomial is followed by one with smaller leader, after finitely many steps either L and M will be empty or the element which is chosen in step 2 in Algorithm 2.2.21 will be an equation. Termination of Algorithm 2.2.20 now follows from the fact that modifications of type b), d), e), f), and h) do not change the vector $(d_1, \ldots, d_n)$. $\quad\sqcap$

Remark 2.2.33. In order to prevent a large growth of expressions and to simplify the final result, two strategies should be included at certain stages of the above algorithms. The left hand side of each equation and inequation should be divided by its numerical content, i.e., to obtain a primitive (multivariate) polynomial. Moreover, and also more generally, if all coefficients of the left hand side of an equation $p = 0$ or inequation $p \neq 0$ are divisible by a non-trivial factor r of the left hand side of an inequation $q \neq 0$, then they should be replaced with their quotients by r. In particular, it is worthwhile to apply this simplification, if possible, after $\text{init}(p) \neq 0$ has been inserted into the first component of a triple (cf. steps 4, 13, and 17 in Alg. 2.2.21).

For instance, the following two simple algebraic systems over $\mathbb{Q}[x,y]$ are equivalent, where $x > y$.

$$\boxed{\begin{array}{c} 2y\underline{x}+4y^2 = 0 \\[2mm] 2y \neq 0 \end{array}} \qquad \Longleftrightarrow \qquad \boxed{\begin{array}{c} \underline{x}+2y = 0 \\[2mm] y \neq 0 \end{array}}$$

In Algorithms 2.2.25 (*GCDSplit*), 2.2.27 (*LCMSplit*), and 2.2.30 (*SquarefreeSplit*) it is not specified in step 9, 7, and 6, respectively, which power a_i of $\mathrm{init}(r_{i+1})$ should be chosen. The power with exponent $\deg_v(r_i) - \deg_v(r_{i+1}) + 1$ allows a polynomial division without fractions in any case because the polynomial division involves at most $\deg_v(r_i) - \deg_v(r_{i+1}) + 1$ subtractions, but a proper divisor of this power may allow this as well for a particular pair r_i, r_{i+1} of polynomials. Using subresultant polynomial remainder sequences (cf., e.g., [Mis93]) is a considerable improvement (cf. also [BGL$^+$12, Sect. 2]).

Furthermore, in order to avoid repeated computations, for each equation and inequation information about whether the initial and discriminant of its left hand side are ensured not to vanish on the solution set of the algebraic system should be recorded. If the equation or inequation is inserted into M and its initial is known not to vanish, the insertion of $\mathrm{init}(p) \neq 0$ can be neglected in steps 4, 13, and 17 in Algorithm 2.2.21 (*ProcessInitial*) and step 5, 14, or 18, respectively, can be skipped. Similarly, in Algorithm 2.2.23 (*ProcessDiscriminant*) step 5 or 7 can be applied to an equation or inequation, respectively, which is chosen in step 2 and which is known to have non-vanishing discriminant.

We demonstrate Algorithm 2.2.20 (*AlgebraicThomasDecomposition*) on two examples.

Example 2.2.34. We revisit Example 2.2.12, p. 65, where $R = \mathbb{Q}[x,a,b,c]$ and the total ordering $>$ on the set of variables is given by $x > c > b > a$. In step 1 we have

$$Q = \{(\{ax^2+bx+c = 0\}, \emptyset, \emptyset)\}.$$

Steps 4 and 5 in Algorithm 2.2.21 (*ProcessInitial*) insert two triples into Q whose solution sets form a partition of the solution set of the initial triple:

$$Q = \{(\{a = 0, ax^2+bx+c = 0\}, \emptyset, \emptyset), (\{a \neq 0\}, \{ax^2+bx+c = 0\}, \emptyset)\}.$$

The first triple in this enumeration is dealt with by Algorithm 2.2.21, which moves the equation $a = 0$ to the second component. We omit both the inequation $1 \neq 0$ and the inconsistent algebraic system containing the equation $1 = 0$, which arise from the splitting in steps 4 and 5. Similarly, the inequation in the second triple is moved to the second component, and we omit inequations with constant left hand sides and inconsistent systems here:

$$Q = \{(\{ax^2+bx+c = 0\}, \{a = 0\}, \emptyset), (\emptyset, \{a \neq 0, ax^2+bx+c = 0\}, \emptyset)\}.$$

The left hand side of the first equation in the first triple is replaced with $bx+c$ by Algorithm 2.2.17 (*Reduce*), and steps 4 and 5 in Algorithm 2.2.21 split this triple according to the vanishing or non-vanishing of the initial of the modified equation. Algorithm 2.2.23 (*ProcessDiscriminant*) is applied to the second triple, which moves the inequation to the third component in step 7:

$$Q = \{(\{b=0, bx+c=0\}, \{a=0\}, \emptyset), (\{b\neq0\}, \{a=0, bx+c=0\}, \emptyset),$$
$$(\emptyset, \{ax^2+bx+c=0\}, \{a\neq0\})\}.$$

The equation $b = 0$ in the first triple is moved to the second component and a subsequent reduction replaces $bx+c$ with c. The inequation $b \neq 0$ in the second triple is moved to the second component. Steps 15 and 16 in Algorithm 2.2.30 split the third triple and add the inequation $4ac-b^2 \neq 0$ and the equation $4ac-b^2 = 0$, respectively:

$$Q = \{(\{c=0\}, \{a=0, b=0\}, \emptyset), (\emptyset, \{a=0, b\neq0, bx+c=0\}, \emptyset),$$
$$(\{4ac-b^2\neq0\}, \emptyset, \{a\neq0, ax^2+bx+c=0\}), \qquad\qquad (2.34)$$
$$(\{4ac-b^2=0, 2ax+b=0\}, \emptyset, \{a\neq0\})\}.$$

The equation $c = 0$ in the first triple is moved to the second component and subsequently all three equations are moved to the third component. Similarly, all elements of the second component of the second triple are moved to the third component in steps 5 and 7 in Algorithm 2.2.23. These two triples give rise to the following simple algebraic systems (cf. also the end of Ex. 2.2.12):

$b\underline{x}+c = 0$	
	$c = 0$
$b \neq 0$	$b = 0$
$a = 0$	$a = 0$

Algorithm 2.2.21 (*ProcessInitial*) splits the third triple in (2.34) according to the vanishing or non-vanishing initial of $4ac-b^2$ (steps 17 and 18):

$$(\{a\neq0\}, \{4ac-b^2\neq0\}, \{a\neq0, ax^2+bx+c=0\}),$$
$$(\{a=0, 4ac-b^2\neq0\}, \emptyset, \{a\neq0, ax^2+bx+c=0\}).$$

The equation $a = 0$ in the first component of the second triple is moved to the second component and the inequation $a \neq 0$ from the third to the first one. A subsequent reduction shows that this triple defines an inconsistent algebraic system.

The first triple is dealt with by applying Algorithm 2.2.27 (*LCMSplit*) to the pair of inequations $a \neq 0$, $a \neq 0$. Step 15 produces the triple

$$(\emptyset, \{a\neq0, 4ac-b^2\neq0\}, \{ax^2+bx+c=0\}),$$

which yields the simple system

$$
\boxed{
\begin{aligned}
&a\underline{x}^2 + b\underline{x} + c = 0 \\
&4a\underline{c} - b^2 \neq 0 \\[2em]
&a \neq 0
\end{aligned}
}
$$

after the inequations have been moved from the second to the third component by Algorithm 2.2.23 (*ProcessDiscriminant*).

The fourth triple in (2.34) is split into two triples by steps 4 and 5 in Algorithm 2.2.21 (*ProcessInitial*):

$$
\begin{aligned}
&(\{4a \neq 0, 2ax + b = 0\}, \{4ac - b^2 = 0\}, \{a \neq 0\}), \\
&(\{4a = 0, 4ac - b^2 = 0, 2ax + b = 0\}, \emptyset, \{a \neq 0\}).
\end{aligned}
\tag{2.35}
$$

Again, a reduction reveals that the second triple has an empty solution set. After another application of Algorithm 2.2.27 (*LCMSplit*), we obtain the remaining simple algebraic system of the Thomas decomposition (cf. also the end of Ex. 2.2.12):

$$
\boxed{
\begin{aligned}
&2a\underline{x} + b = 0 \\
&4a\underline{c} - b^2 = 0 \\[2em]
&a \neq 0
\end{aligned}
}
$$

We give an outline of a computation of a Thomas decomposition which is a little bit more involved. Advantage is taken of simplifications as described in Remark 2.2.33.

Example 2.2.35. Let $R = \mathbb{Q}[x, y, z]$. The *Steiner quartic surface* (cf., e.g., [Bak10, p. 221]) is defined by the equation

$$
x^2 y^2 + x^2 z^2 + y^2 z^2 - xyz = 0.
\tag{2.36}
$$

We choose the total ordering $x > y > z$ on the set of variables.

Algorithm 2.2.20 (*AlgebraicThomasDecomposition*) starts by splitting the original algebraic system according to vanishing or non-vanishing of the initial $y^2 + z^2$. In the former case the original equation is reduced to $zyx + z^4 = 0$. The updated system is split again according to vanishing or non-vanishing of the initial zy. Again, in the former case the analogous case distinction yields, after application of Algorithm 2.2.30 (*SquarefreeSplit*), the simple system

$$
\{z = 0, y = 0\}
$$

and produces only inconsistent algebraic systems otherwise. In case of the algebraic system containing the inequation $zy \neq 0$, Algorithm 2.2.25 (*GCDSplit*) is applied to $p_1 = y^2 + z^2$ and $p_2 = zy$. Three new algebraic systems are generated, one in step 6 containing the equation $z = 0$, which is inconsistent, one in step 13 containing $z \neq 0$ and $z^4 = 0$, which is also inconsistent, and one in step 14, which after applying Algorithms 2.2.27 (*LCMSplit*) and 2.2.30 (*SquarefreeSplit*) yields the simple system

$$\{z \neq 0,\ \underline{y}^2 + z^2 = 0,\ y\underline{x} + z^3 = 0\}.$$

The branch emerging from the case $y^2 + z^2 \neq 0$ remains to be dealt with. Application of *SquarefreeSplit* to this inequation splits the algebraic system into one containing $z^2 = 0$ and $y \neq 0$ and another one containing $z^2 \neq 0$ and $y^2 + z^2 \neq 0$. In the former case (2.36) is reduced to $y^2 x^2 - zyx = 0$, which simplifies to $yx^2 - zx = 0$ because of $y \neq 0$. After computing the square-free part of z^2, the reduced form of the simplified equation modulo z, and the square-free part of x^2, we obtain the simple system

$$\{z = 0,\ y \neq 0,\ x = 0\}.$$

In the latter case *SquarefreeSplit* is applied to $z^2 \neq 0$ and then to (2.36). This generates two new branches to which the equation or inequation with left hand side

$$4z^2 y^4 + (4z^4 - z^2) y^2 \tag{2.37}$$

is added, respectively. The first branch, where the original equation is replaced with

$$2(y^2 + z^2) x - zy = 0, \tag{2.38}$$

produces two simple systems. The essential steps amount to applying *SquarefreeSplit* to $z^2 \neq 0$ and to the equation with left hand side (2.37). Thus, the cases of vanishing or non-vanishing of $4z^2 - 1$ and in the latter case that of $16z^4 - 8z^2 + 1$ are investigated. If $4z^2 - 1 = 0$ is imposed, then (2.37) is reduced to y^4 and (2.38) is reduced to $(4y^2 + 1)x - 2zy = 0$. Applying *GCDSplit* to $4z^2 - 1 = 0$ and $z \neq 0$, *SquarefreeSplit* to $y^4 = 0$, and reduction modulo y yields the simple system

$$\{4z^2 - 1 = 0,\ y = 0,\ x = 0\}.$$

Since $4z^2 - 1$ divides $16z^4 - 8z^2 + 1$, the algebraic system containing $4z^2 - 1 \neq 0$ and $16z^4 - 8z^2 + 1 = 0$ does not contribute to the Thomas decomposition. In the case of the algebraic system containing $4z^2 - 1 \neq 0$ and $16z^4 - 8z^2 \mid 1 \neq 0$ the equation with left hand side (2.37) is replaced with $4y^3 + (4z^2 - 1)y = 0$. After application of *SquarefreeSplit* to the least common multiple $16z^5 - 8z^3 + z \neq 0$ of the inequations with leader z which have been encountered before, we obtain the simple system

$$\{4z^3 - z \neq 0,\ 4\underline{y}^3 + (4z^2 - 1)\underline{y} = 0,\ 2(y^2 + z^2)\underline{x} - zy = 0\}.$$

The branch addressing the inequation with left hand side (2.37) yields the rest of the Thomas decomposition. It is treated by first applying *LCMSplit* to this inequation and $y^2 + z^2 \neq 0$, which essentially reveals that, in presence of the inequation $z \neq 0$, the least common multiple of their left hand sides is

$$4z^2 y^6 + (8z^4 - z^2) y^4 + (4z^6 - z^4) y^2.$$

Again, due to the inequation $z \neq 0$, only the case of non-vanishing initial $4z^2$ is relevant. Application of *SquarefreeSplit* to the simplified inequation

$$4y^6 + (8z^2 - 1) y^4 + (4z^4 - z^2) y^2 \neq 0 \qquad (2.39)$$

produces five algebraic systems. One of them contains $8z^2 - 1 = 0$ and the others contain the corresponding inequation. Among the latter systems one contains $16z^4 - 4z^2 + 1 = 0$ and the others contain the complementary condition. The equation $4z^4 - z^2 = 0$ is imposed in exactly one of the complementary systems and the two remaining ones incorporate the inequation with the same left hand side. One of these two contains $z^4 (2z+1)^2 (2z-1)^2 (8z^2 - 1) = 0$ and the other one the corresponding inequation.

In the very first case (2.36) is reduced to $(8y^2 + 1) x^2 - 8zyx + y^2 = 0$ and (2.39) is reduced to $64y^6 - y^2 \neq 0$. After applying *SquarefreeSplit* to $8z^2 - 1 = 0$ and to $64y^6 - y^2 \neq 0$, we obtain the simple system

$$\{8z^2 - 1 = 0, 64y^5 - y \neq 0, (8y^2 + 1)\underline{x}^2 - 8zy\underline{x} + y^2 = 0\}.$$

The algebraic system containing $16z^4 - 4z^2 + 1 = 0$ leads after the application of *GCDSplit* to this equation and the least common multiple of $8z^2 - 1 \neq 0$ and $z \neq 0$ and the application of *SquarefreeSplit* to the equation $16z^4 - 4z^2 + 1 = 0$ and to the inequation $16y^6 + (32z^2 - 4) y^4 - y^2 \neq 0$ to the simple system

$$\{16z^4 - 4z^2 + 1 = 0, 16\underline{y}^5 + 4(8z^2 - 1)\underline{y}^3 - \underline{y} \neq 0, (y^2 + z^2)\underline{x}^2 - zy\underline{x} + z^2 y^2 = 0\}.$$

The third of the five systems mentioned previously is dealt with by first applying *LCMSplit* to $8z^2 - 1 \neq 0$ and $z \neq 0$. The equation $4z^4 + z^2 = 0$ is used to reduce (2.39) to $4y^6 + (8z^2 - 1) y^4 \neq 0$. Then *GCDSplit* is applied to $4z^4 + z^2 = 0$ and $8z^3 - z \neq 0$, which replaces the equation with $4z^3 - z = 0$, and the inequation is reduced to $z \neq 0$. Subsequently, *GCDSplit* replaces the equation with $4z^2 - 1 = 0$ and removes $z \neq 0$. Then (2.36) is reduced to $(4y^2 + 1) x^2 - 4zyx + y^2 = 0$ and $4y^6 + (8z^2 - 1) y^4 \neq 0$ to $4y^6 + y^4 \neq 0$. After application of *SquarefreeSplit*, we get

$$\{4z^2 - 1 = 0, 4y^3 + y \neq 0, (4y^2 + 1)\underline{x}^2 - 4zy\underline{x} + y^2 = 0\}.$$

Finally, the inequations with leader z in the last of the five systems are combined by a series of calls of *LCMSplit* resulting in the inequation

$$z^4 (2z+1)^2 (2z-1)^2 (8z^2 - 1) (16z^4 - 4z^2 + 1) \neq 0. \qquad (2.40)$$

The inequation (2.39) had been replaced by *SquarefreeSplit* with an inequation whose left hand side has degree five in y. After imposing the condition that its initial does not vanish, this inequation simplifies to $4y^5 + (8z^2 - 1)y^3 + (4z^4 - z^2)y \neq 0$. The square-free part of the inequation with leader z is determined next. Then *SquarefreeSplit* is applied to the inequation with leader y, distinguishing the cases of vanishing or non-vanishing of $8z^2 - 1$, of $32z^4 - 8z^2 + 3$, of $4z^4 - z^2$, and of $z^4(2z+1)^2(2z-1)^2(8z^2-1)$. After further applications of *GCDSplit*, *LCMSplit*, and *SquarefreeSplit*, the algebraic system containing $32z^4 - 8z^2 + 3 = 0$ and the one containing $z^4(2z+1)^2(2z-1)^2(8z^2-1) \neq 0$ each yield one simple system.

We conclude by displaying the constructed Thomas decomposition of (2.36), listing the simple systems in order of increasing dimension of their solution sets.

$(2z+1)(2z-1) = 0$	$z(2z+1)(2z-1) \neq 0$
$y = 0$	$\underline{y}(4\underline{y}^2 + 4z^2 - 1) = 0$
$x = 0$	$2(y^2 + z^2)\underline{x} - zy = 0$

$z = 0$	$z = 0$	$8z^2 - 1 = 0$
$y = 0$	$y \neq 0$	$\underline{y}(8\underline{y}^2 + 1)(8\underline{y}^2 - 1) \neq 0$
	$x = 0$	$(8y^2 + 1)\underline{x}^2 - 8zy\underline{x} + y^2 = 0$

$(2z+1)(2z-1) = 0$	$16z^4 - 4z^2 + 1 = 0$
$\underline{y}(4\underline{y}^2 + 1) \neq 0$	$\underline{y}(16\underline{y}^4 + 4(8z^2 - 1)\underline{y}^2 - 1) \neq 0$
$(4y^2 + 1)\underline{x}^2 - 4zy\underline{x} + y^2 = 0$	$(y^2 + z^2)\underline{x}^2 - zy\underline{x} + z^2y^2 = 0$

$z \neq 0$	$32z^4 - 8z^2 + 3 = 0$
$\underline{y}^2 + z^2 = 0$	$\underline{y}(32\underline{y}^4 + 8(8z^2 - 1)\underline{y}^2 - 3) \neq 0$
$y\underline{x} + z^3 = 0$	$(y^2 + z^2)\underline{x}^2 - zy\underline{x} + z^2y^2 = 0$

$z(2z+1)(2z-1)(8z^2-1)(16z^4-4z^2+1)(32z^4-8z^2+3) \neq 0$
$\underline{y}(4\underline{y}^4 + (8z^2-1)\underline{y}^2 + z^2(4z^2-1)) \neq 0$ (2.41)
$(y^2 + z^2)\underline{x}^2 - zy\underline{x} + z^2y^2 = 0$

The simple system (2.41) is the generic simple system for the prime ideal of R generated by (2.36) as discussed in Subsect. 2.2.3 (cf. also Ex. 2.2.68, p. 107).

Note that some unnecessary case distinctions are avoided when using subresultants. For more details about this technique, we refer to [BGL$^+$12].

2.2.2 Simple Differential Systems

This subsection gives a modern description of the method of J. M. Thomas of decomposing systems of finitely many partial differential equations and inequations into finitely many so-called simple systems. The set of solutions of the given system is thereby partitioned into the solution sets of the simple systems, and using the simple systems, e.g., an effective membership test for the radical differential ideal defined by the given system is made possible. We restrict our attention to analytic solutions on connected open subsets of $\mathbb{C}^n$. Before stating the definition of a simple differential system, we elaborate on certain formal manipulations of differential polynomials, on which Thomas' algorithm is based.

Let $R := K\{u_1,\ldots,u_m\}$ be the differential polynomial ring in u_1, ..., u_m with commuting derivations ∂_1, ..., ∂_n, where K is a computable differential field of characteristic zero (with derivations $\partial_1|_K, \ldots, \partial_n|_K$). We define the set

$$\Delta := \{\partial_1,\ldots,\partial_n\}$$

and the (commutative) monoid $\text{Mon}(\Delta)$ consisting of the monomials in $\partial_1, \ldots, \partial_n$. For $\theta \in \text{Mon}(\Delta)$ we denote by $\deg(\theta)$ the total degree of the monomial θ. If L is a subset of R, then $\langle L \rangle$ is defined to be the differential ideal of R generated by L.

In what follows, we fix a ranking $>$ on $K\{u_1,\ldots,u_m\}$ (i.e., a total ordering on

$$\text{Mon}(\Delta)u := \{\,(u_k)_J \mid 1 \leq k \leq m, J \in (\mathbb{Z}_{\geq 0})^n\,\} \tag{2.42}$$

which respects the action of the derivations and which is a well-ordering; cf. Subsect. A.3.2, p. 249, for more details). Then for every non-constant differential polynomial $p \in R - K$, the *leader* $\text{ld}(p)$, the *initial* $\text{init}(p)$, and the *discriminant* $\text{disc}(p)$ are defined as in Definition 2.2.1 by considering p as polynomial in the finitely many indeterminates $(u_k)_J$ which occur in it, totally ordered by the ranking $>$. For any subset $P \subseteq R$ we define

$$\text{ld}(P) := \{\,\text{ld}(p) \mid p \in P, p \notin K\,\}.$$

Remark 2.2.36. For any non-constant differential polynomial $p \in R - K$ and any $i \in \{1,\ldots,n\}$, the defining properties of a ranking imply that the leader of $\partial_i\, p$ equals $\partial_i\,\text{ld}(p)$. In fact, $\partial_i\, p$ is a polynomial of degree one in $\partial_i\,\text{ld}(p)$, i.e., every proper partial derivative of a differential polynomial is *quasi-linear*. This observation implies important relations among differential polynomials in terms of polynomial division, which we discuss next.

Definition 2.2.37. Let $p \in R - K$. The *separant* of p is defined to be the differential polynomial

$$\text{sep}(p) := \frac{\partial p}{\partial\,\text{ld}(p)},$$

i.e., the formal partial derivative of p with respect to its leader. It is the coefficient of the leader $\partial_i \operatorname{ld}(p)$ of the derivative $\partial_i p$ for any $i \in \{1, \ldots, n\}$.

Remark 2.2.38. Let $p_1 \in R$ and $p_2 \in R - K$. Proper partial derivatives of the leader of p_2 can be eliminated from p_1 by applying Euclidean pseudo-division in an appropriate way, using the fact that any proper derivative of p_2 is quasi-linear (cf. Rem. A.3.6, p. 250, for details). In order to avoid to deal with a partial derivative of $\operatorname{ld}(p_2)$ twice, these derivatives should be processed in decreasing order with respect to the ranking. Apart from this *differential reduction*, which multiplies p_1 by the separant of p_2 to realize the desired cancelation, Euclidean pseudo-division modulo p_2 eliminates powers of $\operatorname{ld}(p_2)$ in p_1 whose exponents are greater than or equal to the degree of p_2 in $\operatorname{ld}(p_2)$. This *algebraic reduction* multiplies p_1 by the initial of p_2. In both cases, the computation is performed in such a way that no fractions of non-constant differential polynomials are involved.

Remark 2.2.39. Using the algebraic reduction technique from the previous remark, we apply the algebraic version of Thomas' algorithm (cf. Rem. 2.2.11, Alg. 2.2.20) to a finite differential system S over R, i.e., a finite set of equations and inequations whose left hand sides are elements of R and whose right hand sides are zero. This set is viewed as an algebraic system in the finitely many indeterminates $(u_k)_J$ which occur in it. Let us consider (m-tuples of) F-valued analytic functions as candidates for solutions, where F is an extension field of the subfield of constants of K. A solution of the system consists of one analytic function f_k of $z_1, \ldots, z_n$ for each differential indeterminate u_k such that every equation and inequation of the system is satisfied upon substitution of $\frac{\partial^{|J|} f_k}{\partial z^J}$ for $(u_k)_J$, $J \in (\mathbb{Z}_{\geq 0})^n$. Taylor expansion translates the problem into algebraic equations and inequations for the Taylor coefficients of a solution. It is therefore convenient to assume that F is algebraically closed. The defining properties of a simple algebraic system (cf. Def. 2.2.4) ensure that a sequence of Taylor coefficients defining a solution of the *algebraic* system corresponding to $S_{<(u_k)_J}$, i.e., the equations and inequations in S with leader smaller than $(u_k)_J$, can be adjusted to be a sequence defining a solution of the algebraic system corresponding to $S_{\leq(u_k)_J}$. However, differential consequences of S must also be taken into account, which may again be equations with a smaller leader (cf. also the discussion leading to Remark 2.1.67).

Recall that Thomas' algorithm splits systems according to vanishing or non-vanishing initials so that pseudo-divisions do not change the total solution set. It also splits systems according to the possible square-free parts until every left hand side in each system is a square-free polynomial (for every possible specialization). Let us assume that a differential system is a simple algebraic system in the above sense. Then the discriminant of each equation is non-zero when evaluated at any solution of the system. Note that in the differential algebra context the discriminant is essentially the resultant of the differential polynomial and its separant (cf. Def. 2.2.1). Moreover, the initial of any partial derivative of a differential polynomial is equal to the separant. Therefore, pseudo-division modulo partial derivatives of equations of a system that is simple in the above sense transforms a differential polynomial into an equivalent one. (For comments about singular solutions of a differential system, cf. Remark 2.2.59.)

We adopt the following piece of notation from the case of algebraic systems. For any differential system

$$S = \{p_i = 0, q_j \neq 0 \mid i \in I, j \in J\}, \qquad p_i, q_j \in R,$$

where I and J are index sets, we denote by

$$S^= := \{p_i \mid i \in I\}, \qquad S^{\neq} := \{q_j \mid j \in J\}$$

the set of left hand sides of equations and inequations in S, respectively.

Given a set of differential polynomials which are the left hand sides of the equations of a simple algebraic system, the following algorithm performs differential reductions in order to eliminate leaders which are proper partial derivatives of other leaders in the system. This is a preparatory step for computing a cone decomposition of the multiple-closed set (with respect to the action of $\mathrm{Mon}(\Delta)$) generated by the leaders, which is discussed afterwards.

Algorithm 2.2.40 (*Auto-reduce* **for differential polynomials**).

Input: $L \subset R - K$ finite and a ranking $>$ on R such that $L = S^=$ for some finite differential system S which is simple as an algebraic system (in the finitely many indeterminates $(u_k)_J$ which occur in it, totally ordered by $>$)

Output: $a \in \{\textbf{true}, \textbf{false}\}$ and $L' \subset R - K$ finite such that

$$\langle L' \rangle : q^\infty = \langle L \rangle : q^\infty, \qquad q := \prod_{p \in L} \mathrm{sep}(p), \qquad (2.43)$$

and, in case $a = \textbf{true}$, there exists no $p_1, p_2 \in L'$, $p_1 \neq p_2$, such that we have $\mathrm{ld}(p_1) \in \mathrm{Mon}(\Delta)\mathrm{ld}(p_2)$

Algorithm:

1: $L' \leftarrow L$
2: **while** there exist $p_1, p_2 \in L'$, $p_1 \neq p_2$ and $\theta \in \mathrm{Mon}(\Delta)$ such that we have $\mathrm{ld}(p_1) = \theta\,\mathrm{ld}(p_2)$ **do**
3:     $L' \leftarrow L' - \{p_1\}$; $v \leftarrow \mathrm{ld}(p_1)$
4:     $r \leftarrow \mathrm{sep}(p_2) \cdot p_1 - \mathrm{init}(p_1) \cdot v^{d-1} \cdot \theta p_2$, where $d := \deg_v(p_1)$
5:     **if** $r \neq 0$ **then**
6:         **return** (**false**, $L' \cup \{r\}$)
7:     **end if**
8: **end while**
9: **return** (**true**, L')

Remarks 2.2.41. a) Since L is the set of left hand sides of equations in a simple algebraic system, we have $L \cap K = \emptyset$. For the same reason, L' is a triangular set with respect to the ranking $>$ in the first round of the loop and, while $r = 0$, also in later rounds. Therefore, $\deg(\theta) > 0$ holds inside the loop, and step 4 eliminates $\operatorname{ld}(p_1)$ from p_1 (cf. also Rem. A.3.6 a), p. 250). Since this process can be understood as replacing the largest term (possibly multiplied by a polynomial with smaller leader) with a sum of terms that are smaller with respect to $>$, and since $>$ is a well-ordering, the algorithm terminates. By Remark 2.2.39, $\operatorname{sep}(p_2)$ does not vanish when evaluated at any solution of the system. Hence, (2.43) holds. Correctness of the algorithm is clear.

b) Note that, once we have $r \neq 0$, the equality (2.43) still holds, but further reductions as in step 4 would not be guaranteed to respect the solution set (when $p_2 = r$ is chosen as a divisor). Therefore, $L' \cup \{r\}$ is returned immediately in this case with the intention that the algebraic consequences of this system are examined by the algebraic version of Thomas' algorithm, which also takes care of the initials and separants of the system.

c) For efficiency reasons it is desirable to find pseudo-remainders in step 4 with least possible leader with respect to the ranking (if not constant), because these lend themselves to be divisors of many other polynomials of the system. Therefore, p_2 and then p_1 should be chosen with least possible leaders[12] in step 2.

Remark 2.2.42. We apply the combinatorics of Janet division (cf. Subsect. 2.1.1) in the present context in order to construct a generating set (in an appropriate sense) of all differential polynomial consequences of a finite system of polynomial partial differential equations (ignoring for a moment necessary splittings of systems). Let $p_1, \ldots, p_s \in R - K$ be non-constant differential polynomials. The chosen ranking on R uniquely determines $\theta_1, \ldots, \theta_s \in \operatorname{Mon}(\Delta)$ and $k_1, \ldots, k_s \in \{1, \ldots, m\}$ such that

$$\operatorname{ld}(p_i) = \theta_i u_{k_i}, \qquad i = 1, \ldots, s.$$

We interpret $p_1, \ldots, p_s$ as left hand sides of PDEs. Then every partial derivative of each p_i is the left hand side of a consequence of the system. Therefore, for each $k \in \{1, \ldots, m\}$, the set of $\theta \in \operatorname{Mon}(\Delta)$ such that θu_k is the leader of an equation that is a consequence of the system is $\operatorname{Mon}(\Delta)$-multiple-closed. Hence, Δ serves as the set X of symbols referred to in Subsect. 2.1.1. We assume that a total ordering on Δ is chosen which is used by Algorithms 2.1.6 and 2.1.8 to construct Janet decompositions of multiple-closed sets of monomials in Δ and their complements, respectively. The choice of the ranking $>$ on R and the choice of the total ordering on Δ are independent, the former one singling out the leader of each non-constant differential polynomial, the latter one determining Janet decompositions. The symbol $>$ will continue to refer to the ranking on R.

---

[12] More information about the polynomials at hand should be taken into account to enhance this basic heuristic strategy, because it turns out that an implementation is slowed down drastically when polynomials get too large (measured in number of terms, say), so that it may be reasonable to trade compactness against rank.

Definition 2.2.43. Let $M \subseteq \mathrm{Mon}(\Delta)u$ (cf. (2.42)). For $k \in \{1,\dots,m\}$ we define $M_k := \{\theta \in \mathrm{Mon}(\Delta) \mid \theta u_k \in M\}$.

a) We call the set M *multiple-closed* if $M_1, \dots, M_m$ are $\mathrm{Mon}(\Delta)$-multiple-closed. A set $G \subseteq \mathrm{Mon}(\Delta)u$ such that $G_1, \dots, G_m$ are generating sets for $M_1, \dots, M_m$, respectively, where

$$G_k := \{\theta \in \mathrm{Mon}(\Delta) \mid \theta u_k \in G\},$$

is called a *generating set* for M. The multiple-closed set generated by G is denoted by

$$[G] := \mathrm{Mon}(\Delta)G = \bigcup_{k=1}^{m} \mathrm{Mon}(\Delta)G_k u_k.$$

b) Let M be multiple-closed. For $k = 1, \dots, m$, let

$$\{(\theta_1^{(k)}, \mu_1^{(k)}), \dots, (\theta_{t_k}^{(k)}, \mu_{t_k}^{(k)})\}$$

be a Janet decomposition of M_k (or of $\mathrm{Mon}(\Delta) - M_k$, cf. Def. 2.1.11, p. 15). Then

$$\bigcup_{k=1}^{m} \{(\theta_1^{(k)}u_k, \mu_1^{(k)}), \dots, (\theta_{t_k}^{(k)}u_k, \mu_{t_k}^{(k)})\}$$

is called a *Janet decomposition* of M (resp. of $\mathrm{Mon}(\Delta)u - M$). The *cones* of the Janet decomposition are given by $\mathrm{Mon}(\mu_i^{(k)})\theta_i^{(k)}u_k$, $i = 1, \dots, t_k$, $k = 1, \dots, m$. If the Janet decomposition is constructed from the generating set G for M, then we call the set of generators $\theta_i^{(k)}u_k$ of the cones the *Janet completion* of G.

For the rest of this section, we fix a total ordering on Δ such that the Janet completion of any set $G \subseteq \mathrm{Mon}(\Delta)u$ is uniquely defined.

Definition 2.2.44. Let $T = \{(p_1,\mu_1),\dots,(p_s,\mu_s)\}$, $p_i \in R - K$, $\mu_i \subseteq \Delta$, $i = 1,\dots,s$.

a) The set T is said to be *Janet complete* if

$$\{\mathrm{ld}(p_1),\mathrm{ld}(p_2),\dots,\mathrm{ld}(p_s)\}$$

equals its Janet completion and, for each $i \in \{1,\dots,s\}$, μ_i is the set of multiplicative variables of the cone with generator $\mathrm{ld}(p_i)$ in the Janet decomposition $\{(\mathrm{ld}(p_1),\mu_1),\dots,(\mathrm{ld}(p_s),\mu_s)\}$ of $[\mathrm{ld}(p_1),\dots,\mathrm{ld}(p_s)]$ (cf. Def. 2.2.43).

b) An element $r \in R$ is said to be *Janet reducible modulo T* if there exist a jet variable $v \in \mathrm{Mon}(\Delta)u$ and $(p,\mu) \in T$ such that v occurs in r and $v \in \mathrm{Mon}(\mu)\mathrm{ld}(p)$. In this case, (p,μ) is called a *Janet divisor* of r. If r is not Janet reducible modulo T, then r is also said to be *Janet reduced modulo T*.

The following algorithm applies differential and algebraic reductions to a given differential polynomial in such a way that the remainder of these pseudo-reductions is Janet reduced modulo a given Janet complete set.

Algorithm 2.2.45 (*Janet-reduce* **for differential polynomials**).

Input: $r \in R$, $T = \{(p_1, \mu_1), (p_2, \mu_2), \ldots, (p_s, \mu_s)\}$, and a ranking $>$ on R, where T is Janet complete (with respect to $>$, cf. Def. 2.2.44)

Output: $r' \in R$ and an element b of the multiplicatively closed set generated by $\bigcup_{i=1}^{s}\{\operatorname{init}(p_i), \operatorname{sep}(p_i)\} \cup \{1\}$ such that r' is Janet reduced modulo T, and such that $r' = r$, $b = 1$ if $T = \emptyset$, and $r' + \langle p_1, \ldots, p_s \rangle = b \cdot r + \langle p_1, \ldots, p_s \rangle$ otherwise

Algorithm:

1: $r' \leftarrow r$
2: $b \leftarrow 1$
3: **if** $r' \notin K$ **then**
4:      $v \leftarrow \operatorname{ld}(r')$
5:      **while** $r' \notin K$ and there exist $(p, \mu) \in T$ and $\theta \in \operatorname{Mon}(\mu)$ with $\deg(\theta) > 0$ such that $v = \theta \operatorname{ld}(p)$ **do**
6:          $r' \leftarrow \operatorname{sep}(p) \cdot r' - \operatorname{init}(r') \cdot v^{d-1} \cdot \theta p$, where $d := \deg_v(r')$
7:          $b \leftarrow \operatorname{sep}(p) \cdot b$
8:      **end while**
9:      **while** $r' \notin K$ and there exists $(p, \mu) \in T$ with $\operatorname{ld}(p) = v$, $\deg_v(r') \geq \deg_v(p)$ **do**
10:          $r' \leftarrow \operatorname{init}(p) \cdot r' - \operatorname{init}(r') \cdot v^{d-d'} \cdot p$, where $d := \deg_v(r')$ and $d' := \deg_v(p)$
11:          $b \leftarrow \operatorname{init}(p) \cdot b$
12:      **end while**
13:      **while** there exists a coefficient c of r' (as a polynomial in v) which is not Janet reduced modulo T **do**
14:          $(r'', b') \leftarrow$ *Janet-reduce*$(c, T, >)$
15:          replace the coefficient $b' \cdot c$ in $b' \cdot r'$ with r'' and replace r' with this result
16:          $b \leftarrow b' \cdot b$
17:      **end while**
18: **end if**
19: **return** (r', b)

The following remarks are analogous to Remarks 2.2.18 for the algebraic case.

Remarks 2.2.46. a) Algorithm 2.2.45 terminates because for the recursive calls in step 14 each coefficient c of r' is either constant or has a leader which is smaller than v with respect to $>$, which is a well-ordering, and the properties of r' which are achieved by steps 5–12 are retained by the recursion. The uniqueness of the Janet divisor of a jet variable implies that the result of Algorithm 2.2.45 is uniquely determined for the given input, so that remarks similar to Remark 2.1.39 a), p. 28, can be made. However, as opposed to Algorithm 2.1.38, the differential and the algebraic reductions are pseudo-reductions in general.

b) Let $r_1, r_2 \in R$ and T be as in the input of Algorithm 2.2.45, and define q to be the product of all $\mathrm{init}(p_i)$ and all $\mathrm{sep}(p_i)$, $i = 1, \ldots, s$. In general, the equality

$$r_1 + \langle p_1, \ldots, p_s \rangle : q^\infty = r_2 + \langle p_1, \ldots, p_s \rangle : q^\infty$$

does not imply that the results of applying *Janet-reduce* to r_1 and r_2, respectively, are equal. However, later on (cf. Prop. 2.2.50) it is shown that, if T is defined by the subset $S^=$ of equations of a simple differential system S (cf. Def. 2.2.49), then the result of applying *Janet-reduce* to $r \in R$ is zero if and only if we have $r \in \langle p_1, \ldots, p_s \rangle : q^\infty$. In this case, we refer to the first component r' of the output of *Janet-reduce* applied to r as the *Janet normal form of r modulo T*. In order to simplify notation, we denote the result r' of *Janet-reduce* applied to r, T, $>$ by $\mathrm{NF}(r, T, >)$, even if T does not have the properties mentioned above.

The polynomial q defined in part b) of the previous remarks will play an important role in what follows.

Remark 2.2.47. The method of J. M. Thomas for treating a differential system S applies the algebraic decomposition technique (cf. Rem. 2.2.11, Alg. 2.2.20), which in general causes a splitting of the system. Restricting attention to one of these simple systems and assuming that this system is not split further, an ascending chain of multiple-closed subsets of $\mathrm{Mon}(\Delta)u$ is produced, which terminates by Lemma 2.1.2, p. 10. The current multiple-closed set is generated by the leaders of the equations of the differential system, from which dispensable equations have been removed by Algorithm 2.2.40 (*Auto-reduce*). If the latter algorithm finds a new differential consequence, Thomas' algorithm for algebraic systems is applied first to the augmented system, and this process is iterated until Algorithm 2.2.40 (*Auto-reduce*) confirms that the leaders of the differential equations form the minimal generating set for the multiple-closed set they generate.

Let G be the set of left hand sides of these equations. The Janet decomposition of the multiple-closed set $[\mathrm{ld}(G)]$ is constructed as described in Subsect. 2.1.1. To this end, Algorithm 2.1.6 (*Decompose*), p. 11, is applied, but in a slightly modified way (cf. also Rem. 2.1.41, p. 28, for the corresponding adaptation of *Decompose* for Janet's algorithm). This algorithm is applied directly to G, in the sense that its run is determined by $\mathrm{ld}(g)$ for $g \in G$, but the application of $d \in \Delta$ to $\mathrm{ld}(g)$ is replaced with the application of the derivation d to g. The result $\{(p_1, \mu_1), \ldots, (p_s, \mu_s)\}$ consists of pairs of a non-constant differential polynomial p_i in R and a subset μ_i of Δ. The elements of μ_i (of $\Delta - \mu_i$) are called the *(non-) admissible derivations* for $p_i = 0$. In the differential version of Thomas' algorithm, *Decompose* will be applied in this adapted version.

Using the Janet decomposition, Thomas' algorithm tries to find new differential consequences by applying derivations to an equation for which they are non-admissible and computing the Janet reductions of these derivatives.

Definition 2.2.48. A Janet complete set $T = \{(p_1, \mu_1), \ldots, (p_s, \mu_s)\}$ (as in Definition 2.2.44 a)) is said to be *passive*, if

$$\mathrm{NF}(d\, p_i, T, >) = 0 \qquad \text{for all} \quad d \in \overline{\mu_i} = \Delta - \mu_i, \quad i = 1, \ldots, s$$

(where we recall that $\mathrm{NF}(r, T, >)$ is the result of Algorithm 2.2.45 (*Janet-reduce*) applied to r, T, $>$). A system of partial differential equations $\{p_1 = 0, \ldots, p_s = 0\}$, where $p_i \in R - K$, $i = 1, \ldots, s$, is said to be *passive* if the Janet completion of $\{p_1, \ldots, p_s\}$ (using the fixed ranking on R and the fixed total ordering on Δ) and the corresponding sets of admissible derivations define a passive Janet complete set.

The result of Thomas' algorithm for differential systems is a finite set of differential systems which are simple, a notion that is defined next.

Definition 2.2.49. Let a ranking $>$ on $K\{u_1, \ldots, u_m\}$ and a total ordering on the set $\Delta = \{\partial_1, \ldots, \partial_n\}$ be fixed. A system S of polynomial partial differential equations and inequations

$$p_1 = 0, \quad \ldots, \quad p_s = 0, \quad q_1 \neq 0, \quad \ldots, \quad q_t \neq 0,$$

where $p_1, \ldots, p_s, q_1, \ldots, q_t \in R - K$, $s, t \in \mathbb{Z}_{\geq 0}$, is said to be *simple* if the following three conditions are satisfied.

a) The system S is simple as an algebraic system (in the finitely many indeterminates $(u_k)_J$ which occur in it, totally ordered by the ranking $>$).
b) The system of partial differential equations $\{p_1 = 0, \ldots, p_s = 0\}$ is passive.
c) The left hand sides of the inequations $q_1 \neq 0, \ldots, q_t \neq 0$ are Janet reduced modulo the left hand sides of the passive system $\{p_1 = 0, \ldots, p_s = 0\}$.

Janet division associates (according to the chosen ordering of Δ) with each equation $p_i = 0$ a set $\mu_i \subseteq \Delta$ of *admissible derivations* in the sense that the monomials in the derivations in μ_i are those elements of $\mathrm{Mon}(\Delta)$ which are potentially applied to p_i for Janet reduction of some differential polynomial. The complement $\overline{\mu_i}$ of μ_i in Δ consists of the *non-admissible derivations* for $p_i = 0$. We refer to $\theta\, p_i$, where $\theta \in \mathrm{Mon}(\mu_i)$, as an *admissible derivative* of p_i.

The next proposition gives a description in terms of a radical differential ideal of all differential equations that are consequences of a simple differential system. Janet reduction modulo the simple differential system decides membership of a differential polynomial to the corresponding radical differential ideal. The statements are analogous to Proposition 2.2.7 in the algebraic case, which is used in the proof.

Proposition 2.2.50. *Let a simple differential system S over R be given by*

$$p_1 = 0, \quad \ldots, \quad p_s = 0, \quad q_1 \neq 0, \quad \ldots, \quad q_t \neq 0.$$

Let $E := \langle P \rangle$ be the differential ideal of R generated by $P := \{p_1, \ldots, p_s\}$ and define the product q of the initials and separants of all elements of P. Then

$$E : q^{\infty} := \{p \in R \mid q^r \cdot p \in E \text{ for some } r \in \mathbb{Z}_{\geq 0}\}$$

is a radical differential ideal. A differential polynomial $p \in R$ is an element of $E : q^{\infty}$ if and only if the Janet normal form of p modulo $p_1, \ldots, p_s$ is zero.

Remark 2.2.51. Similarly to the algebraic case (cf. Rem. 2.2.8, p. 62), the assertion of Proposition 2.2.50 does not depend on the inequations $q_1 \neq 0, \ldots, q_t \neq 0$ because it describes the radical differential ideal of all differential polynomials in R vanishing on the analytic solutions of the given simple differential system, which is not influenced by inequations (cf. also p. 98).

Proof (of Proposition 2.2.50). By definition of the saturation $E : q^{\infty}$, every element $p \in R$ for which Algorithm 2.2.45 (*Janet-reduce*) yields pseudo-remainder zero is an element of $E : q^{\infty}$. Conversely, let $p \in E : q^{\infty}$ be arbitrary. Then there exist $r \in \mathbb{Z}_{\geq 0}$, $k_1, \ldots, k_s \in \mathbb{Z}_{\geq 0}$, and $c_{i,j} \in R - \{0\}$, $\theta_{i,j} \in \mathrm{Mon}(\Delta)$, $j = 1, \ldots, k_i$, $i = 1, \ldots, s$, such that

$$q^r \cdot p = \sum_{i=1}^{s} \left(\sum_{j=1}^{k_i} c_{i,j} \, \theta_{i,j} \right) p_i. \tag{2.44}$$

Our aim is to replace each term $c_{i,j} \, \theta_{i,j} \, p_i$ on the right hand side for which $\theta_{i,j}$ is divisible by a derivation that is non-admissible for $p_i = 0$, with a suitable linear combination of derivatives of $p_1, \ldots, p_s$ not involving any non-admissible derivations (cf. also Rem. 2.1.41, p. 28). Passivity of $\{ p_1 = 0, \ldots, p_s = 0 \}$ (cf. Def. 2.2.49 b)) guarantees that Janet reduction (Alg. 2.2.45) computes such a linear combination if $\theta_{i,j}$ involves only one non-admissible derivation. This computation is a pseudo-reduction in general, so that substitution of the term in question may require multiplying equation (2.44) by a suitable power of q first. Iterating this substitution process and dealing with terms as above in decreasing order with respect to the ranking constructs a representation as in (2.44), possibly with larger r, in which no $\theta_{i,j}$ is divisible by any derivation that is non-admissible for $p_i = 0$. This shows that for every non-zero element p of $E : q^{\infty}$ there exists a Janet divisor of $\mathrm{ld}(p)$ in the passive set defined by $p_1 = 0, \ldots, p_s = 0$. Consequently, the last part of the assertion holds. (Moreover, the uniqueness of the Janet divisor of a jet variable implies the uniqueness of the representation of $q^r \cdot p$ as in (2.44) with admissible derivations only and further conditions on $c_{i,j}$ (for fixed r).)

In order to prove that $E : q^{\infty}$ is a radical differential ideal, let us first define, for any polynomial algebra $K[V] \subset R$, where V is a finite subset of $\mathrm{Mon}(\Delta)u$ such that $S^= \subset K[V]$ and $S^{\neq} \subset K[V]$, the (non-differential) ideal I_V of $K[V]$ which is generated by $p_1, \ldots, p_s$. Since S is simple as an algebraic system (cf. Def. 2.2.49 a)), Proposition 2.2.7, p. 62, implies that $I_V : q^{\infty}$ is a radical ideal of $K[V]$.

Assume that $p \in R$ satisfies $p^k \in E : q^{\infty}$ for some $k \in \mathbb{N}$. Using the first part of the proof, the Janet normal form of p^k modulo $p_1, \ldots, p_s$ is zero. Hence, we obtain an equation of the form (2.44), where p is replaced with p^k. Let p' be the Janet normal form of p modulo $p_1, \ldots, p_s$. Then, using the passivity again, no proper derivative of any $\mathrm{ld}(p_i)$ occurs in p'. We raise the equation

$$q^{r'} \cdot p = p' + \sum_{i=1}^{s} \left(\sum_{j=1}^{k_i'} c_{i,j}' \, \theta_{i,j}' \right) p_i, \tag{2.45}$$

which is constructed by Janet reduction, to the k-th power. After arranging for the left hand sides of this power and of the equation for p^k to be equal by multiplying by a suitable power of q, the difference of the right hand sides expresses $q^l \cdot (p')^k$ for some $l \in \mathbb{Z}_{\geq 0}$ as an R-linear combination of $p_1, \ldots, p_s$ because the proper admissible derivatives of the $\mathrm{ld}(p_i)$ and hence the proper admissible derivatives of the p_i cancel. By defining the polynomial algebra $K[V]$ appropriately (such that $K[V]$ contains all relevant jet variables), we conclude that we have $q^l \cdot (p')^k \in I_V$, thus $(p')^k \in I_V : q^\infty$. It follows that $p' \in I_V : q^\infty \subseteq E : q^\infty$, and therefore, $p \in E : q^\infty$. $\qquad\square$

In order to define a Thomas decomposition for differential systems, we first need to discuss the notion of solution of a differential system.

Recall from Definition 2.2.2 that the set $\mathrm{Sol}_{\overline{K}}(S)$ of solutions of a given algebraic system S is the set of tuples in $\overline{K}^n$ which satisfy the equations and inequations of S. Correspondingly, we are going to define now the set of analytic solutions $\mathrm{Sol}_\Omega(S)$ on a certain subset Ω of $\mathbb{C}^n$ of a differential system S.

From now on we focus on differential equations with (complex) analytic or meromorphic coefficients and we will consider analytic solutions. Let Ω be an open and connected subset of $\mathbb{C}^n$ with coordinates $z_1, \ldots, z_n$ and K the field of meromorphic functions on Ω. The differential polynomial ring $R := K\{u_1, \ldots, u_m\}$ is defined with meromorphic coefficients and with commuting derivations $\partial_1, \ldots, \partial_n$ extending partial differentiation with respect to $z_1, \ldots, z_n$ on K. Let a ranking $>$ on R be fixed. We assume that input to the algorithms is provided in such a way that the arithmetic operations can be carried out effectively when computing with coefficients in K, that equality of such coefficients can be decided, etc.

Given a differential system, an appropriate choice of the set Ω may often be difficult to make before the algebraic and differential consequences of the system have been analyzed. The latter task is achieved by the methods discussed in this section. The defining properties of a simple differential system imply that each PDE of such a system can locally be solved for the highest derivative of some u_k. Therefore, analytic solutions exist in some open neighborhood of any point that is sufficiently generic. (It is sufficient to exclude those points which are poles of meromorphic functions occurring in the given PDEs and those which are zeros of meromorphic functions f for which the resolution process uses division by f, cf. also Rem. 2.1.70, p. 53, for the linear case.) *Usually, we assume that Ω is chosen in such a way that the given system has analytic solutions on Ω.*

The following example shows that a prior choice of Ω may in general exclude certain solutions (depending also on initial or boundary conditions).

Example 2.2.52. The analytic solutions of the ordinary differential equation (ODE) $u' + u^2 = 0$ for an unknown function u of z are uniquely determined by the choice of $u(z_0)$ for any $z_0 \in \mathbb{C}$. Let us choose $z_0 = 0$. For $u(0) = 0$ the solution is identically zero. Given $u(0) = \frac{1}{c}$ with $c \in \mathbb{C} - \{0\}$, the solution $u(z) = \frac{1}{z+c}$ is analytic in an open neighborhood of any point in $\mathbb{C} - \{-c\}$ and has a pole at $z = -c$. The open neighborhood and Ω have to avoid the point $-c$. Alternatively, one may allow meromorphic solutions.

Definition 2.2.53. Let $\Omega \subseteq \mathbb{C}^n$ be open and connected, K the differential field of meromorphic functions on Ω, and $R := K\{u_1, \ldots, u_m\}$. Let

$$S = \{\, p_i = 0, q_j \neq 0 \mid i \in I, j \in J \,\}, \qquad p_i, q_j \in R,$$

where I and J are index sets. We define the *set of (complex analytic) solutions (on Ω)* or *differential variety*[13] *of S (defined on Ω)* by

$$\mathrm{Sol}_\Omega(S) := \{\, f = (f_1, \ldots, f_m) \mid f_k \colon \Omega \to \mathbb{C} \text{ analytic}, k = 1, \ldots, m,$$
$$p_i(f) = 0, q_j(f) \neq 0, i \in I, j \in J \,\},$$

where $p_i(f)$ and $q_j(f)$ are obtained from p_i and q_j, respectively, by substituting f_k for u_k and the partial derivatives of f_k for the corresponding jet variables in u_k. For any set V of m-tuples of analytic functions $\Omega \to \mathbb{C}$ the set

$$\mathscr{I}_R(V) := \{\, p \in R \mid p(v) = 0 \text{ for all } v \in V \,\}$$

is called the *vanishing ideal of V in R*.

Remark 2.2.54. Usually, we assume that I and J are finite index sets. By the Basis Theorem of Ritt-Raudenbush (cf., e.g., Thm. A.3.22, p. 256, or [Kol73, Sect. III.4]), every system of polynomial PDEs is equivalent to a finite one, which shows that the assumption on I is without loss of generality. However, similarly to the algebraic case (cf. Rem. 2.2.3), in general, an infinite set of inequations cannot be reduced to a finite set of inequations with the same solution set. (If F is a differentially closed differential field (cf. [Kol99, pp. 580–583]), then the subsets of F^m which are sets of solutions of systems S of polynomial differential equations defined over F, i.e., $J = \emptyset$, are the closed sets of the *Kolchin topology* on F^m.)

Definition 2.2.55. Let

$$S = \{\, p_i = 0, q_j \neq 0 \mid i \in I, j \in J \,\}, \qquad p_i, q_j \in R,$$

be a system of partial differential equations and inequations, where I and J are index sets and J is finite. A *Thomas decomposition* of S or of $\mathrm{Sol}_\Omega(S)$ is a finite collection of simple differential systems $S_1, \ldots, S_k$ such that

$$\mathrm{Sol}_\Omega(S) = \mathrm{Sol}_\Omega(S_1) \uplus \ldots \uplus \mathrm{Sol}_\Omega(S_k)$$

is a partition of $\mathrm{Sol}_\Omega(S)$.

The following algorithm constructs a Thomas decomposition of a given finite differential system S in finitely many steps. Note that we give a succinct presentation of such an algorithm and ignore efficiency issues. For other variants and details

---

[13] The term *differential variety* is used here mainly in contrast to the term *variety* in Definition 2.2.2 and should not be confused with the solution set of a differential system in an infinite jet space, which is also referred to as a *diffiety*, cf., e.g., [Vin84].

about the latter point we refer to [Ger08], [Ger09], [BGL+10], [BGL+12, Sect. 3].
Some remarks about implementations are also given in Subsect. 2.2.6.

Similarly to the algebraic case, a Thomas decomposition of a differential system
is by no means uniquely determined. Its algorithmic construction may be enhanced
by using factorization of polynomials (cf. also Remark 2.2.11). Since this possibility
depends on the properties of the differential field K, we will not use factorization.

Algorithm 2.2.56 (*DifferentialThomasDecomposition*).

Input: A finite differential system S over R, a ranking $>$ on R, and a total ordering
on Δ (used by *Decompose*)

Output: A Thomas decomposition of S

Algorithm:

1: $Q \leftarrow \{S\}$; $T \leftarrow \emptyset$
2: **repeat**
3:     choose $L \in Q$ and remove L from Q
4:     compute a Thomas decomposition $\{A_1, \ldots, A_r\}$ of L considered as an alge-
    braic system (cf. Rem. 2.2.11 or Alg. 2.2.20, and Rem. 2.2.39)
5:     **for** $i = 1, \ldots, r$ **do**
6:       **if** $A_i = \emptyset$ **then**   *// no equation and no inequation*
7:         **return** $\{\emptyset\}$
8:       **else**
9:         $(a, G) \leftarrow$ *Auto-reduce*$(A_i^=, >)$ *// cf. Alg. 2.2.40*
10:         **if** $a = $ **true then**
11:           $J \leftarrow$ *Decompose*(G) *// cf. Rem. 2.2.47*
12:           $P \leftarrow \{\mathrm{NF}(d\,p, J, >) \mid (p, \mu) \in J, d \in \overline{\mu}\}$ *// cf. Alg. 2.2.45*
13:           **if** $P \subseteq \{0\}$ **then**   *// J is passive*
14:             replace each inequation $q \neq 0$ in A_i with $\mathrm{NF}(q, J, >) \neq 0$
15:             **if** $0 \notin A_i^{\neq}$ **then**
16:               insert $\{p = 0 \mid (p, \mu) \in J\} \cup \{q \neq 0 \mid q \in A_i^{\neq}\}$ into T
17:             **end if**
18:           **else if** $P \cap K \subseteq \{0\}$ **then**
19:             insert $\{p = 0 \mid (p, \mu) \in J\} \cup \{p = 0 \mid p \in P - \{0\}\} \cup$
                            $\{q \neq 0 \mid q \in A_i^{\neq}\}$ into Q
20:           **end if**
21:         **else**
22:           insert $\{p = 0 \mid p \in G\} \cup \{q \neq 0 \mid q \in A_i^{\neq}\}$ into Q
23:         **end if**
24:       **end if**
25:     **end for**
26: **until** $Q = \emptyset$
27: **return** T

Theorem 2.2.57. *a) Algorithm 2.2.56 terminates and is correct.*
b) Let

$$S = \{\, p_1 = 0, \ldots, p_s = 0, q_1 \neq 0, \ldots, q_t \neq 0 \,\}$$

be a simple system in the result T of Algorithm 2.2.56. Define q to be the product of all $\mathrm{init}(p_i)$ *and all* $\mathrm{sep}(p_i)$, *$i = 1, \ldots, s$. Moreover, let $I := \langle p_1, \ldots, p_s \rangle : q^\infty$, and let $\mu_1, \ldots, \mu_s \subseteq \Delta$ be the sets of admissible derivations of $p_1, \ldots, p_s$, respectively, and $J := \{\, (p_i, \mu_i) \mid i = 1, \ldots, s \,\}$. Then we have*

$$\biguplus_{i=1}^{s} \mathrm{Mon}(\mu_i)\, \mathrm{ld}(p_i) = \mathrm{ld}(I).$$

For any $r \in R$ we have

$$r \in I \iff \mathrm{NF}(r, J, >) = 0.$$

c) Let $C_1, \ldots, C_k$ be the cones of a Janet decomposition of the complement of $[\mathrm{ld}(p_1), \ldots, \mathrm{ld}(p_s)]$ *in* $\mathrm{Mon}(\Delta)u$. *Then the cosets in R/I with representatives in the disjoint union $C_1 \uplus \ldots \uplus C_k$ form a maximal subset of R/I that is algebraically independent over K.*

Proof. In order to prove that *DifferentialThomasDecomposition* terminates, it is sufficient to show that we have $Q = \emptyset$ after finitely many steps. Apart from step 1, new elements are inserted into Q in steps 19 and 22.

In case of step 22, differential reduction in Algorithm 2.2.40 (*Auto-reduce*) computed in step 9 a non-zero differential polynomial, which is the left hand side of an equation in the new system that is inserted into Q. The algebraic version of Thomas' algorithm in step 4 will apply algebraic reductions to this system and possibly split this system. Hence, steps 4 and 9 apply algebraic and differential reductions alternately until the result G is simple as an algebraic system and for every pair (p, p') of distinct elements of G, $\mathrm{ld}(p)$ is reduced with respect to p'. Similarly to the auto-reduction method without case distinctions (cf. Rem. A.3.6 c), p. 250), each system constructed in step 4 will, after finitely many steps, either be recognized as inconsistent or be turned into a system G having the above property. During this process, the differentiation order of leaders in such a system is bounded by the maximum of the differentiation orders of the leaders in the system from which the algebraic version of Thomas' algorithm started. The generation of systems is therefore governed by the algebraic splitting method for a polynomial ring in finitely many variables.

In case of step 19, Algorithm 2.2.45 (*Janet-reduce*) returned a non-constant differential polynomial p' in step 12 which is Janet reduced modulo J. Therefore, we have either $\mathrm{ld}(p') \notin [\mathrm{ld}(G)]$, or $\mathrm{ld}(p')$ equals $\mathrm{ld}(p)$ for some $(p, \mu) \in J$ and the degree of p' in $\mathrm{ld}(p')$ is smaller than the corresponding one of p. In the former case, the new system N inserted into Q in step 19 will define a multiple-closed set which properly contains $[\mathrm{ld}(G)]$. In the latter case, we distinguish two kinds of systems that are derived from N by the algebraic version of Thomas' algorithm in step 4 (e.g., by steps 4 and 5 of Algorithm 2.2.21). A system of the first kind contains an equation (e.g., $p' = 0$) with leader $\mathrm{ld}(p) = \mathrm{ld}(p')$, different from $p = 0$, whose initial is guaranteed not to vanish, so that a pseudo-reduction of p is performed. Then the degree of p in $\mathrm{ld}(p)$ decreases. In a system of the second kind such a pseudo-reduction of p (as a polynomial in $\mathrm{ld}(p) = \mathrm{ld}(p')$) is prevented as a result of equating $\mathrm{init}(p')$ with zero. If $\mathrm{init}(p')$ is constant, such a system will be discarded. Otherwise, it contains a new equation with leader $\mathrm{ld}(\mathrm{init}(p'))$. Since $\mathrm{init}(p')$ is Janet reduced modulo J, a pseudo-reduction of $\mathrm{init}(p')$ could at most be performed modulo equations originating from other elements of P. We may assume that $p' \in P$ is chosen such that no pseudo-reduction of $\mathrm{init}(p')$ is possible. Again, we have either $\mathrm{ld}(\mathrm{init}(p')) \notin [\mathrm{ld}(G)]$, or a pseudo-reduction of the left hand side of an equation with leader $\mathrm{ld}(\mathrm{init}(p'))$ is performed, or the initial of the new equation is equated with zero. By iterating this argument, we conclude that either we obtain an equation whose leader is not contained in $[\mathrm{ld}(G)]$ or for some generator v of $[\mathrm{ld}(G)]$ the minimal degree of v as a leader in equations in G is decreased. Since the latter situation can occur only finitely many times, in any case the multiple-closed set $[\mathrm{ld}(G)]$ will be enlarged after finitely many steps.

Hence, termination follows from the termination of the algebraic version of Thomas' algorithm and Dickson's Lemma (cf. Lemma 2.1.2, p. 10).

We are going to show the correctness of *DifferentialThomasDecomposition*. The algebraic version of Thomas' algorithm only performs algebraic pseudo-reductions and splittings while maintaining the total solution set. Hence, the solution sets of the systems $A_1, \ldots, A_r$ in step 4 form a partition of the solution set of L.

In step 9, Algorithm 2.2.40 (*Auto-reduce*) applies differential reductions to equations in a system A_i which is simple as an algebraic system. An equation is replaced with its pseudo-remainder if the latter is non-zero, and Algorithm 2.2.40 stops after the first proper replacement. From the discussion in Remark 2.2.39 we conclude that this transformation does not change the solution set of A_i because the separant which is used for pseudo-division does not vanish on the solution set.

The set J constructed in step 11 is Janet complete. Therefore, $\mathrm{NF}(dp, J, >)$ in step 12 and $\mathrm{NF}(q, J, >)$ in step 14 are well-defined and are realized by applying Algorithm 2.2.45 (*Janet-reduce*). Step 12 computes left hand sides of equations that are consequences of A_i. If some of them are non-zero constants, then these consequences reveal an inconsistent differential system, which is therefore discarded. If some non-constant consequences are obtained and all constant consequences are zero, then the former ones are inserted into the system to be processed again in a later round. If $P \subseteq \{0\}$ holds in step 13, then J is a passive set. Then conditions a) and b) of Definition 2.2.49 are satisfied, and after step 14 condition c) is also ensured if the system is not inconsistent. Thus, every differential system which is inserted into T in step 16 is simple.

In every step of *DifferentialThomasDecomposition* the solution sets of the differential systems in Q and in T form a partition of the solution set of S. Hence, the result T is a Thomas decomposition of S.

The statements in parts b) and c) of the theorem are immediate consequences of Proposition 2.2.50. □

Remark 2.2.58. The result of Algorithm 2.2.56 is empty if and only if the input system S is inconsistent. If it equals $\{\emptyset\}$ (i.e., a set consisting of one empty system), then the input system admits all analytic functions on Ω as solutions.

Remark 2.2.59. The notion of a *singular solution* of a differential equation dates back to the 18th century and research by, e.g., A. C. Clairaut, J.-L. Lagrange, P.-S. Laplace, and S. D. Poisson (cf., e.g., [Inc56, footnote on p. 87], [Kol99, Sect. 1.8]). The intuitive idea of this concept is that the solutions of a differential equation form a number of families, each of which is parametrized by a number of constants (or functions in case of underdetermined systems) which can be chosen arbitrarily. If one family is identified as constituting the *general solution* (obtained by a generic choice of the constants or functions), then all solutions which do not belong to this family are said to be singular. A more rigorous definition was given by J.-G. Darboux [Dar73, p. 158]: A solution of a differential equation is said to be singular if it is also a solution of the separant of the equation.

Thomas' algorithm splits systems according to vanishing or non-vanishing initials of equations and their partial derivatives, the initials of the latter being the separants of the equations (cf. Rem. 2.2.39). A Thomas decomposition of a differential system therefore allows to detect singular solutions of the system. More generally, the problem of determining how singular solutions are distributed among irreducible components of a differential variety was a major motivation for J. F. Ritt to develop differential algebra. For more details and contributions to this problem we refer to [Ham93], [Rit36], [Hub97, Hub99], [Kol99, Sect. 1.8], and the references therein.

Example 2.2.60. In order to investigate the singular solutions of the ordinary differential equation (with non-constant coefficients)

$$\frac{dU}{dt}^2 - 4t\frac{dU}{dt} - 4U + 8t^2 = 0,$$

we are going to compute a Thomas decomposition of this system. We denote by R the differential polynomial ring $\mathbb{Q}(t)\{u\}$ in one differential indeterminate u with derivation ∂_t, which restricts to formal differentiation with respect to t on $\mathbb{Q}(t)$. Let the differential polynomial corresponding to the left hand side be

$$p := u_t^2 - 4t\,u_t - 4u + 8t^2.$$

No case distinction is necessary for the initial. The separant of p equals $2u_t - 4t$. Euclidean division applied to p and $\text{sep}(p)$ (as polynomials in u_t) yields $u - t^2$, which is, up to a constant factor, the discriminant of p as a polynomial in u_t. We obtain the following Thomas decomposition (where the set of admissible derivations is also recorded for each equation):

$\underline{u_t}^2 - 4t\,\underline{u_t} - 4u + 8t^2 = 0 \ \{\ \partial_t\ \}$	
$\underline{u} - t^2 \neq 0$	$\underline{u} - t^2 = 0 \ \{\ \partial_t\ \}$

The analytic solutions of the first system are given by $U(t) = 2((t+c)^2 + c^2)$, where c is an arbitrary (real or complex) constant. The solution $U(t) = t^2$ of the second system is an essential singular solution. Considering all real analytic solutions at the same time, the singular solution is distinguished as an envelope of the general solution.

Fig. 2.3 A visualization of the essential singular solution in Example 2.2.60 as an envelope

Example 2.2.61. We are going to compute a Thomas decomposition of

$$\frac{\partial U}{\partial t} - 6U \frac{\partial U}{\partial x} + \frac{\partial^3 U}{\partial x^3} = 0, \qquad U \frac{\partial^2 U}{\partial t \partial x} - \frac{\partial U}{\partial t} \frac{\partial U}{\partial x} = 0,$$

the differential system given by the Korteweg-de Vries equation (cf., e.g., [BC80]) and another partial differential equation for $U(t,x)$ to be discussed in Sect. 3.3 (cf. also Ex. 3.3.49, p. 228).

Let $R := K\{u\}$ be the differential polynomial ring in one differential indeterminate u with commuting derivations ∂_t, ∂_x over a differential field K of characteristic zero (with derivations $\partial_t|_K$, $\partial_x|_K$). The jet variable $u_{(i,j)}$, $i, j \in \mathbb{Z}_{\geq 0}$, will also be denoted by u_{t^i, x^j}. We set

$$p := u_t - 6u u_x + u_{x,x,x}, \qquad q := u u_{t,x} - u_t u_x$$

and choose the degree-reverse lexicographical ranking on R satisfying $u_t > u_x$ (cf. Ex. A.3.3, p. 250).

We have $\mathrm{ld}(p) = u_{x,x,x}$, $\mathrm{ld}(q) = u_{t,x}$, $\mathrm{init}(p) = 1$, and $\mathrm{init}(q) = u$. Hence,

$$\{p = 0, q = 0\}$$

is a triangular set. We replace this system with two systems

$$\{p = 0, q = 0, u = 0\}, \quad \{p = 0, q = 0, u \neq 0\}$$

according to vanishing or non-vanishing initial of q. (No case distinctions are necessary for the separants.) The first system is equivalent to the simple differential system

$$S_1 := \{u = 0\}.$$

The second system is simple as an algebraic system (cf. Def. 2.2.49 a)), but not passive. We define $\Delta := \{\partial_t, \partial_x\}$ and give ∂_t priority over ∂_x for Janet division (cf. Alg. 2.1.6 and Def. 2.2.43). Then the admissible derivations for p and q are given by $\{\partial_x\}$ and $\{\partial_t, \partial_x\}$, respectively. Janet reduction of $\partial_t p$ modulo $\{(p, \{\partial_x\}), (q, \{\partial_t, \partial_x\})\}$ yields the following non-zero pseudo-remainder:

$$r := u(u\,p_t - q_{x,x}) - u\,u_t\,p + u_x\,q_x = u^2\,u_{t,t} - u(6u^2 - u_{x,x})\,u_{t,x} - u_t\,u_x\,u_{x,x} - u\,u_t^2.$$

The augmented system $\{p = 0, q = 0, r = 0, u \neq 0\}$ is simple as an algebraic system, and the passivity check only involves Janet reduction of $\partial_t q$ modulo $\{(p, \{\partial_x\}), (q, \{\partial_x\}), (r, \{\partial_t, \partial_x\})\}$. The result is:

$$s := u((u\,q_t - r_x) - (6u^2 - u_{x,x})\,q_x + q\,p) + 3u_x\,r + 3(2u^2\,u_x - u\,u_t - u_x\,u_{x,x})\,q$$

$$= 6u^3\,u_t\,u_{x,x}.$$

We have $\mathrm{init}(s) = 6u^3\,u_t$. Now, $\mathrm{init}(s) \neq 0$ implies $u_{x,x} = 0$, which results in the simple system

$$S_2 := \{u_t - 6u\,u_x = 0, u_{x,x} = 0, u \neq 0\}.$$

On the other hand, $\mathrm{init}(s) = 0$ implies $u_t = 0$, hence the simple system

$$S_3 := \{u_t = 0, u_{x,x,x} - 6u\,u_x = 0, u_{x,x} \neq 0, u \neq 0\}.$$

	$\underline{u_t} - 6u\,u_x = 0\ \{\partial_t, \partial_x\}$	$u_t = 0\ \{\partial_t, \partial_x\}$
		$\underline{u_{x,x,x}} - 6u\,u_x = 0\ \{*, \partial_x\}$
	$u_{x,x} = 0\ \{*, \partial_x\}$	$u_{x,x} \neq 0$
$u = 0\ \{\partial_t, \partial_x\}$	$u \neq 0$	$u \neq 0$

For an explicit integration of these simple systems, cf. Example 3.3.49, p. 228.

2.2.3 The Generic Simple System for a Prime Ideal

In this subsection we prove that in every Thomas decomposition of a prime (algebraic or differential) ideal there exists a unique simple system that is in a precise sense the most generic one in the decomposition. Moreover, a corollary to Theorem 2.2.57 of the previous subsection is obtained, which shows how membership to a radical (algebraic or differential) ideal can be decided using a Thomas decomposition.

The statements below will be at the same time about algebraic and differential systems using the following notation.

For the rest of this section, R denotes either the commutative polynomial algebra $K[x_1, \ldots, x_n]$ over a field K of characteristic zero or the differential polynomial ring $K\{u_1, \ldots, u_m\}$, where K is the field of meromorphic functions on a connected open subset Ω of $\mathbb{C}^n$, as in the previous subsection. If S is an algebraic or differential system, we will write $\mathrm{Sol}(S)$ referring to either $\mathrm{Sol}_{\overline{K}}(S)$ (cf. Def. 2.2.2) or $\mathrm{Sol}_\Omega(S)$ (cf. Def. 2.2.53). We will also use $\langle P \rangle$ to denote the ideal and the differential ideal, respectively, generated by the set P depending on the context, and $\mathscr{I}_R$ is a notation for the vanishing ideal in both cases.

Recall that for both algebraic and differential systems

$$S = \{p_1 = 0, \ldots, p_s = 0\}$$

of equations over R a theorem holds, called Nullstellensatz, which states that, if $p \in R$ vanishes on $\mathrm{Sol}(S)$, then some power of p is an element of $E := \langle p_1, \ldots, p_s \rangle$, i.e., we have

$$\mathscr{I}_R(\mathrm{Sol}(S)) = \sqrt{E}. \tag{2.46}$$

(The theorem is due to D. Hilbert in the algebraic case and to J. F. Ritt and H. W. Raudenbush in the differential case; cf. also Thm. A.3.24, p. 258). In particular, if $\langle p_1, \ldots, p_s \rangle$ is a radical ideal, then $\mathscr{I}_R(\mathrm{Sol}(S)) = \langle p_1, \ldots, p_s \rangle$ holds.

The following lemma generalizes (2.46) and will be essential in what follows.

Lemma 2.2.62. *Let*

$$S = \{p_i = 0, q_j \neq 0 \mid i \in I, j \in J\}, \qquad p_i, q_j \in R,$$

be a (not necessarily simple) system, where I and J are index sets and J is finite. Define $E := \langle p_i \mid i \in I \rangle$ and $q := \prod_{j \in J} q_j$. Then we have

$$\mathscr{I}_R(\mathrm{Sol}(S)) = \sqrt{E : q^\infty}.$$

Proof. Let $f \in R$. If $q^r f^s \in E$ for some $r \in \mathbb{Z}_{\geq 0}$ and $s \in \mathbb{N}$, then $f(x)^s = 0$ for all $x \in \mathrm{Sol}(S)$ because $q(x)^r \neq 0$ for all $x \in \mathrm{Sol}(S)$. Since $\mathrm{Sol}(S)$ is a subset of an integral domain, we have $f(x) = 0$ for all $x \in \mathrm{Sol}(S)$, i.e., $f \in \mathscr{I}_R(\mathrm{Sol}(S))$. Conversely, $f(x) = 0$ for all $x \in \mathrm{Sol}(S)$ implies $(qf)(x) = 0$ for all $x \in \mathrm{Sol}(\{p_i = 0 \mid i \in I\})$. By the Nullstellensatz, there exists $s \in \mathbb{N}$ such that $(qf)^s \in E$. It follows that $f \in \sqrt{E : q^\infty}$. $\square$

For any set V of elements of $\overline{K}^n$ or of m-tuples of analytic functions on Ω, we define

$$\overline{V} := \mathrm{Sol}(\{p = 0 \mid p \in \mathscr{I}_R(V)\}),$$

which is a shorthand notation used in the next corollary and in what follows, and which is reminiscent of the closure with respect to the Zariski topology. (We expect no confusion with the notation $\overline{K}$ for an algebraic closure of K.)

Corollary 2.2.63. *Let S be a (not necessarily simple) system as in Lemma 2.2.62, $E := \langle p_i \mid i \in I \rangle$, and $q := \prod_{j \in J} q_j$. Then we have*

$$\overline{\mathrm{Sol}(S)} - \mathrm{Sol}(S) = \mathrm{Sol}(\{ p = 0 \mid p \in \sqrt{E : q^\infty} \} \cup \{ q = 0 \}).$$

In particular, $\overline{\mathrm{Sol}(S)} - \mathrm{Sol}(S)$ is closed (i.e., equals its closure).

Proof. Since on the one hand all polynomials in $\sqrt{E : q^\infty}$ vanish on $\mathrm{Sol}(S)$ and on the other hand $p_i \in \sqrt{E : q^\infty}$ for all $i \in I$, we have

$$\mathrm{Sol}(S) = \mathrm{Sol}(\{ p = 0 \mid p \in \sqrt{E : q^\infty} \} \cup \{ q \neq 0 \}).$$

A reformulation of the statement of the previous lemma is:

$$\overline{\mathrm{Sol}(S)} = \mathrm{Sol}(\{ p = 0 \mid p \in \sqrt{E : q^\infty} \}).$$

Now an elementary observation shows that the claim holds. $\square$

A central result of this subsection is the corollary to the next proposition. The proof of the corollary uses the following lemma.

Lemma 2.2.64. *Let S and S' be systems as in Lemma 2.2.62 satisfying $\mathrm{Sol}(S') \neq \emptyset$, $\mathrm{Sol}(S) \cap \mathrm{Sol}(S') = \emptyset$, and $\overline{\mathrm{Sol}(S)} \subseteq \overline{\mathrm{Sol}(S')}$. Then we have $\overline{\mathrm{Sol}(S)} \neq \overline{\mathrm{Sol}(S')}$.*

Proof. It follows from $\mathrm{Sol}(S) \cap \mathrm{Sol}(S') = \emptyset$ and

$$\mathrm{Sol}(S) \subseteq \overline{\mathrm{Sol}(S')} = \mathrm{Sol}(S') \uplus (\overline{\mathrm{Sol}(S')} - \mathrm{Sol}(S'))$$

that $\mathrm{Sol}(S)$ is a subset of $\overline{\mathrm{Sol}(S')} - \mathrm{Sol}(S')$. By Corollary 2.2.63, $\overline{\mathrm{Sol}(S')} - \mathrm{Sol}(S')$ is closed. Since it is a proper subset of $\overline{\mathrm{Sol}(S')}$, the claim follows. $\square$

Proposition 2.2.65. *Suppose that $p_1, \ldots, p_s \in R$ generate a prime (algebraic or differential) ideal of R and that*

$$V := \mathrm{Sol}(\{ p_1 = 0, \ldots, p_s = 0 \})$$

is the union of finitely many non-empty sets $V_1, \ldots, V_k$ of the form $V_i = \mathrm{Sol}(S_i)$ for (not necessarily simple) systems S_i of equations and inequations over R. Then we have $V = \overline{V_i}$ for some $i \in \{ 1, \ldots, k \}$.

Proof. By the Nullstellensatz and since $p_1, \ldots, p_s$ generate a radical ideal, we have

$$\langle p_1, \ldots, p_s \rangle = \mathscr{I}_R(V) = \mathscr{I}_R(V_1) \cap \ldots \cap \mathscr{I}_R(V_k).$$

Each vanishing ideal $\mathscr{I}_R(V_j)$ has a representation as intersection of finitely many prime ideals, all of which clearly contain $\mathscr{I}_R(V)$. The uniqueness of the minimal representation of a radical ideal in such a form implies that the prime ideal $\langle p_1, \ldots, p_s \rangle$ occurs in the minimal representation of at least one $\mathscr{I}_R(V_i)$, and we have $\mathscr{I}_R(V) = \mathscr{I}_R(V_i)$. It follows that $V = \overline{V_i}$. $\square$

Corollary 2.2.66. *Suppose that* $p_1, \ldots, p_s \in R$ *generate a prime ideal of* R *and let* $S_1, \ldots, S_k$ *be a Thomas decomposition of* $V := \mathrm{Sol}(\{ p_1 = 0, \ldots, p_s = 0 \})$. *Then there exists a unique* $i \in \{1, \ldots, k\}$ *such that* $\overline{\mathrm{Sol}(S_i)} = V$. *Moreover, the prime ideal* $\mathscr{I}_R(\mathrm{Sol}(S_i)) = \langle p_1, \ldots, p_s \rangle$ *is a proper subset of* $\mathscr{I}_R(\mathrm{Sol}(S_j))$ *for every* $j \neq i$.

Proof. The existence of i follows from Proposition 2.2.65 applied to $V_j := \mathrm{Sol}(S_j)$, $j = 1, \ldots, k$. On the other hand, let $j \in \{1, \ldots, k\}$, $j \neq i$. Then we have $\overline{\mathrm{Sol}(S_i)} \neq \emptyset$, $\overline{\mathrm{Sol}(S_j)} \neq \emptyset$, and $\mathrm{Sol}(S_i) \cap \mathrm{Sol}(S_j) = \emptyset$. By Lemma 2.2.64, equality of $\overline{\mathrm{Sol}(S_i)}$ and $\overline{\mathrm{Sol}(S_j)}$ is impossible. This proves the uniqueness.

We have $\mathscr{I}_R(\mathrm{Sol}(S_j)) = \mathscr{I}_R(\overline{\mathrm{Sol}(S_j)})$ for all $j \in \{1, \ldots, k\}$ and furthermore $\mathscr{I}_R(\overline{\mathrm{Sol}(S_i)}) = \mathscr{I}_R(V)$. Since V is the union of $\mathrm{Sol}(S_1), \ldots, \mathrm{Sol}(S_k)$, we have

$$\mathscr{I}_R(\mathrm{Sol}(S_i)) = \mathscr{I}_R(\mathrm{Sol}(S_1)) \cap \ldots \cap \mathscr{I}_R(\mathrm{Sol}(S_k)).$$

Therefore, $\mathscr{I}_R(\mathrm{Sol}(S_i))$ is properly contained in each $\mathscr{I}_R(\mathrm{Sol}(S_j))$, $j \neq i$. $\qquad \square$

Definition 2.2.67. Let $S_1, \ldots, S_k$ be a Thomas decomposition of a system of equations whose left hand sides generate a prime (algebraic or differential) ideal of R. The simple system S_i in Corollary 2.2.66 is called the *generic simple system* of the given Thomas decomposition[14].

Example 2.2.68. A Thomas decomposition of the Steiner quartic surface defined by

$$x^2 y^2 + x^2 z^2 + y^2 z^2 - xyz = 0 \tag{2.47}$$

is constructed in Example 2.2.35, p. 84. This surface is an irreducible variety, i.e., the ideal of $\overline{\mathbb{Q}}[x, y, z]$ generated by the left hand side of (2.47) is prime. The generic simple system of the Thomas decomposition in Example 2.2.35 is system (2.41) because it is the only one whose solution set has Zariski closure of dimension two.

The generic simple system of a Thomas decomposition can be determined if the pairwise inclusion relations among the radical ideals $\mathscr{I}_R(\mathrm{Sol}(S_i)) = \langle S_i^= \rangle : q_i^\infty$ can be checked effectively, where q_i is the product of the initials (and separants in the differential case) of all elements of $S_i^=$. Corollary 2.2.71 below solves this problem.

As before, we deal at the same time with algebraic and differential systems S, implying that pseudo-reductions modulo $S^=$ are understood to be pseudo-reductions modulo the elements of $S^=$ in the algebraic case and pseudo-reductions modulo the elements of $S^=$ and their derivatives in the differential case.

Proposition 2.2.69. *Let* S_1 *and* S_2 *be simple systems over* R. *For* $i = 1, 2$, *we define* $I_i := \langle S_i^= \rangle : q_i^\infty$, *where* q_i *is the product of the initials (and separants in the differential case) of all elements of* $S_i^=$. *If* I_1 *is a prime ideal and if we have* $I_1 \subseteq I_2$ *and* $\mathrm{ld}(I_1) = \mathrm{ld}(I_2)$, *then the equality* $I_1 = I_2$ *holds.*

---

[14] The notion of *generic simple system* should not be confused with the notion of *general component* of an irreducible differential polynomial p (cf. [Kol73, Sect. IV.6] or [Rit50, p. 167]), which is a certain prime differential ideal not containing the separant of p.

Proof. First of all, we have the equality $I_1 : q_2^\infty = I_1$. In fact, if $q_2^k \cdot p \in I_1$ for some $k > 0$ and $p \in R - I_1$, then we have $q_2 \in I_1$ because I_1 is a prime ideal; thus, q_2 vanishes on $\mathrm{Sol}(I_1)$ and hence on $\mathrm{Sol}(I_2)$, which is a contradiction to the fact that $q_2 \notin \mathscr{I}_R(\mathrm{Sol}(S_2))$ because S_2 is a simple system.

In order to prove the proposition, we show that the pseudo-remainder of every $p_2 \in S_2^=$ modulo the equations in S_1 (and, in the differential case, their consequences obtained by applying admissible derivations) is zero. Then it follows $\langle S_2^= \rangle \subseteq I_1$ (by Thm. 2.2.57 b)) and $I_2 = \langle S_2^= \rangle : q_2^\infty \subseteq I_1 : q_2^\infty = I_1$.

Let us assume that the pseudo-remainder modulo $S_1^=$ of some element of $S_2^=$ is non-zero, and let $p_2 \in S_2^=$ be such an element with minimal possible $\mathrm{ld}(p_2)$.

The hypothesis $\mathrm{ld}(I_1) = \mathrm{ld}(I_2)$ implies that there exists (an admissible derivative of) an element of $S_1^=$ with the same leader as p_2. Let us denote this algebraic pseudo-divisor by p_1 and the variable $\mathrm{ld}(p_1) = \mathrm{ld}(p_2)$ by x. In case $\deg_x(p_1) \leq \deg_x(p_2)$ algebraic pseudo-division of p_2 modulo p_1 is possible resulting in an element of I_2 which is either zero or has leader smaller than x or has leader x and smaller degree in x than p_1. The first two cases are contradictions to the choice of p_2. Hence, it is sufficient to consider the case $\deg_x(p_1) > \deg_x(p_2)$.

The pseudo-remainder of p_1 modulo $S_2^=$ is zero because $I_1 \subseteq I_2$. Let q be the product of the initials of (the admissible derivatives of) the elements of $S_2^=$ which are involved in the pseudo-reduction of p_1 modulo $S_2^=$. Then we have $q \in K$ or $\mathrm{ld}(q) < x$. There exist $k \in \mathbb{Z}_{\geq 0}$, $c \in R$, and $r \in I_2$ with $\mathrm{ld}(r) < x$ such that $q^k \cdot p_1 = c \cdot p_2 + r$. Now $\deg_x(p_1) > \deg_x(p_2)$ implies $\mathrm{ld}(c) = x$. By the choice of p_2, the pseudo-remainder modulo $S_1^=$ of (every admissible derivative of) every element of $S_2^=$ with leader smaller than x is zero. Therefore, we have $q^k \cdot p_1 - r = c \cdot p_2 \in I_1$. The assumption $p_2 \notin I_1$ and the fact that I_1 is prime imply $c \in I_1$. But the pseudo-remainder of c modulo $S_1^=$ is non-zero due to $\deg_x(c) < \deg_x(p_1)$, which is a contradiction. $\qquad\square$

Remark 2.2.70. In the differential case the pseudo-divisor p_1 in the proof of the previous proposition is actually not a proper derivative of an element of $S_1^=$ because every partial derivative of a differential polynomial has degree one in its leader.

Corollary 2.2.71. *Let $S_1, \ldots, S_k$ be a Thomas decomposition of a system of equations whose left hand sides generate a prime ideal of R. Set inclusion defines a partial order on $L := \{\mathrm{ld}(I_1), \ldots, \mathrm{ld}(I_k)\}$, where $I_i := \langle S_i^= \rangle : q_i^\infty$ and q_i is defined to be the product of the initials (and separants in the differential case) of all elements of $S_i^=$, $i = 1, \ldots, k$. Then L has a unique least element $\mathrm{ld}(I_i)$. It determines the generic simple system S_i among $S_1, \ldots, S_k$.*

Proof. Let $i \in \{1, \ldots, k\}$ be such that S_i is the generic simple system of the given Thomas decomposition. By Corollary 2.2.66, I_i is a proper subset of I_j for every $j \neq i$. Hence, we have $\mathrm{ld}(I_i) \subseteq \mathrm{ld}(I_j)$ for every $j \neq i$. By Proposition 2.2.69, each of these inclusions is strict. $\qquad\square$

In the differential case, each set $\mathrm{ld}(I_j)$ is infinite. An effective method which determines the generic simple system of a Thomas decomposition of a prime differential ideal will be given in Proposition 2.2.82.

We draw another important conclusion from Lemma 2.2.62.

Proposition 2.2.72. *Let a (not necessarily simple) system be given by*

$$S = \{\, p_i = 0, q_j \neq 0 \mid i \in I, j \in J \,\}, \qquad p_i, q_j \in R,$$

where I and J are index sets and J is finite. Define $E := \langle\, p_i \mid i \in I \,\rangle$ and $q := \prod_{j \in J} q_j$. Moreover, let $S_1, \ldots, S_k$ be a Thomas decomposition of S, and for $i = 1, \ldots, k$, define $E^{(i)} := \langle\, S_i^= \,\rangle$ and the product $q^{(i)}$ of the initials (and separants in the differential case) of all elements of $S_i^=$. Then we have

$$\sqrt{E : q^\infty} = (E^{(1)} : (q^{(1)})^\infty) \cap \ldots \cap (E^{(k)} : (q^{(k)})^\infty).$$

Proof. The Thomas decomposition defines a partition $\mathrm{Sol}(S_1) \uplus \ldots \uplus \mathrm{Sol}(S_k)$ of $\mathrm{Sol}(S)$. As a consequence of the Nullstellensatz and Propositions 2.2.7 and 2.2.50, we have $\mathscr{I}_R(\mathrm{Sol}(S_i)) = E^{(i)} : (q^{(i)})^\infty$ for all $i = 1, \ldots, k$. Now, Lemma 2.2.62 implies

$$\begin{aligned}
\sqrt{E : q^\infty} &= \mathscr{I}_R(\mathrm{Sol}(S)) = \mathscr{I}_R(\mathrm{Sol}(S_1) \uplus \ldots \uplus \mathrm{Sol}(S_k)) \\
&= \mathscr{I}_R(\mathrm{Sol}(S_1)) \cap \ldots \cap \mathscr{I}_R(\mathrm{Sol}(S_k)) \\
&= (E^{(1)} : (q^{(1)})^\infty) \cap \ldots \cap (E^{(k)} : (q^{(k)})^\infty).
\end{aligned}$$

$\square$

Membership to a radical (algebraic or differential) ideal can therefore be decided by computing the pseudo-remainder modulo every simple system in a Thomas decomposition.

Corollary 2.2.73. *Let $p_1, \ldots, p_s \in R$ and let $S_1, \ldots, S_k$ be a Thomas decomposition of $\{\, p_1 = 0, \ldots, p_s = 0 \,\}$ (with respect to any total ordering of the indeterminates or any ranking on R). For $r \in R$ let r_i be the pseudo-remainder of r modulo (the equations of) S_i, $i = 1, \ldots, k$. Then we have*

$$r \in \sqrt{\langle p_1, \ldots, p_s \rangle} \quad \Longleftrightarrow \quad r_i = 0 \quad \text{for all } i = 1, \ldots, k.$$

Proof. The claim follows from Theorem 2.2.57 b) and Proposition 2.2.72. $\square$

Finally, the disjointness of the solution sets of the simple systems in a Thomas decomposition implies a statement about the corresponding prime ideals which is more general than the one given in Corollary 2.2.66.

Proposition 2.2.74. *Let S_1 and S_2 be two different simple systems in a Thomas decomposition. Moreover, for $i = 1, 2$, let $\mathscr{I}_R(\mathrm{Sol}(S_i)) = P_{i,1} \cap \ldots \cap P_{i,r_i}$ be the minimal representation of the vanishing ideal as intersection of prime (differential) ideals. If a prime (differential) ideal P of R satisfies*

$$\mathscr{I}_R(\mathrm{Sol}(S_1)) \subseteq P \quad \text{and} \quad \mathscr{I}_R(\mathrm{Sol}(S_2)) \subseteq P,$$

then some $P_{i,j}$ is properly contained in P. In particular, the sets of prime ideals $\{P_{1,1}, \ldots, P_{1,r_1}\}$ and $\{P_{2,1}, \ldots, P_{2,r_2}\}$ are disjoint.

Proof. Let us define $V := \mathrm{Sol}(\{\, p = 0 \mid p \in P\,\})$ and $V_{i,j} := \mathrm{Sol}(\{\, p = 0 \mid p \in P_{i,j}\,\})$, $i \in \{1,2\}$, $j \in \{1,\ldots,r_i\}$. First of all, we have $V_{i,j} \cap \mathrm{Sol}(S_i) \neq \emptyset$ for each j. Otherwise, $\mathrm{Sol}(\{\, p = 0 \mid p \in \bigcap_{k \neq j} P_{i,k}\,\})$ would be a closed set containing $\mathrm{Sol}(S_i)$ and contained in $\overline{\mathrm{Sol}(S_i)}$, and $P_{i,j}$ would be redundant in $P_{i,1} \cap \ldots \cap P_{i,r_i}$.

By the hypothesis of the proposition, we have $V \subseteq \overline{\mathrm{Sol}(S_i)}$ for $i = 1, 2$. According to the definition of a Thomas decomposition, $\mathrm{Sol}(S_1)$ and $\mathrm{Sol}(S_2)$ are disjoint. Using the notation $C_i := \overline{\mathrm{Sol}(S_i)} - \mathrm{Sol}(S_i)$, $i = 1, 2$, we therefore have

$$V \subseteq \overline{\mathrm{Sol}(S_1)} \cap \overline{\mathrm{Sol}(S_2)} \subseteq C_1 \cup C_2$$

and

$$(V \cap C_1) \cup (V \cap C_2) = V.$$

Now, $V \cap C_1$ and $V \cap C_2$ are closed sets by Corollary 2.2.63 and because V is closed. Since the vanishing ideal of the closed set V is a prime ideal, we conclude that we have $V \cap C_1 = V$ or $V \cap C_2 = V$. Hence, there exists $i \in \{1,2\}$ such that

$$V \subseteq C_i = \overline{\mathrm{Sol}(S_i)} - \mathrm{Sol}(S_i).$$

Since C_i and $V_{i,j}$ are closed, $C_i \cap V_{i,j}$ is closed. Moreover, $C_i \cap V_{i,j}$ is a proper subset of $V_{i,j}$, because otherwise $V_{i,j} \cap \mathrm{Sol}(S_i) \neq \emptyset$ implies $C_i \cap \mathrm{Sol}(S_i) \neq \emptyset$, which is a contradiction. Now $P_{i,1} \cap \ldots \cap P_{i,r_i} \subseteq P$ implies that there exists $j \in \{1,\ldots,r_i\}$ such that $P_{i,j} \subseteq P$, because $P_{i,k}$ and P are prime. Then $V \subseteq C_i \cap V_{i,j} \subsetneq V_{i,j}$ implies $P_{i,j} \subsetneq P$. The last claim of the proposition follows from the minimality of the representation of $\mathscr{I}_R(\mathrm{Sol}(S_i))$, $i = 1, 2$. $\square$

2.2.4 Comparison and Complexity

Some comments about the relationship of simple systems in the sense of Thomas and other types of triangular sets and about the complexity of computing a Thomas decomposition are given in this subsection.

Remark 2.2.75. Conditions a) and b) of Definition 2.2.4, p. 61, of a simple algebraic system imply that the set $S^=$ of equations of such a system S is a consistent triangular set, i.e., a triangular set admitting solutions (cf. Rem. 2.2.5). The radical ideal generated by $S^=$ is a characterizable ideal, as membership to it can be decided effectively by applying pseudo-reductions (cf. [Hub03a, Hub03b]). Moreover, the set $S^=$ of equations of a simple system is a regular chain (cf. [ALMM99]).

In addition to these properties, the solution sets of the simple systems in a Thomas decomposition are pairwise disjoint. In practice, achieving this requirement often is at the cost of a more involved computation. However, this strong geometric property may be used both in theory (cf., e.g., the previous subsection) and for concrete applications (e.g., for counting solutions; cf. [Ple09a] for the case of algebraic systems).

Recall that the Nullstellensatz for analytic functions (cf. Thm. A.3.24, p. 258) states that a differential polynomial q which vanishes on all analytic solutions of a system $\{p_1 = 0, \ldots, p_s = 0\}$ of polynomial partial differential equations is an element of the radical differential ideal I generated by $p_1, \ldots, p_s$. Now by Corollary 2.2.73, based on Theorem 2.2.57 b), p. 100, membership to I can be decided using a Thomas decomposition. In order to rate the complexity of this membership problem, we would like to know estimates for the degrees and differential orders of differential polynomials occurring in a representation of q as element of I.

Let us assume that an upper bound is known for the maximum number of differentiations which need to be applied to any of the p_i such that some power of q is in the algebraic ideal generated by the p_i and their derivatives of bounded order (in a polynomial ring with finitely many indeterminates). Then the (constructive) membership problem reduces to an effective version of Hilbert's Nullstellensatz. A well-known result in this direction can be stated as follows (cf. [Bro87]).

Theorem 2.2.76. *Let $p_1, \ldots, p_s$ be non-zero elements of a commutative polynomial algebra over $\mathbb{Q}$ in r indeterminates, d the maximum degree of $p_1, \ldots, p_s$, and let $\mu := \min\{r, s\}$. If q is in the radical ideal generated by $p_1, \ldots, p_s$, then there exist $e \in \mathbb{N}$ and elements $c_1, \ldots, c_s$ of the same polynomial algebra such that*

$$q^e = \sum_{i=1}^{s} c_i\, p_i, \quad e \le (\mu+1)(r+2)(d+1)^{\mu+1}, \quad \deg(c_j) \le (\mu+1)(r+2)(d+1)^{\mu+2}$$

for all $j = 1, \ldots, s$ such that $c_j \ne 0$.

An upper bound for the number of differentiations necessary for the above reduction was obtained in [GKOS09]. We may assume that $p_1, \ldots, p_s, q$ are non-zero.

Theorem 2.2.77. *Let d be the maximum of the degrees of $p_1, \ldots, p_s, q$, and let h be the maximum of the differential orders (of jet variables in any differential indeterminate) of the same polynomials. If q is an element of the radical differential ideal generated by $p_1, \ldots, p_s$, then some power of q is a linear combination of $p_1, \ldots, p_s$ and their derivatives up to order at most*

$$A(n+8, \max\{m, h, d\}),$$

where A is the Ackermann function recursively defined by

$$A(0, m) = m+1, \quad A(n+1, 0) = A(n, 1), \quad A(n+1, m+1) = A(n, A(n+1, m)).$$

This bound, of course, allows an extreme growth of the polynomials involved in a computation of a Thomas decomposition. However, on the other hand, it applies to every other differential elimination method of this kind, e.g., the Rosenfeld-Gröbner algorithm, and no smaller bound is known up to now for the general case of partial differential equations.

For systems of ordinary differential equations, Seidenberg's elimination method (cf. [Sei56]) was improved by D. Grigoryev in [Gri89], where upper bounds for the time complexity, the differential orders, the degrees, and the bit sizes of the resulting differential polynomials are given in terms of the corresponding data for the input.

In [DJS14] the statement of Theorem 2.2.77 was improved for the case of ordinary differential equations with constant coefficients in a field of characteristic zero. An upper bound L for the order of derivatives is given by (mhd) raised to the power $2^{c(mh)^3}$ for some universal constant $c > 0$, and the exponent of q may be bounded by $d^{m(h+L+1)}$, where h is the maximum of 2 and the differential orders as above.

2.2.5 Hilbert Series for Simple Differential Systems

In this subsection we define the generalized Hilbert series for simple differential systems. It allows, in particular, to determine effectively the generic simple system in a Thomas decomposition of a prime differential ideal, which is our main application in Sect. 3.3. More benefits of the generalized Hilbert series (e.g., as indicated by the corresponding notion in the case of linear differential polynomials, cf. Subsect. 2.1.5) will be studied in the future.

We mention that alternative notions capturing in some sense the dimension of the solution set of a system of polynomial differential equations were developed by, e.g., E. R. Kolchin (cf. [Kol64]), J. Johnson (cf. [Joh69a]), A. Levin (cf., e.g., [Lev10]), and recently by M. Lange-Hegermann (cf. [LH14]).

Let Ω be an open and connected subset of $\mathbb{C}^n$ with coordinates $z_1, \ldots, z_n$ and K the differential field of meromorphic functions on Ω. We denote by R the differential polynomial ring $K\{u_1, \ldots, u_m\}$ in the differential indeterminates $u_1, \ldots, u_m$ with commuting derivations $\partial_1, \ldots, \partial_n$ extending partial differentiation with respect to $z_1, \ldots, z_n$ on K. We define

$$\mathrm{Mon}(\Delta)u := \{\partial^J u_i \mid 1 \le i \le m, J \in (\mathbb{Z}_{\ge 0})^n\}$$

and fix a ranking $>$ on R.

Definition 2.2.78. For any subset M of $\mathrm{Mon}(\Delta)u$ the *generalized Hilbert series* of M is defined by

$$H_M(\partial_1, \ldots, \partial_n) := \sum_{\partial^J u_i \in M} \partial^J u_i \in \bigoplus_{i=1}^m \mathbb{Z}[[\partial_1, \ldots, \partial_n]] u_i,$$

where for simplicity the differential indeterminates $u_1, \ldots, u_m$ are used here also as generators of a free $\mathbb{Z}[[\partial_1, \ldots, \partial_n]]$-module of rank m. For $i = 1, \ldots, m$, we define $H_{M,i}(\partial_1, \ldots, \partial_n)$ by

$$H_M(\partial_1, \ldots, \partial_n) = \sum_{i=1}^m H_{M,i}(\partial_1, \ldots, \partial_n) u_i,$$

and we identify $H_M(\partial_1, \ldots, \partial_n)$ with $H_{M,1}(\partial_1, \ldots, \partial_n)$ in case $m = 1$.

Remark 2.2.79. Let S be a simple differential system with set of equations

$$\{p_1 = 0, \ldots, p_s = 0\}$$

and corresponding sets of admissible derivations $\mu_1, \ldots, \mu_s \subseteq \Delta$ (cf. Def. 2.2.48). We define q to be the product of all $\mathrm{init}(p_i)$ and all $\mathrm{sep}(p_i)$, $i = 1, \ldots, s$. The simple differential system provides a Janet decomposition of the multiple-closed set $[\mathrm{ld}(p_1), \ldots, \mathrm{ld}(p_s)]$, which defines a partition of $M := \mathrm{ld}(\langle p_1, \ldots, p_s \rangle : q^\infty)$ by Theorem 2.2.57 b):

$$M = \biguplus_{i=1}^{s} \mathrm{Mon}(\mu_i) \, \mathrm{ld}(p_i).$$

Accordingly, the generalized Hilbert series of M is obtained from the simple system S as

$$H_M(\partial_1, \ldots, \partial_n) = \sum_{i=1}^{s} \left(\prod_{d \in \mu_i} \frac{1}{1-d} \right) \mathrm{ld}(p_i).$$

The simple differential system S defines a partition of $\mathrm{Mon}(\Delta)u$ into the set M of jet variables which occur as leaders of equations that are consequences of S and the set $\mathrm{Mon}(\Delta)u - M$ of remaining jet variables. We call the elements of M the *principal jet variables* and the elements of $\mathrm{Mon}(\Delta)u - M$ the *parametric jet variables* of the simple differential system S.

In a similar manner, a Janet decomposition of $\overline{M} := \mathrm{Mon}(\Delta)u - M$ with cones $C_j = \mathrm{Mon}(v_j)\theta_j u_{i_j}$, $j = 1, \ldots, k$, allows to write the generalized Hilbert series of $\overline{M}$ in the form

$$H_{\overline{M}}(\partial_1, \ldots, \partial_n) = \sum_{j=1}^{k} \left(\prod_{d \in v_j} \frac{1}{1-d} \right) \theta_j u_{i_j},$$

which enumerates the parametric jet variables of S via expansion of the (formal) geometric series.

Definition 2.2.80. Let S be a simple differential system and $M := \mathrm{ld}(\langle S^= \rangle : q^\infty)$, where q is the product of the initials and separants of all elements of the set $S^=$ and $\overline{M} := \mathrm{Mon}(\Delta)u - M$. Then the *Hilbert series counting the principal jet variables* or the *parametric jet variables* of S is the formal power series in λ with non-negative integer coefficients defined by

$$H_S(\lambda) := \sum_{i=1}^{m} H_{M,i}(\lambda, \ldots, \lambda), \quad H_{\overline{S}}(\lambda) := \sum_{i=1}^{m} H_{\overline{M},i}(\lambda, \ldots, \lambda), \text{ respectively.}$$

Remark 2.2.81. In the same way as a Janet basis for a system of linear partial differential equations allows to determine all (formal) power series solutions for the system (cf. Rem. 2.1.67, p. 50), a simple differential system is a formally integrable system of PDEs. Whereas in the linear case the Taylor coefficients corresponding to the principal derivatives are uniquely determined by any choice of values for the Taylor coefficients corresponding to the parametric derivatives, the equations of a simple differential system allow a finite number of values for the Taylor coefficients that are associated with the principal jet variables. If a principal jet variable is the leader of an equation of the simple system, then its degree in that equation coincides with this number because the left hand side is a square-free polynomial. Taylor coefficients for proper derivatives of such leaders are uniquely determined due to quasi-linearity (cf. Rem. 2.2.36).

By part c) of Theorem 2.2.57, p. 100, values for all Taylor coefficients corresponding to the parametric jet variables can be chosen independently. Any choice yields finitely many (formal) power series solutions of the simple system.

The following proposition describes a method which determines the generic simple system in a Thomas decomposition of a system of differential equations whose left hand sides generate a prime differential ideal (cf. Def. 2.2.67).

Proposition 2.2.82. *Suppose that $p_1, \ldots, p_s \in R$ generate a prime differential ideal and let $S_1, \ldots, S_k$ be a Thomas decomposition of the system $\{p_1 = 0, \ldots, p_s = 0\}$. For $i = 1, \ldots, k$, let H_{S_i} (and $H_{\overline{S_i}}$) be the Hilbert series counting the principal jet variables (the parametric jet variables, respectively) of S_i. Comparing the sequences of Taylor coefficients lexicographically, the unique index $i \in \{1, \ldots, k\}$ for which H_{S_i} (or $H_{\overline{S_i}}$) is the least (the greatest, respectively) among these Hilbert series determines the generic simple system S_i of this Thomas decomposition.*

Proof. For every $j \in \mathbb{Z}_{\geq 0}$, the coefficient of λ^j in the Taylor expansion of H_{S_i} equals the number of elements of $\mathrm{ld}(\langle S_i^= \rangle : q_i^\infty)$ which are jet variables of differentiation order j, where q_i is the product of the initials and separants of all elements of $S_i^=$. Obviously, the set inclusions referred to in Corollary 2.2.71 can be checked by comparing the sequences of these coefficients lexicographically. $\square$

Example 2.2.83. Let $R = K\{u\}$ be the differential polynomial ring in one differential indeterminate u with commuting derivations $\partial_w, \partial_x, \partial_y, \partial_z$ over a differential field K of characteristic zero (with derivations $\partial_v|_K$, $v \in \{w,x,y,z\}$). We choose the degree-reverse lexicographical ranking $>$ on R which extends the ordering $\partial_w u > \partial_x u > \partial_y u > \partial_z u$ (cf. Ex. A.3.3, p. 250).

Let us consider the differential system given by

$$
\begin{cases}
u_{w,y} = u_{w,z} = u_{x,y} = u_{x,z} = 0, & \begin{vmatrix} u & u_w & u_y \\ u_x & u_{w,x} & 0 \\ u_z & 0 & u_{y,z} \end{vmatrix} = 0, \\[2em]
\begin{vmatrix} u_w & u_{w,x} \\ u_{w,w} & u_{w,w,x} \end{vmatrix} = \begin{vmatrix} u_x & u_{x,x} \\ u_{w,x} & u_{w,x,x} \end{vmatrix} = \begin{vmatrix} u_y & u_{y,z} \\ u_{y,y} & u_{y,y,z} \end{vmatrix} = \begin{vmatrix} u_z & u_{z,z} \\ u_{y,z} & u_{y,z,z} \end{vmatrix} = 0.
\end{cases}
\tag{2.48}
$$

For reasons that will become clear in Sect. 3.3, the differential ideal of R which is generated by the left hand sides of these equations is prime. (In fact, the left hand sides of the equations in (2.48) generate the prime differential ideal of $K\{u\}$ consisting of all differential polynomials in u which vanish under substitution of $f_1(w) \cdot f_2(x) + f_3(y) \cdot f_4(z)$ for u, where $f_1, \ldots, f_4$ are analytic functions.)

Using the Maple package `DifferentialThomas` (cf. Subsect. 2.2.6), we obtain a Thomas decomposition of (2.48) consisting of simple systems $S_1, \ldots, S_8$. Their Hilbert series counting the parametric jet variables are

$$H_{\overline{S_1}}(\lambda) = 1 + 4\lambda + 5\lambda^2 + \frac{4\lambda^3}{1-\lambda},$$

$$H_{\overline{S_2}}(\lambda) = H_{\overline{S_3}}(\lambda) = 1 + 3\lambda + 4\lambda^2 + \frac{3\lambda^3}{1-\lambda},$$

$$H_{\overline{S_4}}(\lambda) = H_{\overline{S_5}}(\lambda) = H_{\overline{S_6}}(\lambda) = H_{\overline{S_7}}(\lambda) = H_{\overline{S_8}}(\lambda) = 1 + 2\lambda + \frac{2\lambda^2}{1-\lambda}.$$

Proposition 2.2.82 implies that S_1 is the generic simple system of the Thomas decomposition. This system is the following one:

$$\begin{cases} (u\,u_{y,z} - u_y\,u_z)\,u_{w,x} - u_w\,u_x\,u_{y,z} = 0, \; \{\partial_w, \partial_x, \partial_y, \partial_z\}, \\ u_{w,y} = 0, \; \{\partial_w, *, \partial_y, \partial_z\}, \\ u_{w,z} = 0, \; \{\partial_w, *, *, \partial_z\}, \\ u_{x,y} = 0, \; \{*, \partial_x, \partial_y, \partial_z\}, \\ u_{x,z} = 0, \; \{*, \partial_x, *, \partial_z\}, \\ u_y\,u_{y,y,z} - u_{y,y}\,u_{y,z} = 0, \; \{*, *, \partial_y, \partial_z\}, \\ u_z\,u_{y,z,z} - u_{y,z}\,u_{z,z} = 0, \; \{*, *, *, \partial_z\}, \\ u \neq 0, \\ u_y \neq 0, \\ u_z \neq 0, \\ u\,u_{y,z} - u_y\,u_z \neq 0. \end{cases} \qquad (2.49)$$

For illustrative purposes we also give the generalized Hilbert series for S_1. We denote by E the differential ideal of R which is generated by the left hand sides of the equations in (2.49) and we define $q := u_y\,u_z\,(u\,u_{y,z} - u_y\,u_z)$. Then the generalized Hilbert series $H_M(\partial_w, \partial_x, \partial_y, \partial_z)$ of $M := \mathrm{ld}(E : q^\infty)$ is determined along the lines of Remark 2.2.79:

$$\frac{\partial_w\,\partial_x\,u}{(1-\partial_w)(1-\partial_x)(1-\partial_y)(1-\partial_z)} + \frac{\partial_w\,\partial_y\,u}{(1-\partial_w)(1-\partial_y)(1-\partial_z)} + \frac{\partial_w\,\partial_z\,u}{(1-\partial_w)(1-\partial_z)}$$

$$+ \frac{\partial_x\,\partial_y\,u}{(1-\partial_x)(1-\partial_y)(1-\partial_z)} + \frac{\partial_x\,\partial_z\,u}{(1-\partial_x)(1-\partial_z)} + \frac{\partial_y^2\,\partial_z\,u}{(1-\partial_y)(1-\partial_z)} + \frac{\partial_y\,\partial_z^2\,u}{1-\partial_z}.$$

A Janet decomposition of the complement $\overline{M}$ of M in $\mathrm{Mon}(\{\partial_w, \partial_x, \partial_y, \partial_z\})u$ yields the generalized Hilbert series of $\overline{M}$:

$$\frac{1}{1-\partial_z}+\partial_y+\frac{\partial_x}{1-\partial_x}+\frac{\partial_w}{1-\partial_w}+\partial_y\,\partial_z+\frac{\partial_y^2}{1-\partial_y}. \qquad (2.50)$$

Expansion of this geometric series enumerates the parametric jet variables of S_1. The representation (2.50) of the generalized Hilbert series as a rational function indicates that (the essential part of) the set of analytic solutions of (2.49) can be parametrized by four arbitrary analytic functions $f_1(w)$, $f_2(x)$, $f_3(y)$, $f_4(z)$. The conditions $f_3' \neq 0$, $f_4' \neq 0$, which are implied by the inequations $u_y \neq 0$, $u_z \neq 0$ in S_1, are not reflected by the generalized Hilbert series (2.50). A direct comparison of the set of parametric jet variables (enumerated in (2.50)) with the Taylor coefficients of $f_1, \ldots, f_4$, which may be chosen arbitrarily to define a solution of (2.49) of the form

$$u(w,x,y,z) = f_1(w) \cdot f_2(x) + f_3(y) \cdot f_4(z),$$

is hindered, e.g., because the Taylor coefficients of $f_1, \ldots, f_4$ of order zero add up to the Taylor coefficient of u of order zero. (Note also that (2.49) admits further solutions, e.g., certain analytic functions of the form $f_1(w) + f_2(x) + f_3(y) + f_4(z)$.)

2.2.6 Implementations

This subsection is devoted to implementations of Thomas' algorithm and also refers to related packages.

Thomas' algorithm has been implemented in the packages `AlgebraicThomas` and `DifferentialThomas` for the computer algebra system Maple by T. Bächler and M. Lange-Hegermann, respectively, at Lehrstuhl B für Mathematik, RWTH Aachen [BLH] (with the help of V. P. Gerdt and the author of this monograph). As important efficiency issues have been ignored in the above presentation of Thomas' algorithm, the implementation of these packages is not along the lines of the algorithms in the previous subsections. For instance, the packages avoid to apply pseudo-reduction to the same pair of polynomials repeatedly (which occurs in different branches of the splittings of systems). In the algebraic case, handling of the square-free part of a polynomial, being often a very expensive part of the computation, may be postponed. The growth of polynomials during the computation of a Thomas decomposition is especially severe in the differential case due to the product rule of differentiation. Strategies which counteract this growth and heuristics for choosing the equation or inequation to be treated next are needed for an efficient implementation. Factorization of polynomials (whenever possible) leads to further splittings, but the gain in simplification is usually significant. The remarks in the beginning of Subsect. 2.1.6 concerning Janet division also apply to the differential version of Thomas' algorithm. For more details, we refer to [BGL⁺10, BGL⁺12].

The package `AlgebraicThomas` computes Thomas decompositions of algebraic systems. It includes procedures which determine counting polynomials of quasi-affine or quasi-projective varieties as defined in [Ple09a]. Set-theoretic constructions can be applied to solution sets, e.g., forming complements and inter-sections. Moreover, comprehensive Thomas decompositions can be computed, i.e., Thomas decompositions of parametric systems such that specialization of parameters respects the structure of the Thomas decomposition.

The package DifferentialThomas implements Thomas' algorithm for differential systems. The field of coefficients for the differential polynomial ring can be any differential field supported by Maple, and the ranking may be chosen from a list of standard rankings or may be specified in terms of a matrix as discussed in Remark 3.1.39, p. 142. Building on the concept of Janet division (cf. Rem. 2.2.42 and Rem. 2.2.47), the package provides combinatorial data given by the constructed Janet decomposition, such as, e.g., Hilbert series. Moreover, it includes procedures that solve differential systems in terms of (truncated) power series or via the built-in solvers of Maple.

A very useful feature is the possibility to stop the computation of a Thomas decomposition as soon as a given number of simple systems have been constructed by the program and to output these simple systems. Options given to the program determine whether special branches or the generic branch of the splittings of systems should be preferred. The computation of the Thomas decomposition may then be continued, starting from the point where the previous computation stopped. This feature is used in applications presented in Subsect. 3.3.5.

The Maple package epsilon (by Dongming Wang) [Wan04] is a collection of several implementations of triangular decomposition methods. In particular, computation of Thomas decompositions of algebraic systems is possible using epsilon.

The packages RegularChains (by F. Lemaire, M. Moreno Maza, and Y. Xie) [LMMX05] and DifferentialAlgebra (by F. Boulier and E. S. Cheb-Terrab), formerly diffalg (by F. Boulier and E. Hubert, cf. also [Hub00]), are part of the standard Maple library. The latter one is now based on the BLAD libraries (written by F. Boulier in the programming language C) [Bou]. These packages compute decompositions of algebraic varieties and systems of polynomial ordinary and partial differential equations, respectively. In the former case, the decomposition is constructed in terms of regular chains [ALMM99], in the latter case the Rosenfeld-Gröbner algorithm [BLOP09] is applied, which computes finitely many characterizable ideals [Hub03a, Hub03b] whose intersection equals the radical differential ideal generated by the input. In both cases, the decomposition of the solution set is not a disjoint one, in general, but can be made disjoint in principle.

For a further comparison of these packages including timings for benchmark examples, we refer to [BGL$^+$12, Sect. 4].

Chapter 3
Differential Elimination for Analytic Functions

Abstract The implicitization problem for certain parametrized sets of complex analytic functions is solved in this chapter by developing elimination methods based on Janet's and Thomas' algorithms. The first section discusses important elimination problems in detail, in particular, how to compute the intersection of a left ideal of an Ore algebra and a subalgebra which is generated by certain indeterminates, the intersection of a submodule of a finitely generated free module over an Ore algebra and the submodule which is generated by certain standard basis vectors, and the intersection of a radical differential ideal of a differential polynomial ring and a differential subring which is generated by certain differential indeterminates. These techniques allow, e.g., to determine all consequences of a given PDE system involving only certain of the unknown functions. Compatibility conditions for inhomogeneous linear systems are also addressed. The second section treats sets of complex analytic functions given by linear parametrizations, whereas the third section develops differential elimination techniques for multilinear parametrizations. Applications to symbolic solving of PDE systems are given. For instance, a family of exact solutions of the Navier-Stokes equations is computed.

3.1 Elimination

From a constructive point of view, the term "elimination" can be understood as the process of finding among given relations between objects some or all relations not involving certain of these objects. We focus on algebraic equations, which we assume, for algorithmic purposes, to be finite in number. Eliminating a specified set of variables from these equations means determining a generating set for the consequences which are independent of these variables. Geometrically speaking, the elimination process shifts attention from the whole solution set to its projection onto the range of values for the remaining unknowns.

Iteration of this technique may allow to solve the system of equations successively, i.e., to find all values admitted for one of the unknowns depending on the

© Springer International Publishing Switzerland 2014

D. Robertz, *Formal Algorithmic Elimination for PDEs*,
Lecture Notes in Mathematics 2121, DOI 10.1007/978-3-319-11445-3_3

information already determined for other unknowns. For linear equations with co-efficients in a field this is accomplished by Gaussian elimination and back substitution. More generally, triangular sets, as for example given by simple systems in the sense of J. M. Thomas, realize the same idea for systems of algebraic equations, cf. Remark 2.2.5, p. 61.

In later sections of this chapter, the intention of performing elimination is to construct an implicit description of the projection rather than explicitly solving the given system. A very simple example in the context of algebraic geometry is given by the unit circle in the real affine plane with coordinates x, y, where the passage from the rational parametrization

$$t \longmapsto (x,y) = \left(\frac{2t}{t^2+1}, \frac{t^2-1}{t^2+1} \right)$$

to the equation

$$x^2 + y^2 - 1 = 0$$

is an instance of the implicitization problem. Note that the first representation of the circle admits an easy construction of points on the variety, while the second one allows a straightforward check whether a given point belongs to the circle. In this chapter we study the analogous implicitization problem for algebraic differential equations and their analytic solutions.

For lack of space, we will not consider here the important concept of resultant, which is classical in elimination theory (cf., e.g., [Jou91], [EM99]). The (multivariate) resultant of $n+1$ polynomials $p_0, \ldots, p_n$ in variables $x_1, \ldots, x_n$ is an irreducible polynomial in the coefficients of $p_0, \ldots, p_n$ that is zero if these coefficients are chosen such that $p_0, \ldots, p_n$ have a common root (cf., e.g., [GKZ94], also for variants for homogeneous polynomials). This concept has been extended to polynomial differential equations, and the development of elimination methods using differential resultants is ongoing work (cf., e.g., [CF97], [RS10]).

Another approach to polynomial elimination is the characteristic set method by J. F. Ritt and Wen-tsün Wu. The techniques of this approach are similar to J. M. Thomas', but usually waive the disjointness of the decomposition of the solution set. We refer to, e.g., [Rit50], [Wu00], [Wan01], [CLO07], and Subsect. A.3.2.

In Subsect. 3.1.1 we discuss methods for the elimination of variables, i.e., more precisely, ways to compute the intersection of a (left) ideal of an Ore algebra with the subalgebra which is generated by the remaining indeterminates. Such an intersection is also called an elimination ideal. The approach that is pursued here is to construct Janet bases with respect to specific term orderings, which rank the variables to be eliminated higher than the remaining ones. It is well-known that such a computation can be extremely time-consuming. A much more efficient method in this context computes Janet bases for the same ideal repeatedly while varying the term ordering (cf. Rem. 3.1.9). We refer to this method as "degree steering". An analogous technique which uses Buchberger's algorithm is called Gröbner walk.

One immediate application of the possibility to find generating sets for elimination ideals is determining the kernel of a homomorphism between finitely presented commutative algebras (cf. Rem. 3.1.10).

Being a hard computational problem in general, the construction of an elimination ideal profits from utilizing additional structure at hand. Subsection 3.1.2 concentrates on the case where the variables to be eliminated occur in relatively few monomials in the generators. Using the concept of a Veronese embedding from projective algebraic geometry, elimination problems that arise in the context of the later Sect. 3.3 can be solved much more efficiently (cf. Ex. 3.1.20).

Subsection 3.1.3 deals with the problem of determining the intersection of a given submodule of a finitely generated free (left) module of tuples with another submodule which is generated by certain of the standard basis vectors, i.e., standard basis vectors are eliminated instead of variables. Correspondingly, the methods developed in this subsection allow to compute the kernel of a homomorphism between finitely presented (left) modules over Ore algebras. An application to systems of linear partial differential equations allows to eliminate certain of the unknown functions. This technique is generalized to polynomial partial differential equations in Subsect. 3.1.4 using Thomas' algorithm. It effects a projection of the solution set onto the components which define the remaining unknown functions. The method of degree steering can be carried over to the settings of Subsects. 3.1.3 and 3.1.4 in a straightforward way.

The topics of Subsect. 3.1.5 are syzygies and compatibility conditions for inhomogeneous systems of linear functional equations. Determining the latter amounts to a syzygy computation (cf. also Subsect. A.1.3). This is an elimination problem because all consequences of the given equations which do not involve the unknown functions, and hence only the inhomogeneous part, are to be found. We show that a Janet basis almost includes a generating set for its syzygies, which is again a Janet basis with respect to a suitable term ordering, and iteration of this procedure yields a free resolution of the presented module. Moreover, we demonstrate the relevance of injective modules for solving systems of linear functional equations and outline a formal algebraic approach to systems theory, which builds on these ideas.

3.1.1 Elimination of Variables using Janet Bases

In this subsection Janet's algorithm is used to compute the intersection of a left ideal of an Ore algebra D with a certain subalgebra. The subalgebra is generated by a subset of the indeterminates of D. The task is therefore to eliminate the other indeterminates. In principle, computing a Janet basis with respect to a suitable lexicographical term ordering solves this task. Since this method is not feasible in practice, we generalize a technique called "degree steering", introduced in [PR08], to Ore algebras. The approach discussed below will be enhanced in the next subsection for generating sets with special monomial structure.

In what follows, we consider an Ore algebra D as defined in the beginning of Subsect. 2.1.3, i.e., D is obtained as iterated Ore extension with (commuting) indeterminates $\partial_1, \ldots, \partial_l$ of either a field K, in which case we set $n := 0$, or a commutative polynomial algebra $K[z_1, \ldots, z_n]$, where K is a field (of any characteristic) or $K = \mathbb{Z}$, $n \geq 0$.

As in Sect. 2.1, we assume that the operations in D which are necessary for executing the algorithms described below can be carried out effectively, e.g., arithmetic in D and deciding equality of elements in D.

Let $q \in \mathbb{N}$. We denote by $e_1, \ldots, e_q$ the standard basis vectors of the free left D-module $D^{1 \times q}$. Throughout this subsection we suppose that D satisfies Assumption 2.1.23 (p. 22) and that $>$ is an admissible term ordering on $\mathrm{Mon}(D^{1 \times q})$ (i.e., satisfies Assumption 2.1.29, p. 23), so that Janet bases for submodules of $D^{1 \times q}$ (with respect to $>$) can be computed as described in Subsect. 2.1.3.

Definition 3.1.1. Given any subset $\{y_1, \ldots, y_r\}$ of $\mathrm{Indet}(D)$, we denote the smallest subalgebra of D which contains $y_1, \ldots, y_r$ by $K\langle y_1, \ldots, y_r \rangle$. Let Y be the smallest subset of $\mathrm{Indet}(D)$ such that $K\langle y_1, \ldots, y_r \rangle$ is contained in the free left K-module which is generated by $\mathrm{Mon}(Y)$ (cf. Def. 2.1.16; for $Y = \emptyset$ this is supposed to be K). Then we define

$$\mathrm{Mon}(K\langle y_1, \ldots, y_r \rangle) := \mathrm{Mon}(Y),$$

and, denoting by $f_1, \ldots, f_q$ the standard basis vectors of $K\langle y_1, \ldots, y_r \rangle^{1 \times q}$, we set

$$\mathrm{Mon}(K\langle y_1, \ldots, y_r \rangle^{1 \times q}) := \bigcup_{k=1}^{q} \mathrm{Mon}(K\langle y_1, \ldots, y_r \rangle) f_k.$$

Example 3.1.2. Let $A_2(K)$ be the Weyl algebra over K (cf. Ex. 2.1.18 b)) with generators $z_1, z_2, \partial_1, \partial_2$ satisfying $z_1 z_2 = z_2 z_1$, $\partial_1 \partial_2 = \partial_2 \partial_1$, and $\partial_i z_j = z_j \partial_i + \delta_{i,j}$, $i, j = 1, 2$, where $\delta_{i,j}$ is the Kronecker symbol. For every subset $\{y_1, \ldots, y_r\}$ of $\{z_1, z_2, \partial_1, \partial_2\}$ we have

$$\mathrm{Mon}(K\langle y_1, \ldots, y_r \rangle) = \mathrm{Mon}(\{y_1, \ldots, y_r\}). \tag{3.1}$$

On the other hand, let $D = \mathbb{Z}[z_1, z_2][\partial_1; \sigma_1, \delta_1]$ be the Ore algebra defined by the commutation rules

$$z_1 z_2 = z_2 z_1, \quad \partial_1 z_1 = z_1 \partial_1 + 2 z_2, \quad \partial_1 z_2 = z_2 \partial_1.$$

Then D satisfies Assumption 2.1.23, and the degree-reverse lexicographical ordering on D (extending any ordering of the indeterminates) is admissible (cf. Ex. 2.1.30, p. 23). We have

$$\mathrm{Mon}(K\langle z_1, \partial_1 \rangle) = \mathrm{Mon}(\{z_1, z_2, \partial_1\})$$

because the representation of an element of $K\langle z_1, \partial_1 \rangle$ as left $\mathbb{Z}$-linear combination of monomials of D in general involves z_1, ∂_1, and z_2. However, (3.1) holds for every subset $\{y_1, \ldots, y_r\}$ of $\{z_1, z_2, \partial_1\}$ which is different from $\{z_1, \partial_1\}$.

The following lemma is a generalization to the case of the Ore algebra D of a criterion developed in [PR08, Lemma 4.1].

Lemma 3.1.3. *Let J be a Janet basis for a submodule M of the free left D-module $D^{1 \times q}$ with respect to any admissible term ordering $>$ on $D^{1 \times q}$. Let $\{y_1, \ldots, y_r\}$ be a subset of $\mathrm{Indet}(D)$ and $R := K\langle y_1, \ldots, y_r \rangle$. If*

$$\{ p \in J \mid p \in R^{1 \times q} \} = \{ p \in J \mid \mathrm{lm}(p) \in \mathrm{Mon}(R^{1 \times q}) \},$$

then $J \cap R^{1 \times q}$ is a generating set for the left R-module $M \cap R^{1 \times q}$.

Proof. If $M \cap R^{1 \times q} = \{0\}$, there is nothing to prove. Let $0 \neq p \in M \cap R^{1 \times q}$. Since J is a Janet basis for M, there exists a Janet divisor of $\mathrm{lm}(p)$ in J, i.e., $p_1 \in J$ such that $\mathrm{lm}(p)$ is the leading monomial of a left multiple of p_1. Then $\mathrm{lm}(p) \in \mathrm{Mon}(R^{1 \times q})$ implies $\mathrm{lm}(p_1) \in \mathrm{Mon}(R^{1 \times q})$ by the definition of the subalgebra R of D and Assumption 2.1.23. By the hypothesis of the lemma we have $p_1 \in R^{1 \times q}$. Hence, subtracting a suitable left multiple of p_1 from p yields $p' \in M \cap R^{1 \times q}$ satisfying $\mathrm{lm}(p) > \mathrm{lm}(p')$. Janet reduction of p modulo J is therefore a computation in $R^{1 \times q}$. $\square$

Definition 3.1.4. Let $I_1, \ldots, I_k$ form a partition of $\{1, \ldots, n+l\}$ such that $i_1 \in I_{j_1}$, $i_2 \in I_{j_2}$, $i_1 \leq i_2$ implies $j_1 \leq j_2$. Let $\pi \colon \{1, \ldots, n+l\} \to \mathrm{Indet}(D)$ be a bijection and set $B_j := \pi(I_j)$, where $j = 1, \ldots, k$. For every $m \in \mathrm{Mon}(D)$ and $j \in \{1, \ldots, k\}$ we denote by $m^{(j)}$ the result of substituting in m each indeterminate in $\bigcup_{i \neq j} B_i$ with 1. The *block ordering* on $\mathrm{Mon}(D)$ *with blocks of variables* $B_1, \ldots, B_k$ (extending the total ordering $\pi(1) > \pi(2) > \ldots > \pi(n+l)$ of the indeterminates and the degree-reverse lexicographical ordering on the blocks) is defined for $m_1, m_2 \in \mathrm{Mon}(D)$ by

$$m_1 > m_2 \quad :\Longleftrightarrow \quad \begin{cases} m_1 \neq m_2 \quad \text{and} \\ m_1^{(j)} > m_2^{(j)} \quad \text{for} \quad j = \min\{\, 1 \leq i \leq k \mid m_1^{(i)} \neq m_2^{(i)} \,\}, \end{cases}$$

where monomials on the right hand side are compared with respect to the degree-reverse lexicographical ordering (induced by π, cf. Ex. 2.1.27, p. 23).

For $q \in \mathbb{N}$, we call any extension of a block ordering on $\mathrm{Mon}(D)$ to a term ordering on $\mathrm{Mon}(D^{1 \times q})$ a *block ordering on* $\mathrm{Mon}(D^{1 \times q})$ (with the same blocks).

Example 3.1.5. The coarsest partition of $\mathrm{Indet}(D)$ defines a degree-reverse lexicographical ordering on $\mathrm{Mon}(D)$, whereas the finest partition defines a lexicographical ordering on $\mathrm{Mon}(D)$. In general, the contributions of the blocks are compared lexicographically. (Extensions to $\mathrm{Mon}(D^{1 \times q})$ for $q > 1$ as term-over-position or position-over-term orderings are defined as in Example 2.1.28, p. 23.)

Remark 3.1.6. Clearly, only the block orderings that are admissible in the sense of Assumption 2.1.29, p. 23, are relevant for our purposes. Let $B_1, \ldots, B_k$ be blocks of variables as defined above. If for all $i = 1, \ldots, n$ and $j = 1, \ldots, l$ the σ_j-derivation δ_j satisfies that $\delta_j(z_i)$ is a polynomial of total degree at most one and involves only indeterminates in $\{z_1, \ldots, z_n\} \cap (B_r \cup B_{r+1} \cup \ldots \cup B_k)$, where $z_i \in B_r$, then the block ordering with blocks of variables $B_1, \ldots, B_k$ is admissible.

The next proposition shows that computing a Janet basis with respect to a block ordering solves elimination problems along the lines of Lemma 3.1.3. In what follows, we write $K\langle B_i,\ldots,B_k\rangle$ for $K\langle y \mid y \in B_i \cup B_{i+1} \cup \ldots \cup B_k\rangle$. Janet decompositions of multiple-closed subsets of $\mathrm{Mon}(K\langle B_i,\ldots,B_k\rangle^{1\times q})$ are defined with respect to the total ordering on $B_i \cup B_{i+1} \cup \ldots \cup B_k$ that is the restriction of the chosen total ordering on $\mathrm{Indet}(D)$.

Proposition 3.1.7. *Let J be a Janet basis for a submodule M of the free left D-module $D^{1\times q}$ with respect to an admissible block ordering $>$ with blocks of variables $B_1, \ldots, B_k$. Then, for every $i \in \{1,\ldots,k\}$, the elements of $J \cap K\langle B_i,\ldots,B_k\rangle^{1\times q}$ form a Janet basis for the left $K\langle B_i,\ldots,B_k\rangle$-module $M \cap K\langle B_i,\ldots,B_k\rangle^{1\times q}$ with respect to the restriction of $>$ to $\mathrm{Mon}(K\langle B_i,\ldots,B_k\rangle^{1\times q})$, which is an admissible block ordering with blocks of variables $B_i, \ldots, B_k$.*

Proof. By Definition 3.1.4 of a block ordering, the hypothesis of Lemma 3.1.3 is satisfied (with $K\langle y_1,\ldots,y_r\rangle = K\langle B_i,\ldots,B_k\rangle$). Hence, $J \cap K\langle B_i,\ldots,B_k\rangle^{1\times q}$ is a generating set for the left $K\langle B_i,\ldots,B_k\rangle$-module $M \cap K\langle B_i,\ldots,B_k\rangle^{1\times q}$.

Note that the intersection of a $\mathrm{Mon}(X)$-multiple-closed subset S of $\mathrm{Mon}(X)$ with $\mathrm{Mon}(Y)$ for any subset Y of X is a $\mathrm{Mon}(Y)$-multiple-closed subset of $\mathrm{Mon}(Y)$, and that by selecting from a cone decomposition of S those cones whose generators involve only variables in Y and intersecting the corresponding sets of multiplicative variables with Y, we obtain a cone decomposition of $S \cap \mathrm{Mon}(Y)$. Therefore, associating with each element of $J \cap K\langle B_i,\ldots,B_k\rangle^{1\times q}$ the intersection of its set of multiplicative variables with $B_i \cup \ldots \cup B_k$ defines a Janet complete subset of $K\langle B_i,\ldots,B_k\rangle^{1\times q}$ (with respect to the restriction of the total ordering of the indeterminates to $B_i \cup \ldots \cup B_k$).

The check that $J \cap K\langle B_i,\ldots,B_k\rangle^{1\times q}$ is a Janet basis thus amounts to verifying that Janet reduction (modulo this set) of every left multiple of each generator by a non-multiplicative variable yields zero. However, these verifications are actually the same as the corresponding ones performed by Janet's algorithm over D because the block ordering with blocks of variables $B_1, \ldots, B_k$ used on $\mathrm{Mon}(D^{1\times q})$ restricts to the relevant one on $\mathrm{Mon}(K\langle B_i,\ldots,B_k\rangle^{1\times q})$. The fact that J is a Janet basis with respect to the former term ordering therefore implies the claim.   □

The properties of a block ordering, which are described in the previous proposition, suggest *elimination ordering* as an alternative name for a block ordering.

Remark 3.1.8. As mentioned in the introduction to this section, elimination of variables can be used to compute equations for the (Zariski closure of the) projection of a variety onto some coordinate subspace. Let D be the coordinate ring of the ambient (affine or projective) space. The intersection of the vanishing ideal I of the variety with the subalgebra R of D generated by the coordinates of the subspace is the vanishing ideal of the projection. According to Proposition 3.1.7 it can be determined by computing a Janet basis J for I with respect to a block ordering with blocks of variables B_1 and B_2, where B_2 consists of the variables of R, and by extracting the subset of elements of J involving only variables in B_2.

Note that the projection of an affine variety onto a coordinate subspace is not closed in the Zariski topology in general (as, e.g., for the hyperbola $x \cdot y = 1$ in the affine plane). The *Main Theorem of Elimination Theory* ensures, however, that the projection is closed in the Zariski topology if the space along which the projection is defined is projective (cf., e.g., [Eis95]).

If the variety is given by a rational parametrization, then the previous method can be applied to compute an implicit description of the variety by projecting the graph of this map along the parameter space. For the example of the unit circle given in the introduction, we consider the ideal I of $\mathbb{Q}[t,x,y]$ which is generated by $\{x \cdot (t^2 + 1) - 2t, \, y \cdot (t^2 + 1) - (t^2 - 1)\}$. We compute the intersection of I with $\mathbb{Q}[x,y]$ by removing from the Janet basis for I with respect to the block ordering with blocks of variables $\{t\}$, $\{x,y\}$ the polynomials which involve t. The remaining polynomial is $x^2 + y^2 - 1$.

Conversely, not every variety admits a rational parametrization. For algorithmic solutions of the corresponding problem for algebraic curves, we refer to [SWPD08].

Remark 3.1.9. For reasons of efficiency, the partition of the set of indeterminates which is formed by $B_1, \ldots, B_k$ should be chosen as coarse as possible for the given elimination problem.

Let us assume that the blocks form the finest partition of $\mathrm{Mon}(D)$, i.e., we assume that the block ordering is an admissible lexicographical ordering on $\mathrm{Mon}(D)$. Let J be a Janet basis for a left ideal I of D with respect to this term ordering. Then, for all $i = 1, \ldots, n + l$, the intersection of J with the subalgebra R of D which is obtained by removing the indeterminates in $B_1, B_2, \ldots, B_{i-1}$ as generators, is a generating set for $I \cap R$ (cf. Prop. 3.1.7). In the case of a commutative polynomial algebra D, this allows to find the solutions to the system of algebraic equations given by I recursively, one variable after the other (cf. Ex. 2.1.47, p. 34, for an example and also Remark 2.2.5, p. 61, for the corresponding property of algebraic systems that are simple in the sense of J. M. Thomas).

A Janet basis computation which establishes this strong property is very inefficient in practice. This can be understood by analyzing the polynomial reductions which are performed by Janet's algorithm. A suitable left multiple of a Janet divisor (cf. Def. 2.1.37 b), p. 27) is subtracted from a polynomial in order to eliminate its leading monomial, and every monomial occurring in the result is less than the eliminated one with respect to the term ordering. Using a term ordering which is compatible with the total degree ensures that the result has total degree less than or equal to the total degree of the given polynomial, whereas a lexicographical ordering allows an arbitrarily high total degree of the remainder.

Choosing a coarser partition as mentioned above may often result in a feasible computation for problems of moderate size. However, a block ordering is not adapted to more difficult problems. In [PR08] a strategy, called *degree steering*, is proposed which is akin to the idea of Gröbner walk (cf. [CKM97], [MR88]). Solving an elimination problem by degree steering computes Janet bases for the same ideal or module repeatedly with respect to the degree-reverse lexicographical ordering which is being adjusted to the elimination problem. The comparison of monomials with respect to degrevlex is changed slightly by considering instead of the total degree of a monomial a weighted sum of its exponents. Starting with standard weights 1 for all the indeterminates (or, preferably, weights with respect to which the ideal or module has a generating set consisting of homogeneous elements), the weights of the indeterminates to be eliminated are increased gradually until the elimination is completed, which is detected by Lemma 3.1.3.

Elimination of variables allows to compute a Janet basis for the kernel of a homomorphism of finitely presented commutative algebras.

Remark 3.1.10. Let K be a computable field or $K = \mathbb{Z}$ and define the commutative polynomial algebras $D_1 := K[x_1, \ldots, x_n]$ and $D_2 := K[y_1, \ldots, y_m]$ over K. Moreover, let $\varphi \colon D_1 \to D_2$ be a K-algebra homomorphism[1]. In order to compute the kernel of φ, we consider D_1 and D_2 as subalgebras of $D_{1,2} := K[x_1, \ldots, x_n, y_1, \ldots, y_m]$, define the ideal I of $D_{1,2}$ which is generated by $\{x_1 - \varphi(x_1), \ldots, x_n - \varphi(x_n)\}$, and compute the intersection $I \cap D_1$. The correctness of this approach can be understood as follows. Let p be an n-variate polynomial with coefficients in K. Then $p(x_1, \ldots, x_n)$ and $p(\varphi(x_1), \ldots, \varphi(x_n)) = \varphi(p(x_1, \ldots, x_n))$ represent the same residue class of $D_{1,2}$ modulo I. Hence, the kernel of φ equals the set of representatives in D_1 of the zero residue class in $D_{1,2}/I$. This shows that we have $\ker(\varphi) = I \cap D_1$. Proposition 3.1.7 implies that the subset of polynomials in $x_1, \ldots, x_n$ in a Janet basis for I with respect to a block ordering with blocks of variables $\{y_1, \ldots, y_m\}$, $\{x_1, \ldots, x_n\}$ is a Janet basis for $\ker(\varphi)$.

More generally, if $\varphi \colon D_1/I_1 \to D_2/I_2$ is a homomorphism of K-algebras, where I_i is an ideal of D_i, $i = 1$, 2, then $\varphi(x_j)$ in the above generating set for I is to be replaced with any representative of $\varphi(x_j + I_1)$, $j = 1, \ldots, n$. Now $\ker(\varphi)$ consists of the residue classes in D_1/I_1 which have representatives in $(I + I_1 + I_2) \cap D_1$, and a Janet basis for $I + I_1 + I_2$ with respect to a block ordering yields a Janet basis for $\ker(\varphi)$.

Remark 3.1.11. Let K be a computable field or $K = \mathbb{Z}$ and let I be an ideal of the commutative polynomial algebra $D := K[x_1, \ldots, x_n]$ and $h \in D$. The saturation

$$I : h^\infty := \{d \in D \mid h^r \cdot d \in I \text{ for some } r \in \mathbb{Z}_{\geq 0}\}$$

can be computed using a technique known as Rabinowitsch's trick in algebraic geometry. We may assume that $I \neq D$ and that $h \neq 0$. Let $\{g_1, \ldots, g_m\}$ be a generating set for the ideal I. We consider D as a subalgebra of $D' := K[x_1, \ldots, x_n, u]$, where u is a new indeterminate, and the ideal I' of D' which is generated by $g_1, \ldots, g_m$ and $u \cdot h - 1$. Then $h + I'$ is invertible in D'/I' with inverse $u + I'$. Every element d of $I : h^\infty$ belongs to the zero residue class in D'/I' because $h^r \cdot d \equiv 0$ modulo I' implies $d \equiv 0$ modulo I'. Conversely, if $d \in I' \cap D$, then there exist $a_1, \ldots, a_m, b \in D'$ such that

---

[1] Recall that all algebra homomorphisms are assumed to map the multiplicative identity element to the multiplicative identity element.

$$d = \sum_{i=1}^{m} a_i \cdot g_i + b \cdot (u \cdot h - 1).$$

Substituting $1/h$ for u and multiplying both sides by h^r with $r \in \mathbb{Z}_{\geq 0}$ sufficiently large yields an expression of $h^r \cdot d$ as D-linear combination of the generators of I. Therefore, if J is a Janet basis for I' with respect to a block ordering with blocks of variables $\{u\}, \{x_1, \ldots, x_n\}$, then by Proposition 3.1.7, $J \cap D$ is a Janet basis for the saturation $I : h^\infty$.

Example 3.1.12. We would like to determine all polynomial relations that are satisfied by given rational functions $f_i \in K(x_1, \ldots, x_n)$, $i = 1, \ldots, m$, where K is a computable field or $K = \mathbb{Z}$. Let $f_i = g_i/h_i$, where $g_i, h_i \in K[x_1, \ldots, x_n]$, $h_i \neq 0$, $i = 1, \ldots, m$, and let h be the least common multiple of $h_1, \ldots, h_m$. We consider $D := K[u_1, \ldots, u_m]$ as a subalgebra of $D' := K[x_1, \ldots, x_n, v, u_1, \ldots, u_m]$, where $u_1, \ldots, u_m, v$ are new indeterminates, and the ideal I' of D' which is generated by $h_1 \cdot u_1 - g_1, \ldots, h_m \cdot u_m - g_m$, and $v \cdot h - 1$. Then $h_i + I'$ is invertible in D'/I' with inverse $v \cdot h/h_i + I'$. If $p \in D$ satisfies $p(f_1, \ldots, f_m) = 0$, then $p(u_1, \ldots, u_m)$ and $p(f_1, \ldots, f_m)$ represent the same residue class of D' modulo I', which is the zero residue class by assumption. Therefore, we have $p \in I' \cap D$. Conversely, if $d \in I' \cap D$, then there exist $a_1, \ldots, a_m, b \in D'$ such that

$$d = \sum_{i=1}^{m} a_i \cdot (h_i \cdot u_i - g_i) + b \cdot (v \cdot h - 1).$$

Similarly to the previous remark, substituting $1/h$ for v and multiplying both sides by h^r with $r \in \mathbb{Z}_{\geq 0}$ sufficiently large yields an expression of $p := h^r \cdot d$ as linear combination of $h_i \cdot u_i - g_i$, $i = 1, \ldots, m$, with coefficients in $K[x_1, \ldots, x_n, u_1, \ldots, u_m]$, and we have $p(f_1, \ldots, f_m) = 0$. Since $h \neq 0$, this implies $d(f_1, \ldots, f_m) = 0$. This shows that $I' \cap D$ equals the kernel of the K-algebra homomorphism

$$\psi : K[u_1, \ldots, u_m] \longrightarrow K(x_1, \ldots, x_n) : u_i \longmapsto f_i, \quad i = 1, \ldots, m.$$

Hence, if J is a Janet basis for I' with respect to a block ordering with blocks of variables $\{x_1, \ldots, x_n, v\}, \{u_1, \ldots, u_m\}$, then by Proposition 3.1.7, $J \cap D$ is a Janet basis for $\ker(\psi)$.

For instance, if $f_1 = x/(y+z)$, $f_2 = y/(x+z)$, $f_3 = z/(x+y) \in \mathbb{Q}(x, y, z)$, then a Janet basis for $\ker(\psi)$ consists of the polynomial $2u_1u_2u_3 + u_1u_2 + u_1u_3 + u_2u_3 - 1$.

The following problem, dealt with in [PR08] (cf. also [PR07]), is motivated by the matrix group recognition project (cf., e.g., [O'B06]).

Example 3.1.13. The polynomial relations which are satisfied by the coefficients of the characteristic polynomial of the Kronecker product of two general matrices of square shape are to be determined (cf. also [LGO97]). Some instances of this problem were treated in [PR08] using the technique of degree steering (cf. Rem. 3.1.9). Let K be a computable field or $K = \mathbb{Z}$. We write monic polynomials in t with coefficients in K in the form

$$t^n + \sum_{i=1}^{n} (-1)^i a_i t^{n-i}.$$

Let p_1 and p_2 be two such polynomials of degree n and m and with coefficients a_i and b_j, respectively. Then the characteristic polynomial of the Kronecker product of the companion matrices of p_1 and p_2 is a polynomial of degree $n \cdot m$, whose coefficients c_k can be expressed as polynomials in a_i and b_j:

$$c_1 = a_1 b_1, c_2 = a_1^2 b_2 + a_2 b_1^2 - 2a_2 b_2, \ldots, c_{nm-1} = a_n^{m-1} a_{n-1} b_m^{n-1} b_{m-1}, c_{nm} = a_n^m b_m^n$$

(cf. also [Gla01, Thm. 2.1] for general statements about these polynomials). The task is to eliminate all a_i and b_j, in order to obtain a generating set for the polynomial relations satisfied by the c_k, i.e., to compute the intersection of the corresponding ideal I of $K[a_i, b_j, c_k]$ with the subalgebra $K[c_k]$.

While for the case $(n,m) = (2,2)$ the generating relation $c_3^2 - c_1^2 c_4 = 0$ can be obtained by hand calculation, the case $(n,m) = (2,3)$ was solved in [Sch99] using the computer algebra system Magma [BCP97].

If the grading of the commutative polynomial algebra $K[a_i, b_j, c_k]$ is defined by $\deg(a_i) = i$, $\deg(b_j) = j$, $\deg(c_k) = 2k$, then I is generated by homogeneous polynomials. Clearly, the intersection $I \cap K[c_k]$ is homogeneous with respect to this grading as well, but for $K[c_k]$ we may also choose $\deg(c_k) = k$.

The cases $(n,m) \in \{(2,4),(3,3)\}$ were addressed in [PR08], giving a complete answer for $(n,m) = (2,4)$. It turns out that a good strategy is to deal first with the case $a_n = b_m = c_{nm} = 1$ of matrices of determinant one. Homogenization of generators of the elimination ideal for this special case yields many generators for the general problem already. Starting with the weights defined above (ensuring homogeneity) and increasing the weights of the variables to be eliminated step by step in repeated Janet basis computations for the ideal I with respect to a degree-reverse lexicographical ordering is a flexible technique to arrive at a Janet basis for the elimination ideal, if computationally feasible. We refer to [PR08] for more details and for a report on computations for the cases $(n,m) \in \{(2,4),(3,3)\}$ using the C++ software `ginv` (cf. also Subsect. 2.1.6).

3.1.2 Complexity Reduction by means of a Veronese Embedding

Motivated by the computational difficulty of eliminating variables, we are going to describe a method which reduces the complexity of this problem in cases where the variables to be eliminated occur in relatively few terms of the given generators. This technique will be applied to differential elimination problems in Remark 3.3.17, p. 204, and Remark 3.3.18.

Let K be a field of arbitrary characteristic and denote by $\mathbb{A}^n$ and $\mathbb{P}^n$ affine and projective n-space, respectively, defined over K.

Recall that a *Veronese variety* is the set of points in a projective space which is parametrized by a morphism that maps the homogeneous coordinates of another projective space to the tuple of all monomials in these coordinates of a fixed degree. More precisely, such a parametrization is defined as follows (cf., e.g., [Har95, p. 23]).

Definition 3.1.14. Given $n, d \in \mathbb{N}$, set $N := \binom{n+d}{d} - 1$, and let $m_0, m_1, \ldots, m_N$ be the monomials of degree d in the homogeneous coordinate ring of $\mathbb{P}^n$, in some fixed ordering. The *Veronese map* $v_{n,d}: \mathbb{P}^n \to \mathbb{P}^N$ of degree d (with respect to the chosen ordering) is defined as evaluation of $(m_0 : m_1 : \ldots : m_N)$ at the homogeneous coordinates of points of $\mathbb{P}^n$.

The following well-known proposition lists homogeneous equations whose solution set in $\mathbb{P}^N$ is the image of the Veronese map $v_{n,d}$. A proof can be found, e.g., in [Grö70, III § 3]. (For a study of the Gröbner bases for the toric ideals which are generated by these equations, cf. also [Stu96, Chap. 14].)

Proposition 3.1.15. *Let $K[X_0, \ldots, X_n]$ and $K[M_0, \ldots, M_N]$ be the homogeneous coordinate rings of $\mathbb{P}^n$ and $\mathbb{P}^N$, respectively, and let*

$$v_{n,d}^*: K[M_0, \ldots, M_N] \longrightarrow K[X_0, \ldots, X_n]: M_i \longmapsto m_i, \qquad i = 0, \ldots, N,$$

be the K-algebra homomorphism corresponding to the Veronese map $v_{n,d}$. Moreover, we define $\varepsilon: \{0, \ldots, N\} \to (\mathbb{Z}_{\geq 0})^{n+1}$ by $m_i = X^{\varepsilon(i)}$. Then the kernel of $v_{n,d}^*$ is generated by

$$\{ M_i M_j - M_k M_l \mid 0 \leq i, j, k, l \leq N, \ \varepsilon(i) + \varepsilon(j) = \varepsilon(k) + \varepsilon(l) \},$$

and the image of $v_{n,d}$ is the set of solutions in $\mathbb{P}^N$ of this homogeneous ideal.

Example 3.1.16. Let $d = 2$. Using the notation of the previous proposition and the Kronecker symbol, we let $a(i,j) \in \{0, 1, \ldots, N\}$ for $i, j \in \{1, 2, \ldots, n+1\}$ be such that $\varepsilon(a(i,j))$ is the multi-index in $(\mathbb{Z}_{\geq 0})^{n+1}$ whose k-th entry equals $\delta_{i,k} + \delta_{j,k}$ for $k = 1, \ldots, n+1$. Then the 2×2 minors of the symmetric $(n+1) \times (n+1)$ matrix whose entry at position (i,j) equals $M_{a(i,j)}$ form a generating set for the kernel of $v_{n,2}^*$. For $n = 2$, the image of $v_{2,2}$ is the *Veronese surface* in $\mathbb{P}^5$ (cf., e.g., [Har95, p. 23]), whose vanishing ideal is generated by the 2×2 minors of

$$\begin{pmatrix} M_0 & M_1 & M_2 \\ M_1 & M_3 & M_4 \\ M_2 & M_4 & M_5 \end{pmatrix},$$

if $(m_0, m_1, \ldots, m_5) = (X_0^2, X_0 X_1, X_0 X_2, X_1^2, X_1 X_2, X_2^2)$.

Remark 3.1.17. The homogeneous coordinates of a point in the image of $v_{n,d}$ determine the homogeneous coordinates of a preimage of this point under $v_{n,d}$ uniquely. In fact, the Veronese map $v_{n,d}$ is a homeomorphism from $\mathbb{P}^n$ onto its image in $\mathbb{P}^N$

with respect to the Zariski topology. This is due to the fact that the set of solutions in $\mathbb{P}^n$ of a system of homogeneous equations is not altered if each equation is replaced with its multiples by all monomials of a fixed degree. Hence, every closed subset V of $\mathbb{P}^n$ is the set of solutions of a set of homogeneous polynomials in $m_0, \ldots, m_N$, i.e., the image of V under $v_{n,d}$ is closed.

Remark 3.1.18. Let $K[X_0, \ldots, X_n]$ be the homogeneous coordinate ring of $\mathbb{P}^n$. We would like to construct an implicit description of the image of the morphism of projective varieties

$$\varphi \colon \mathbb{P}^n \longrightarrow \mathbb{P}^r \colon x = (x_0 : \ldots : x_n) \longmapsto (p_0(x) : \ldots : p_r(x)),$$

where $p_0, \ldots, p_r \in K[X_0, \ldots, X_n]$ are homogeneous of the same degree d. Note that the morphism φ factors over a Veronese map. In fact, let $m_0, \ldots, m_N$ be the monomials in $X_0, \ldots, X_n$ of degree d in any fixed ordering, and choose homogeneous polynomials $q_0, \ldots, q_r \in K[M_0, \ldots, M_N]$ such that we have $p_i = q_i(m_0, \ldots, m_N)$ for all $i = 0, \ldots, r$. Then we have $\varphi = \psi \circ v_{n,d}$, where ψ is defined by

$$\psi \colon v_{n,d}(\mathbb{P}^n) \longrightarrow \mathbb{P}^r \colon m = (m_0(x) : \ldots : m_N(x)) \longmapsto (q_0(m) : \ldots : q_r(m))$$

and $v_{n,d}$ is the Veronese map corresponding to the chosen ordering of monomials. In general, $q_0, \ldots, q_r$ are not uniquely determined.

Clearly, this construction can also be applied to the map $\widehat{\varphi} \colon \mathbb{A}^{n+1} \to \mathbb{A}^{r+1}$ between the affine cones of the projective spaces $\mathbb{P}^n$ and $\mathbb{P}^r$ which is defined by φ. If $p_0, \ldots, p_r$ can be expressed as polynomials in less than $N+1$, say t, monomials listed among $m_0, \ldots, m_N$, then $\widehat{\varphi}$ factors over a morphism of affine varieties $\widehat{v} \colon \mathbb{A}^{n+1} \to \mathbb{A}^t$, whose components are defined by those monomials.

The problem to eliminate $X_0, \ldots, X_n$ is simplified, e.g., if it is possible to replace $X_0, \ldots, X_n$ with a smaller number of variables $M_{i_1}, \ldots, M_{i_t}$, which represent monomials $m_{i_1}, \ldots, m_{i_t}$ in $X_0, \ldots, X_n$ as above. In case $t > n$, a reduction of the number of variables may also be realized if some of the chosen polynomials $q_0, \ldots, q_r$ can be solved for certain of the variables $M_{i_1}, \ldots, M_{i_t}$, which allows to remove the latter from the variable list (cf. Ex. 3.1.19 below). Moreover, substituting monomials in $X_0, \ldots, X_n$ of degree greater than one by new variables also reduces the degree of the polynomials under consideration. Of course, the monomials which are represented by $M_{i_1}, \ldots, M_{i_t}$ are not algebraically independent over K in general. However, their relations are provided by Proposition 3.1.15. Adding the relevant relations to the input of the elimination problem, where the above substitutions have been performed, can be understood as a preprocessing for the elimination at comparably little cost (cf. Ex. 3.1.20 below).

Example 3.1.19. We would like to determine all algebraic relations that are satisfied by the partial derivatives up to order three of functions of the form

$$u(x,y,z) = f_1(x) \cdot f_2(y) + f_3(y) \cdot f_4(z), \tag{3.2}$$

where f_1, f_2, f_3, and f_4 are arbitrary analytic functions of one argument. To this end, we compute symbolically all partial derivatives of the equation (3.2) up to order three, form the difference of the left hand side and the right hand side for each equation, and replace the derivatives of u, f_1, ..., f_4 with the corresponding jet variables in U, F_1, ..., F_4, respectively. Let I be the ideal which is generated by these expressions in the commutative polynomial algebra

$$D := \mathbb{Q}[U_{(i_1,i_2,i_3)}, (F_j)_k \mid i_1, i_2, i_3 \in \mathbb{Z}_{\geq 0}, i_1 + i_2 + i_3 \leq 3, j = 1, \ldots, 4, k = 0, \ldots, 3],$$

whose 36 indeterminates are the jet variables up to order three. We will also use

$$U_{\underbrace{x,\ldots,}_{i_1}\underbrace{x,y,\ldots,}_{i_2}\underbrace{y,z,\ldots,z}_{i_3}}$$

as a synonym for $U_{(i_1,i_2,i_3)}$ and F_i', F_i'' etc. as synonyms for the jet variables representing the first and second derivatives of f_i and f_j, respectively.

Our aim is to compute a Janet basis for the intersection of I with the subalgebra $R := \mathbb{Q}[U_{(i_1,i_2,i_3)} \mid i_1, i_2, i_3 \in \mathbb{Z}_{\geq 0}, i_1 + i_2 + i_3 \leq 3]$, which is a polynomial ring in 20 variables. We readily obtain $U_{x,z}$, $U_{x,x,z}$, $U_{x,y,z}$, $U_{x,z,z}$ as generators of this intersection without any further computation, and we may exclude these from the list of variables, if we substitute each of them with zero whenever occurring elsewhere.

Using the previous remark, we consider the subalgebra $\mathbb{Q}[(F_j)_k]$ with 16 indeterminates as coordinate ring of $\mathbb{Q}^{16}$, the algebra $\mathbb{Q}[M_1, \ldots, M_{20}]$ as coordinate ring of $\mathbb{Q}^{20}$, and define the morphism of affine varieties $v \colon \mathbb{Q}^{16} \to \mathbb{Q}^{20}$ with corresponding $\mathbb{Q}$-algebra homomorphism $v^* \colon \mathbb{Q}[M_1, \ldots, M_{20}] \to \mathbb{Q}[(F_j)_k]$ mapping M_1, ..., M_{20} to

$$F_1 F_2, \ F_3 F_4, \ F_1' F_2, \ F_3' F_4, \ F_1 F_2', \ F_3 F_4', \ F_1'' F_2, \ F_1' F_2', \ F_3' F_4, \ F_1 F_2'', \ F_3' F_4', \ F_3 F_4'',$$

$$F_1''' F_2, \ F_1'' F_2', \ F_1' F_2'', \ F_1 F_2''', \ F_3''' F_4, \ F_3'' F_4', \ F_3' F_4'', \ F_3 F_4''',$$

respectively. By Proposition 3.1.15, the kernel of v^* is generated by

$$
\begin{aligned}
G := \{ & M_1 M_8 - M_3 M_5, \quad M_1 M_{14} - M_5 M_7, \quad M_1 M_{15} - M_3 M_{10}, \\
& M_2 M_{11} - M_4 M_6, \quad M_2 M_{18} - M_6 M_9, \quad M_2 M_{19} - M_4 M_{12}, \\
& M_3 M_{14} - M_7 M_8, \quad M_4 M_{18} - M_9 M_{11}, \quad M_5 M_{15} - M_8 M_{10}, \\
& M_6 M_{19} - M_{11} M_{12} \}.
\end{aligned}
\tag{3.3}
$$

We extend v^* to a $\mathbb{Q}$-algebra homomorphism

$$v' \colon D' := \mathbb{Q}[U_{(i_1,i_2,i_3)}, M_1, \ldots, M_{20} \mid i_1, i_2, i_3 \in \mathbb{Z}_{\geq 0}, i_1 + i_2 + i_3 \leq 3] \longrightarrow D$$

by $U_{(i_1,i_2,i_3)} \mapsto U_{(i_1,i_2,i_3)}$ and consider R as a subalgebra of D'. By the homomorphism theorem, we have $D'/\ker(v') \cong \operatorname{im}(v')$. Instead of $I \subseteq \operatorname{im}(v')$ we investigate the ideal I' of D' which is generated by G and

$$\begin{cases} U - M_1 - M_2, & U_x - M_3, & U_y - M_4 - M_5, & U_z - M_6, \\ U_{x,x} - M_7, & U_{x,y} - M_8, & U_{y,y} - M_9 - M_{10}, & U_{y,z} - M_{11}, \\ U_{z,z} - M_{12}, & U_{x,x,x} - M_{13}, & U_{x,x,y} - M_{14}, & U_{x,y,y} - M_{15}, \\ U_{y,y,y} - M_{16} - M_{17}, & U_{y,y,z} - M_{18}, & U_{y,z,z} - M_{19}, & U_{z,z,z} - M_{20}. \end{cases} \tag{3.4}$$

Since those polynomials in (3.4) which depend on a jet variable in U whose index involves x or z define trivial rewriting rules, we remove these polynomials from the list of generators, apply the inverses of these rules to G, and only deal with

$$\{ U - M_1 - M_2, \ U_y - M_4 - M_5, \ U_{y,y} - M_9 - M_{10}, \ U_{y,y,y} - M_{16} - M_{17} \} \cup G,$$

which is a set of homogeneous polynomials in 24 variables. Now we compute a Janet basis for $I' \cap R$ using, e.g., the block ordering with blocks of variables $\{M_1, \ldots, M_{20}\}$, $\{U, U_y, U_{y,y}, U_{y,y,y}\}$ on D' (cf. Prop. 3.1.7), or using the degree steering method (cf. Rem. 3.1.9). The minimal Janet basis consists of the determinants

$$\begin{vmatrix} U_x & U_{x,y} \\ U_{x,x} & U_{x,x,y} \end{vmatrix}, \qquad \begin{vmatrix} U_z & U_{y,z} \\ U_{z,z} & U_{y,z,z} \end{vmatrix}, \qquad \begin{vmatrix} U & U_y & U_{y,y} \\ U_x & U_{x,y} & U_{x,y,y} \\ U_z & U_{y,z} & U_{y,y,z} \end{vmatrix}$$

and Janet normal forms of certain multiples of these which are added by Janet completion.

The previous example was chosen to balance size and relevance of the elimination problem. When applying computer algebra systems that are specialized in polynomial computations to the above elimination problem, we cannot measure any difference in time spent either on the original problem or on the simplified problem using the Veronese map. Therefore, we summarize in the following remark another experiment dealing with a larger problem of the same kind.

Example 3.1.20. We treat the elimination problem of Example 3.1.19 for analytic functions of the form

$$u(v, w, x, y, z) = f_1(v) \cdot f_2(w) + f_3(w) \cdot f_4(x) + f_5(x) \cdot f_6(y) + f_7(y) \cdot f_8(z), \tag{3.5}$$

where $f_1, \ldots, f_8$ are arbitrary analytic functions of one argument. Let I be the ideal of the appropriate polynomial ring D of jet variables (with $56 + 32$ indeterminates[2]) which is generated by the differences of the left hand sides and right hand sides of equation (3.5) and all its derivatives up to order 3.

We compute a Gröbner basis (with respect to a block ordering) using Singular (version 3-1-6) [DGPS12] and Macaulay2 (version 1.4) [GS], in order to determine the intersection of I with the subalgebra of D which is generated by the jet variables representing the partial derivatives of u up to order 3. The computations were

---

[2] After removing the jet variables representing vanishing derivatives of u (analogous to $U_{x,z}$, $U_{x,x,z}$, $U_{x,y,z}$, $U_{x,z,z}$ in Example 3.1.19), $28 + 32$ variables remain, but this has negligible effect on the performance of the Gröbner basis computations in Singular and Macaulay2.

performed on Linux x86-64 using an AMD Opteron 2.3 GHz processor. The reduced Gröbner basis was computed in about 45 minutes by Singular and in about 40 minutes by Macaulay2.

Using Remark 3.1.18, 18 seconds of computation time in Maple were spent for a preprocessing. Using the notation of the previous example, a Janet (or Gröbner) basis J with respect to a degree-reverse lexicographical ordering is computed for the ideal which is generated by the polynomials in which the monomials in $(F_j)_k$ have been substituted by the corresponding variables M_i. This actually amounts to Gaussian elimination. The variables M_i are ranked higher than the jet variables in U with respect to the term ordering. Then the normal form modulo J of each element of the generating set G for the kernel of v^* is computed, in order to enable the removal of trivial rewriting rules from the elimination problem. (In the previous example, this step corresponds to a reduction of (3.3) modulo (3.4), considering variables M_i in (3.4) as leading monomials.) In the present case, the input for the final elimination step consists of the rewritten differences arising from the derivatives u, u_w, u_x, u_y, $u_{w,w}$, $u_{x,x}$, $u_{y,y}$, $u_{w,w,w}$, $u_{x,x,x}$, $u_{y,y,y}$ and the normal forms of the generators for the kernel of v^*. Now we proceed with a block ordering which ranks the 22 remaining variables M_i higher than the remaining 26 jet variables in U.

The Gröbner basis computations in Singular and in Macaulay2 finish instantly. The resulting Gröbner bases for the intersection are easily checked to be the same in every case (up to removal of the jet variables representing vanishing derivatives of u, which may be memorized in an auxiliary list). We have chosen the degree-reverse lexicographical ordering on the jet variables in U extending $U > U_v > U_w > \ldots > U_{y,z,z} > U_{z,z,z}$ (i.e., sorted according to the alphabet for fixed differential order). The reduced Gröbner basis then consists of

$$U_{v,x}, \quad U_{v,y}, \quad U_{v,z}, \quad U_{w,y}, \quad U_{w,z}, \quad U_{x,z}, \quad \begin{vmatrix} U_v & U_{v,w} \\ U_{v,v} & U_{v,v,w} \end{vmatrix}, \quad \begin{vmatrix} U_z & U_{y,z} \\ U_{z,z} & U_{y,z,z} \end{vmatrix},$$

$$
\begin{aligned}
U \cdot &\begin{vmatrix} U_{v,w} & U_{w,x} \\ U_{v,w,w} & U_{w,w,x} \end{vmatrix} \cdot \begin{vmatrix} U_{w,x} & U_{x,y} \\ U_{w,x,x} & U_{x,x,y} \end{vmatrix} \cdot \begin{vmatrix} U_{x,y} & U_{y,z} \\ U_{x,y,y} & U_{y,y,z} \end{vmatrix} - \\
U_v \cdot &\begin{vmatrix} U_w & U_{w,x} \\ U_{w,w} & U_{w,w,x} \end{vmatrix} \cdot \begin{vmatrix} U_{w,x} & U_{x,y} \\ U_{w,x,x} & U_{x,x,y} \end{vmatrix} \cdot \begin{vmatrix} U_{x,y} & U_{y,z} \\ U_{x,y,y} & U_{y,y,z} \end{vmatrix} - \\
U_{w,x} \cdot &\begin{vmatrix} U_{v,w} & U_w \\ U_{v,w,w} & U_{w,w} \end{vmatrix} \cdot \begin{vmatrix} U_x & U_{x,y} \\ U_{x,x} & U_{x,x,y} \end{vmatrix} \cdot \begin{vmatrix} U_{x,y} & U_{y,z} \\ U_{x,y,y} & U_{y,y,z} \end{vmatrix} - \\
U_{x,y} \cdot &\begin{vmatrix} U_{v,w} & U_{w,x} \\ U_{v,w,w} & U_{w,w,x} \end{vmatrix} \cdot \begin{vmatrix} U_{w,x} & U_x \\ U_{w,x,x} & U_{x,x} \end{vmatrix} \cdot \begin{vmatrix} U_y & U_{y,z} \\ U_{y,y} & U_{y,y,z} \end{vmatrix} - \\
U_z \cdot &\begin{vmatrix} U_{v,w} & U_{w,x} \\ U_{v,w,w} & U_{w,w,x} \end{vmatrix} \cdot \begin{vmatrix} U_{w,x} & U_{x,y} \\ U_{w,x,x} & U_{x,x,y} \end{vmatrix} \cdot \begin{vmatrix} U_{x,y} & U_y \\ U_{x,y,y} & U_{y,y} \end{vmatrix},
\end{aligned}
\tag{3.6}
$$

and a reduced S-polynomial of degree 8 of two of these polynomials. (For an application of the generator (3.6), cf. Example 3.3.38, p. 219.)

3.1.3 Elimination of Unknown Functions using Janet Bases

Let D be an Ore algebra as in Subsect. 3.1.1, $q \in \mathbb{N}$, and $e_1, \ldots, e_q$ the standard basis vectors of the free left D-module $D^{1 \times q}$.

Intersecting a submodule M of $D^{1 \times q}$ with the left D-module which is generated by a certain subset of $\{e_1, \ldots, e_q\}$ is an elimination problem. Given a finite generating set for M, we would like to compute a generating set for the submodule of M whose elements do not involve the basis vectors in the complement. Computing a Janet basis with respect to a suitable term ordering on $\mathrm{Mon}(D^{1 \times q})$ achieves this aim. In the context of systems of linear partial differential equations this construction corresponds to determining the linear differential consequences of the system which involve only certain of the unknown functions (cf. Ex. 3.1.28).

If S is a submodule of $D^{1 \times q}$ (which in the following will typically be generated by a certain subset of $\{e_1, \ldots, e_q\}$), then we denote by $\mathrm{Mon}(S)$ the intersection of $\mathrm{Mon}(D^{1 \times q})$ and S.

The next lemma states a sufficient condition for a Janet basis to solve the elimination problem for standard basis vectors. Its analogy to Lemma 3.1.3 is apparent.

Lemma 3.1.21. *Let J be a Janet basis for a submodule M of $D^{1 \times q}$ with respect to any admissible term ordering $>$ on $D^{1 \times q}$. Let P be a subset of $\{1, \ldots, q\}$ and S the submodule of $D^{1 \times q}$ which is generated by $\{e_i \mid i \in P\}$. If*

$$\{p \in J \mid p \in S\} = \{p \in J \mid \mathrm{lm}(p) \in \mathrm{Mon}(S)\},$$

then $J \cap S$ is a generating set for the left D-module $M \cap S$.

Proof. If $M \cap S = \{0\}$, there is nothing to prove. Otherwise, let $0 \neq p \in M \cap S$. Since J is a Janet basis for M, there exists a Janet divisor $p_1 \in J$ of $\mathrm{lm}(p)$. The leading monomial of p is of the form $m \cdot e_i$ for some $m \in \mathrm{Mon}(D)$ and $i \in P$ because $p \in S$. Hence, $\mathrm{lm}(p_1)$ is a left multiple of e_i, too. Due to the assumption of the lemma we have $p_1 \in J \cap S$. Subtracting a suitable left multiple of p_1 from p yields $p' \in M \cap S$ satisfying $\mathrm{lm}(p) > \mathrm{lm}(p')$. We conclude that Janet reduction of p modulo J is a computation only involving left multiples of e_j, $j \in P$.   □

Similar to the elimination problem for variables (cf. Subsect. 3.1.1), a lexicographic comparison of tuple entries according to a partition of $\{e_1, \ldots, e_q\}$ navigates Janet's algorithm to a Janet basis satisfying the condition in Lemma 3.1.21.

Definition 3.1.22. Let $>_1$, $>_2$, $\ldots$, $>_k$ be admissible term orderings on $\mathrm{Mon}(D)$ and let $I_1, I_2, \ldots, I_k$ form a partition of $\{1, \ldots, q\}$ such that for $i_1 \in I_{j_1}$, $i_2 \in I_{j_2}$, $i_1 \leq i_2$ implies $j_1 \leq j_2$. Moreover, let $\pi \colon \{1, \ldots, q\} \to \{e_1, \ldots, e_q\}$ be a bijection and set $B_j := \pi(I_j)$, $j = 1, \ldots, k$. The *block ordering* on $\mathrm{Mon}(D^{1 \times q})$ *with blocks of standard basis vectors* $B_1, \ldots, B_k$ (extending the term orderings $>_1, >_2, \ldots, >_k$ and the total ordering $\pi(1) > \pi(2) > \ldots > \pi(q)$ of the standard basis vectors) is defined for $m_1 e_{i_1}$, $m_2 e_{i_2} \in \mathrm{Mon}(D^{1 \times q})$, where $e_{i_1} \in B_{j_1}$, $e_{i_2} \in B_{j_2}$, by

$$m_1 e_{i_1} > m_2 e_{i_2} \quad :\Longleftrightarrow \quad \begin{cases} j_1 < j_2 \quad \text{or} \quad \Big(\ j_1 = j_2 \quad \text{and} \quad \big(m_1 >_{j_1} m_2 \quad \text{or} \\ \\ (m_1 = m_2 \quad \text{and} \quad \pi^{-1}(e_{i_1}) < \pi^{-1}(e_{i_2})) \big) \ \Big). \end{cases}$$

Example 3.1.23. Let us assume for simplicity that $>_1, \ldots, >_k$ are given by the same admissible term ordering $>$ on $\mathrm{Mon}(D)$. The coarsest partition of $\{1, \ldots, q\}$ extends $>$ to a term-over-position ordering on $\mathrm{Mon}(D^{1 \times q})$, whereas the finest partition defines the position-over-term ordering on $\mathrm{Mon}(D^{1 \times q})$ extending $>$ (cf. Ex. 2.1.28, p. 23).

Remark 3.1.24. The assumption that $>_1, \ldots, >_k$ are admissible term orderings implies that every block ordering extending $>_1, \ldots, >_k$ is admissible.

The next proposition shows how a Janet basis for the intersection of a submodule M of $D^{1 \times q}$ with another submodule, which is generated by certain of the standard basis vectors $e_1, \ldots, e_q$, can be computed.

Proposition 3.1.25. *Let J be a Janet basis for a submodule M of the free left D-module $D^{1 \times q}$ with respect to an admissible block ordering $>$ with blocks of standard basis vectors $B_1, \ldots, B_k$. For $i = 1, \ldots, k$, let $S_{\geq i}$ be the submodule of $D^{1 \times q}$ which is generated by $B_i \cup \ldots \cup B_k$. Then, for every $i \in \{1, \ldots, k\}$, the elements of $J \cap S_{\geq i}$ form a Janet basis for the left D-module $M \cap S_{\geq i}$ with respect to the restriction of $>$ to $\mathrm{Mon}(S_{\geq i})$, which is an admissible block ordering with blocks of standard basis vectors $B_i, \ldots, B_k$.*

Proof. The hypothesis of Lemma 3.1.21 for $P = B_i \cup \ldots \cup B_k$ is satisfied by Definition 3.1.22 of a block ordering. Therefore, $J \cap S_{\geq i}$ is a generating set for $M \cap S_{\geq i}$. Since the construction of a cone decomposition for a multiple-closed set handles the sets of monomials which are multiples of distinct standard basis vectors of $D^{1 \times q}$ separately (cf. Def. 2.1.32), associating with each element of $J \cap S_{\geq i}$ its set of multiplicative variables defined by J yields a Janet complete subset of $S_{\geq i}$. Passivity of this set now follows from the passivity of J. □

Remark 3.1.26. According to the previous proposition, block orderings defined by partitions $B_1 \cup \ldots \cup B_k = \{e_1, \ldots, e_q\}$ with large k lead to Janet bases which solve several elimination problems at the same time. As an extreme case, a Janet basis with respect to a position-over-term ordering (cf. Ex. 3.1.23) admits a trivial way of computing the intersection of the module it generates with the submodule of $D^{1 \times q}$ generated by the last i standard basis vectors, for every $i = 1, \ldots, q$. Thus, establishing an analogy between term orderings with blocks of standard basis vectors and term orderings with blocks of variables via the coarseness of the associated partitions, a position-over-term ordering corresponds to a lexicographical ordering (cf. Ex. 3.1.5).

A term ordering with blocks of standard basis vectors can be approximated by a term-over-position ordering on $\mathrm{Mon}(D^{1 \times q})$ which compares standard basis vectors according to certain weights. Computing Janet bases for the same module repeatedly, while changing these weights appropriately, is an elimination technique analogous to degree steering (cf. Rem. 3.1.9).

Block orderings as above allow, for instance, to compute kernels of homomorphisms of finitely presented left D-modules.

Example 3.1.27. Let $\varphi\colon D^{1\times q} \to D^{1\times r}$ be a homomorphism of left D-modules, denote by $e_1, \ldots, e_q$ the standard basis vectors of $D^{1\times q}$, and define the submodule M of $D^{1\times q} \oplus D^{1\times r}$ which is generated by $\{\,(e_1,\varphi(e_1)),\ldots,(e_q,\varphi(e_q))\,\}$. The fact that φ is a D-linear map implies that the projection of $M \cap (D^{1\times q} \oplus \{0\})$ onto the first q components equals the kernel of φ.

Let $f_1, f_2, \ldots, f_{q+r}$ be the standard basis vectors of $D^{1\times(q+r)} = D^{1\times q} \oplus D^{1\times r}$ and define $>$ to be the block ordering on $\mathrm{Mon}(D^{1\times(q+r)})$ with blocks of standard basis vectors $B_1 := \{f_{q+1},\ldots,f_{q+r}\}$ and $B_2 := \{f_1,\ldots,f_q\}$ extending any given admissible term orderings on $\mathrm{Mon}(D)$. Then, by Proposition 3.1.25, the intersection of a Janet basis for M with respect to $>$ with $D^{1\times q} \oplus \{0\}$ yields, after projection onto the first q components, a Janet basis for the kernel of φ.

More generally, if $\varphi\colon D^{1\times q}/M_1 \to D^{1\times r}/M_2$ is a homomorphism of left D-modules, where M_1 and M_2 are submodules of $D^{1\times q}$ and $D^{1\times r}$ with finite generating sets G_1 and G_2, respectively, then we define the submodule M of $D^{1\times q} \oplus D^{1\times r}$ which is generated by

$$\{\,(e_1,m_1),\ldots,(e_q,m_q)\,\}\cup\{\,(g_1,0) \mid g_1 \in G_1\,\}\cup\{\,(0,g_2) \mid g_2 \in G_2\,\},$$

where $m_i \in D^{1\times r}$ is any representative of $\varphi(e_i+M_1)$, $i=1,\ldots,q$. Now an analogous Janet basis computation for M as above yields a Janet basis for $\ker(\varphi)$.

Example 3.1.28. In the context of systems of linear partial differential equations, elimination of standard basis vectors amounts to elimination of the corresponding dependent variables (or unknown functions) and all their partial derivatives. We illustrate this technique on the system of linear PDEs

$$\begin{cases} \dfrac{\partial a_1}{\partial z}+y\dfrac{\partial a_2}{\partial z}=0, \quad 2y\dfrac{\partial a_1}{\partial x}-\dfrac{\partial a_1}{\partial y}+2y^2\dfrac{\partial a_2}{\partial x}-y\dfrac{\partial a_2}{\partial y}-a_2=0, \\[2ex] x\dfrac{\partial a_2}{\partial y}-x\dfrac{\partial a_2}{\partial z}=0, \\[2ex] \dfrac{\partial u}{\partial x}+\dfrac{\partial u}{\partial z}-x\dfrac{\partial a_1}{\partial x}-x\dfrac{\partial a_1}{\partial z}-a_1=0, \end{cases} \tag{3.7}$$

which comes up in Example 3.2.31, p. 178 (cf. (3.60) and also Example 3.2.33). Here a_1, a_2, u are unknown analytic functions of the coordinates x, y, z of $\mathbb{C}^3$.

Let K be the field of meromorphic functions on $\mathbb{C}^3$, and let $D = K\langle\partial_x,\partial_y,\partial_z\rangle$ be the skew polynomial ring of differential operators with coefficients in K. Elements of $D^{1\times 3}$ are triples (d_1,d_2,d_3) of differential operators representing linear PDEs $d_1\,a_1(x,y,z)+d_2\,a_2(x,y,z)+d_3\,u(x,y,z)=0$. We define $\Delta := \{\partial_x,\partial_y,\partial_z\}$ and choose a block ordering on $\mathrm{Mon}(\Delta)a_1 \cup \mathrm{Mon}(\Delta)a_2 \cup \mathrm{Mon}(\Delta)u$ with blocks $\{a_1,a_2\}$, $\{u\}$ such that $a_1 > a_2 > u$, extending the degree-reverse lexicographical ordering with $\partial_x > \partial_y > \partial_z$. A Janet basis for the PDE system (3.7) with respect to this block ordering can be computed, which consists of eight differential operators. The subset of elements involving neither a_1 nor a_2 has cardinality five. Three of these elements, which already form a generating set for the same left ideal of D, are given in terms of linear PDEs for u as follows:

$$y\left(u_{x,y,z} - u_{x,z,z} + u_{y,z,z} - u_{z,z,z}\right) - \left(u_{x,z} + u_{z,z}\right) = 0,$$

$$4y^2\left(8y^4 - 4xy^2 - x^2\right)\left(y\left(u_{x,x,x,y} - u_{x,x,x,z} + u_{y,z,z,z} - u_{z,z,z,z}\right) - \left(u_{x,x,x} + u_{z,z,z}\right)\right)$$
$$-4y\left(2y^2 + x\right)\left(4y^2 - x\right)\left(y\left(u_{x,x,y,y} - u_{x,x,z,z} - u_{y,y,z,z} + u_{z,z,z,z}\right) - 2\left(u_{x,x,z} - u_{z,z,z}\right)\right)$$
$$+\left(8y^4 + 8xy^2 - x^2\right)\left(y\left(u_{x,y,y,y} + u_{y,y,y,z} - u_{x,z,z,z} - u_{z,z,z,z}\right) - 3\left(u_{x,z,z} + u_{z,z,z}\right)\right)$$
$$+16y^2\left(y^2 - x\right)\left(y\left(u_{x,x,y} - u_{x,x,z} - u_{y,z,z} + u_{z,z,z}\right) + u_{x,y,y} + u_{y,y,z} - u_{x,z,z} - u_{z,z,z}\right.$$
$$\left. - u_{x,x} + u_{z,z}\right) - 32y\left(y^2 - x\right)\left(u_{x,z} + u_{z,z}\right) = 0,$$

$$xy\left(8y^4 - 4xy^2 - x^2\right)\left(2y\left(u_{x,x,y,y,y} - u_{y,y,y,z,z} - u_{x,x,z,z,z} + u_{z,z,z,z,z}\right)\right.$$
$$\left. - \left(u_{x,y,y,y,y} + u_{y,y,y,y,z} - u_{x,z,z,z,z} - u_{z,z,z,z,z}\right) - 6\left(u_{x,x,y,y} - u_{y,y,z,z}\right)\right)$$
$$-4y\left(y^2 - x\right)\left(8y^4 + 12xy^2 + 3x^2\right)\left(u_{x,x,z,z} - u_{x,x,y,y} + u_{y,y,z,z} - u_{z,z,z,z}\right)$$
$$-4y^2\left(8y^4 - 12xy^2 + x^2\right)\left(u_{x,y,y,y} + u_{y,y,y,z}\right) + 6\left(8xy^4 - x^3\right)\left(u_{x,x,z} - u_{z,z,z}\right)$$
$$+4\left(y^2 - x\right)\left(8y^4 + 4xy^2 + x^2\right)\left(u_{x,z,z,z} + u_{z,z,z,z} - 3\left(u_{x,x,z} - u_{z,z,z}\right)\right)$$
$$+2\left(8y^4 - 12xy^2 + 3x^2\right)\left(2y^2\left(u_{x,x,y} - u_{y,z,z}\right) + 2y\left(u_{x,y,y} + u_{y,y,z} - u_{x,x} + u_{z,z}\right)\right.$$
$$\left. + x\left(u_{x,x,z} - u_{z,z,z}\right) - 4\left(u_{x,z} + u_{z,z}\right)\right) + 32y^3\left(2y^2 - 3x\right)\left(u_{x,z,z} + u_{z,z,z}\right) = 0.$$

Remark 3.1.29. Eliminating dependent variables in a system of linear PDEs determines all linear differential consequences of the system involving only the remaining unknown functions. Since the equations are linear, the solution space of the resulting system of linear PDEs equals the projection of the whole solution space onto the components which define the remaining unknown functions.

Example 3.1.30. Let us consider the Cauchy-Riemann equations

$$\begin{cases} \dfrac{\partial u}{\partial x} - \dfrac{\partial v}{\partial y} = 0, \\[2mm] \dfrac{\partial u}{\partial y} + \dfrac{\partial v}{\partial x} = 0, \end{cases}$$

i.e., we may interpret u and v as real and imaginary part of a complex function depending on $x + iy$. They form a Janet basis with respect to a term-over-position ordering. Computation of a Janet basis with respect to a position-over-term ordering satisfying $u > v$ adds the Laplace equation

$$\frac{\partial^2 v}{\partial x^2} + \frac{\partial^2 v}{\partial y^2} = 0$$

to the system. Exchanging the roles played by u and v also yields the Laplace equation in u. Hence, the construction of Janet bases with respect to block orderings recovers the well-known fact that the real and imaginary part of a holomorphic function are harmonic functions.

As another application of term orderings with blocks of standard basis vectors we extend a method by H. Zassenhaus [Zas66] to compute sums and intersections of vector spaces or $\mathbb{Z}$-modules to left modules over Ore algebras. Given two submodules M_1 and M_2 of the free left D-module $D^{1 \times q}$, the aim is to compute Janet bases for $M_1 \cap M_2$ and $M_1 + M_2$. To this end, we use a certain term ordering on $\mathrm{Mon}(D^{1 \times q} \oplus D^{1 \times q})$ which extends given admissible term orderings on both copies of $D^{1 \times q}$.

Algorithm 3.1.31 (Zassenhaus' trick).

Input: Generating sets $G_1 = \{m_{1,1}, \ldots, m_{1,s}\}$ and $G_2 = \{m_{2,1}, \ldots, m_{2,t}\}$ for submodules M_1 and M_2 of $D^{1 \times q}$, respectively, and admissible term orderings $>_1$, $>_2$ on $D^{1 \times q}$

Output: A Janet basis J_1 for $M_1 + M_2$ with respect to $>_1$ and a Janet basis J_2 for $M_1 \cap M_2$ with respect to $>_2$

Algorithm:

1: define the term ordering $>$ on the free left D-module $D^{1 \times 2q}$ by

$$(r_1, r_2) > (r_1', r_2') \quad :\Longleftrightarrow \quad r_1 >_1 r_1' \quad \text{or} \quad (r_1 = r_1' \quad \text{and} \quad r_2 >_2 r_2').$$

2: compute a Janet basis J for the submodule N of $D^{1 \times 2q}$ generated by

$$G := \{ (m_{1,i}, m_{1,i}) \mid i = 1, \ldots, s \} \cup \{ (m_{2,j}, 0) \mid j = 1, \ldots, t \}$$

with respect to $>$

3: return $J_1 := \{ p_1 \mid (p_1, p_2) \in J, \ p_1 \neq 0 \}$ and $J_2 := \{ p_2 \mid (0, p_2) \in J \}$

Proof. Since the first q components of the elements of G are defined to be the elements of the generating set G_1 of M_1 and of the generating set G_2 of M_2, and since Janet's algorithm computes a generating set for the module to which it is applied, J_1 is a generating set for $M_1 + M_2$. Moreover, J_1 is a Janet basis with respect to $>_1$ because, by definition of the term ordering $>$, the leading monomial of $(p_1, p_2) \in J$ with respect to $>$ appears in an entry of p_1, whenever $p_1 \neq 0$, so that a Janet basis check for J_1 is part of the Janet basis computation of J if the last q components in each $2q$-tuple are neglected.

Every element of N of the form $(0, g)$, $g \in D^{1 \times q}$, has a (unique) representation $(0, g) = (g_1, g_1) + (g_2, 0)$, where $g_1 \in M_1$ and $g_2 \in M_2$, and hence we have $g = g_1$ and $g_1 = -g_2 \in M_1 \cap M_2$. Conversely, any element $g \in M_1 \cap M_2$ yields an element $(0, g)$ of N in this way. In fact, J_2 is a generating set for $M_1 \cap M_2$ because, for any $g \in M_1 \cap M_2$, the element $(0, g)$ of N has normal form $(0, 0)$ modulo J, and the reduction to normal form is a computation in $\{0\} \oplus D^{1 \times q}$ by definition of the term ordering $>$, and therefore involves only the elements $(0, p_2)$, $p_2 \in J_2$, of J. For the same reason, J_2 is a Janet basis with respect to $>_2$.   □

3.1.4 Elimination via Thomas Decomposition

In this subsection we explain how to perform a projection of the solution set of a differential system onto the components defining a certain subset of the unknown functions. The elimination techniques to be discussed use simple systems in the sense of J. M. Thomas and generalize the methods of the previous subsections. (For the special case of linear PDEs, cf. Remark 3.1.29.)

Let R be the differential polynomial ring $K\{u_1,\ldots,u_m\}$ in the differential inde-terminates u_1, ..., u_m with commuting derivations ∂_1, ..., ∂_n, where K is a dif-ferential field of characteristic zero (with derivations $\partial_1|_K$, ..., $\partial_n|_K$), in which all arithmetic operations that are needed below can be carried out effectively. We define $\Delta := \{\partial_1,\ldots,\partial_n\}$, the (commutative) monoid $\mathrm{Mon}(\Delta)$ of monomials in $\partial_1, \ldots, \partial_n$, and the set of jet variables in $u_1, \ldots, u_m$

$$\mathrm{Mon}(\Delta)u := \{\,(u_k)_J \mid 1 \leq k \leq m, J \in (\mathbb{Z}_{\geq 0})^n\,\} = \bigcup_{k=1}^{m} \mathrm{Mon}(\Delta)u_k \subset R$$

as before. Moreover, for any subset V of $\{u_1,\ldots,u_m\}$ we denote by $K\{V\}$ the dif-ferential polynomial ring in the differential indeterminates in V (with the corre-sponding restrictions of $\partial_1, \ldots, \partial_n$) and consider it as a subring of R. The set of jet variables in V is defined by

$$\mathrm{Mon}(\Delta)V := \{\,v_J \mid v \in V, J \in (\mathbb{Z}_{\geq 0})^n\,\} \subseteq \mathrm{Mon}(\Delta)u.$$

A differential system over R is understood to be a finite set of equations $p = 0$ and inequations $q \neq 0$ whose left hand sides are differential polynomials in R.

Recall from Sect. 2.2 that, given any differential system S over R, Thomas' al-gorithm constructs a finite partition of the solution set of S, where each set in this partition is defined by a simple differential system (cf. Def. 2.2.49, p. 95).

We start with a lemma which is analogous to Lemmas 3.1.3 and 3.1.21.

Lemma 3.1.32. *Let S be a simple differential system over R with respect to any ranking $>$ on R and define $E := \langle S^= \rangle$ and the product q of the initials and separants of all elements of $S^=$. Let V be a subset of $\{u_1,\ldots,u_m\}$. If*

$$P := \{\,p \in S^= \mid p \in K\{V\}\,\} = \{\,p \in S^= \mid \mathrm{ld}(p) \in \mathrm{Mon}(\Delta)V\,\}, \qquad (3.8)$$

then we have

$$(E : q^\infty) \cap K\{V\} = E' : (q')^\infty,$$

where E' is the differential ideal of $K\{V\}$ generated by P and where q' is the product of the initials and separants of all elements of P.

Proof. Let $p \in E' : (q')^\infty$. By construction, q' divides q. Therefore, every represen-tation of $(q')^r \cdot p$ for some $r \in \mathbb{Z}_{\geq 0}$ as $K\{V\}$-linear combination of the elements of P and their partial derivatives gives rise to a representation of $q^r \cdot p$ as R-linear combination of the same differential polynomials. This proves the inclusion "$\supseteq$".

For the reverse inclusion let $p \in (E : q^\infty) \cap K\{V\}$. Let $S^= = \{p_1,\ldots,p_s\}$. By Proposition 2.2.50, p. 95, the Janet normal form of p modulo $p_1, \ldots, p_s$ is zero. In fact, Algorithm 2.2.45 (*Janet-reduce*), p. 93, constructs a representation of $b \cdot p$ as R-linear combination of $p_1, \ldots, p_s$ and their partial derivatives, where b is a product of powers of the initials and separants of those polynomials among $p_1, \ldots, p_s$ which are used for the pseudo-reductions. If $p \neq 0$, then we have $\mathrm{ld}(p) \in \mathrm{Mon}(\Delta)V$. By assumption (3.8), the Janet divisor of $\mathrm{ld}(p)$ in $S^=$ is an element of $K\{V\}$. Hence, its initial and separant are elements of $K\{V\}$ and so is the result of pseudo-reduction. Janet reduction of p modulo $p_1, \ldots, p_s$ is therefore a computation in $K\{V\}$.   □

Remark 3.1.33. In the situation of the previous lemma, define

$$S' := \{\, p = 0 \mid p \in P \,\} \cup \{\, r \neq 0 \mid r \in S^{\neq} \,\}.$$

Clearly, S' is simple as an algebraic system, $\{\, p = 0 \mid p \in P \,\}$ is a passive system of partial differential equations, and the elements of $S^{\neq}$ are Janet reduced modulo P. In other words, S' is a simple differential system. In general, however, $S^{\neq}$ is not a subset of $K\{V\}$.

Note that inequations $q_1 \neq 0$, $\ldots$, $q_t \neq 0$ of a simple system may be replaced with $q_1 \cdot \ldots \cdot q_t \neq 0$ without violating the simplicity of the system. Conversely, an inequation may be replaced with the inequations defined by the factors of its left hand side, as long as the result is still a triangular set.

In favorable cases, factorization of inequations may allow to extract from S' a simple system over $K\{V\}$, cf. Examples 3.1.40 and 3.1.41 below. Since we do not assume that factorization in R can be carried out effectively, we do not discuss this issue further. Note finally that, even if the hypothesis of Lemma 3.1.32 is fulfilled for each simple system of a Thomas decomposition and the intersection of each simple system with $K\{V\}$ yields a simple system over $K\{V\}$, the resulting set of simple systems is not a Thomas decomposition in general. For examples where the solution sets of these simple systems are not disjoint, cf. again Examples 3.1.40 and 3.1.41.

Definition 3.1.34. Let $I_1, I_2, \ldots, I_k$ form a partition of $\{1,\ldots,m\}$ such that $i_1 \in I_{j_1}$, $i_2 \in I_{j_2}$, $i_1 \leq i_2$ implies $j_1 \leq j_2$. Let $\pi \colon \{1,\ldots,m\} \to \{u_1,\ldots,u_m\}$ be a bijection and set $B_j := \pi(I_j)$, $j = 1, \ldots, k$. Moreover, fix some degree-reverse lexicographical ordering $>$ on $\mathrm{Mon}(\Delta)$. Then the *block ranking* on R *with blocks* $B_1, \ldots, B_k$ (extending $>$ and the total ordering $\pi(1) > \pi(2) > \ldots > \pi(m)$ of the differential indeterminates) is defined for $\theta_1 u_{i_1}, \theta_2 u_{i_2} \in \mathrm{Mon}(\Delta)u$, where $u_{i_1} \in B_{j_1}$, $u_{i_2} \in B_{j_2}$, by

$$\theta_1 u_{i_1} > \theta_2 u_{i_2} \quad :\Longleftrightarrow \quad \begin{cases} j_1 < j_2 \quad \text{or} \quad \Big(j_1 = j_2 \quad \text{and} \quad \big(\theta_1 > \theta_2 \quad \text{or} \\ (\theta_1 = \theta_2 \quad \text{and} \quad \pi^{-1}(u_{i_1}) < \pi^{-1}(u_{i_2})) \big) \Big). \end{cases}$$

Remark 3.1.35. More generally, block rankings may be defined that extend given term orderings $>_1, \ldots, >_k$ on $\mathrm{Mon}(\Delta)$, which are used to compare jet variables involving differential indeterminates of the same block (cf. Def. 3.1.22 for the analogous case of a block ordering with blocks of standard basis vectors).

For blocks $B_1, \ldots, B_k$ of differential indeterminates in R and a set of indices $\{i_1, \ldots, i_r\} \subseteq \{1, \ldots, k\}$, we denote by $K\{B_{i_1}, \ldots, B_{i_r}\}$ the differential polynomial ring in the differential indeterminates in $B_{i_1} \cup \ldots \cup B_{i_r}$, endowed with the corresponding restrictions of $\partial_1, \ldots, \partial_n$, and consider it as a subring of R.

Proposition 3.1.36. *Let S be a simple differential system over R with respect to a block ranking with blocks $B_1, \ldots, B_k$ and define $E := \langle S^= \rangle$ and the product q of the initials and separants of all elements of $S^=$. For every $i = 1, \ldots, k$, let E_i be the differential ideal of $K\{B_i, \ldots, B_k\}$ which is generated by $P_i := S^= \cap K\{B_i, \ldots, B_k\}$, and define the product q_i of the initials and separants of all elements of P_i. Then, for every $i = 1, \ldots, k$, we have*

$$(E : q^\infty) \cap K\{B_i, \ldots, B_k\} = E_i : q_i^\infty.$$

Proof. By Definition 3.1.34 of a block ranking, P_i satisfies condition (3.8) of Lemma 3.1.32 with $V = B_i \cup \ldots \cup B_k$. $\qquad\square$

Corollary 3.1.37. *Let S be a (not necessarily simple) differential system over R and let $S_1, \ldots, S_r$ be a Thomas decomposition of S with respect to a block ranking with blocks $B_1, \ldots, B_k$. Define the differential ideal $E := \langle S^= \rangle$ and the product q of all elements of $S^{\neq}$. Let $i \in \{1, \ldots, k\}$ be fixed. For every $j = 1, \ldots, r$, let $E^{(j)}$ be the differential ideal of $K\{B_i, \ldots, B_k\}$ which is generated by $P_j := S_j^= \cap K\{B_i, \ldots, B_k\}$, and define the product $q^{(j)}$ of the initials and separants of all elements of P_j. Then we have*

$$\sqrt{E : q^\infty} \cap K\{B_i, \ldots, B_k\} = (E^{(1)} : (q^{(1)})^\infty) \cap \ldots \cap (E^{(r)} : (q^{(r)})^\infty).$$

Proof. For $j = 1, \ldots, r$, define the differential ideal $\widehat{E}^{(j)} := \langle S_j^= \rangle$ of R and the product $\widehat{q}^{(j)}$ of the initials and separants of all elements of $S_j^=$. By Proposition 2.2.72, p. 109, we have

$$\sqrt{E : q^\infty} = (\widehat{E}^{(1)} : (\widehat{q}^{(1)})^\infty) \cap \ldots \cap (\widehat{E}^{(r)} : (\widehat{q}^{(r)})^\infty).$$

Intersecting both sides with $K\{B_i, \ldots, B_k\}$ and applying Proposition 3.1.36 to each simple system S_j proves the claim. $\qquad\square$

Remark 3.1.38. Let S be a (not necessarily simple) differential system over R, defined over the differential field K of meromorphic functions on a connected open subset Ω of $\mathbb{C}^n$. Computation of a Thomas decomposition of S with respect to a block ranking with blocks $B_1, \ldots, B_k$ determines, for every $i = 1, \ldots, k$, the vanishing ideal $\mathscr{I}_i$ of the projection of $\mathrm{Sol}_\Omega(S)$ onto the components given by $B_i \cup \ldots \cup B_k$.

Clearly, as the projection of algebraic varieties onto coordinate subspaces is a special case of this remark (cf. also Rem. 3.1.8), the projection of $\mathrm{Sol}_\Omega(S)$ onto the components given by $B_i \cup \ldots \cup B_k$ is a proper subset of $\mathrm{Sol}_\Omega(\mathscr{I}_i)$, in general. For instance, the system $\{u \cdot v - 1 = 0\}$ in two differential indeterminates u, v has no non-zero consequences involving only v. However, the projection of the solution set onto the component defining v, of course, does not contain the zero function. (Thomas' algorithm actually adds the inequation $v \neq 0$, in order to turn the above system into a simple one. Hence, in this case we obtain the correct projection if we take the inequation into account. Due to a certain ambiguity of the set of inequations, which is produced by Thomas' algorithm in general, we will not go into details of this matter here, cf. also Remark 3.1.33.)

Motivated by the result of Proposition 3.1.36, we may call a block ranking also an *elimination ranking*.

The construction of rankings as presented in the following remark allows to adapt the idea of degree steering (cf. Rem. 3.1.9) to the differential algebra context. Rankings which arise in this way can be traced back to C. Riquier, cf. [Riq10, no. 102].

Remark 3.1.39. Let $\varphi \colon \mathrm{Mon}(\Delta)u \to \mathbb{Q}^{(n+m)\times 1} = \mathbb{Q}^{n\times 1} \oplus \mathbb{Q}^{m\times 1}$ be defined by

$$\partial^I u_j \longmapsto (I, e_j)^\top, \qquad I \in (\mathbb{Z}_{\geq 0})^n, \quad 1 \leq j \leq m,$$

where e_1, e_2, ..., e_m are the standard basis vectors of $\mathbb{Q}^{m\times 1}$. Then every matrix $M \in \mathbb{Q}^{r\times(n+m)}$ defines an irreflexive and transitive relation $>$ on $\mathrm{Mon}(\Delta)u$ by

$$v > w \quad :\Longleftrightarrow \quad M\varphi(v) > M\varphi(w), \qquad v, w \in \mathrm{Mon}(\Delta)u, \tag{3.9}$$

where vectors on the right hand side are compared lexicographically. Let us assume that M admits a left inverse (in particular, we have $r \geq n + m$). Then the linear map $\mathbb{Q}^{(n+m)\times 1} \to \mathbb{Q}^{r\times 1}$ induced by M is injective, and $>$ is a total ordering on $\mathrm{Mon}(\Delta)u$. Linearity of matrix multiplication implies that $>$ satisfies condition b) of Definition A.3.1, p. 249, of a ranking. Moreover, condition a) of the same definition holds if and only if, for each $j = 1, \ldots, n$, the first non-zero entry of the j-th column of M is positive. Every ranking $>$ defined by (3.9) is a *Riquier ranking*, i.e.,

$$\theta_1 u_i > \theta_2 u_i \quad \Longleftrightarrow \quad \theta_1 u_j > \theta_2 u_j$$

holds for all θ_1, $\theta_2 \in \mathrm{Mon}(\Delta)$, $1 \leq i, j \leq m$.

Rankings as defined by (3.9) may be used to approximate block rankings and to perform eliminations following the idea of degree steering. Starting with a matrix M of weights with respect to which the computation of a Thomas decomposition of the system under consideration is not too difficult (e.g., the degree-reverse lexicographical ranking with suitable weights), the construction of further Thomas decompositions of the result with respect to rankings tending to a desired block ranking, is a flexible elimination method. A termination criterion is provided by Lemma 3.1.32.

We also refer to [BLMM10], where a conversion routine for differential characteristic sets was developed, which allows to change the ranking.

Example 3.1.40. Let $R = K\{f_1, f_2, f_3, f_4, u\}$ be the differential polynomial ring over K in the five differential indeterminates f_1, f_2, f_3, f_4, u with commuting derivations $\partial_w, \partial_x, \partial_y, \partial_z$. We choose the block ranking $>$ with blocks $\{f_1, f_2, f_3, f_4\}, \{u\}$, where $f_1 > f_2 > f_3 > f_4 > u$, and which extends the degree-reverse lexicographical ordering on $\mathrm{Mon}(\{\partial_w, \partial_x, \partial_y, \partial_z\})$ satisfying $\partial_z > \partial_y > \partial_x > \partial_w$.

Let us consider the following differential system S over R (which is also used to demonstrate an implicitization method in Remark 3.3.15, p. 203):

$$\begin{cases} u - f_1 \cdot f_2 - f_3 \cdot f_4 = 0, \\ (f_1)_x = (f_1)_y = (f_1)_z = 0, \quad (f_2)_w = (f_2)_y = (f_2)_z = 0, \\ (f_3)_w = (f_3)_x = (f_3)_z = 0, \quad (f_4)_w = (f_4)_x = (f_4)_y = 0. \end{cases}$$

We compute a Thomas decomposition of S with respect to $>$ using the package `DifferentialThomas` in Maple (cf. Subsect. 2.2.6). The result consists of 16 simple differential systems. Their intersections with $K\{u\}$ are listed below as systems (3.11) to (3.20) (without indication of the admissible derivations, but with underlined leaders, where not obvious). Both systems (3.14) and (3.15) below arise twice as intersections of simple systems with $K\{u\}$, and the trivial intersection (3.20) arises five times. (Recall also that a Thomas decomposition is by no means uniquely determined.)

The left hand sides of the inequations in the 16 simple systems are irreducible polynomials and, except for the four systems whose intersections with $K\{u\}$ are given by systems (3.16) to (3.19), these polynomials either involve only u or only $f_1, \ldots, f_4$. For instance, in case of (3.17) the inequations are

$$f_2 \neq 0, \quad f_3 \neq 0, \quad f_4 \neq 0, \quad u - f_3 \cdot f_4 \neq 0, \quad u \neq 0, \tag{3.10}$$

which only yield $u \neq 0$ as a consequence[3] for u. The situation for the systems whose intersections with $K\{u\}$ are given by (3.16), (3.18), (3.19) is analogous.

Hence, the intersections with $K\{u\}$ are simple differential systems over $K\{u\}$ (cf. also Rem. 3.1.33). The fact that some of these systems arise more than once under projection of the given Thomas decomposition already shows that the solution sets of these systems are not disjoint in general. (For more details on the pairwise intersections of the solutions to systems (3.11) to (3.20), cf. below.)

In addition to the equations and inequations of the simple systems we also list their analytic solutions defined on a connected open subset Ω of $\mathbb{C}^4$ with coordinates w, x, y, z. Here we assume that K is the field of meromorphic functions on Ω. The sets of solutions are specified in parametrized form in terms of arbitrary analytic functions $f_1, \ldots, f_4$ of one argument. (We refer to the techniques developed in Sect. 3.3 for more details, in particular to Proposition 3.3.7, p. 199.)

Note that every polynomial in (3.11) to (3.20) can be written as a determinant (cf. also Ex. 3.3.6, p. 196).

---

[3] Although $f_1 \neq 0$ is a consequence of $u - f_1 \cdot f_2 - f_3 \cdot f_4$ and (3.10), it is dispensable as a condition in the description of the analytic solutions in (3.17), as is $f_3 \neq 0$.

$$\begin{cases} u_w\,u_{w,w,x} - u_{w,w}\,u_{w,x} = 0, & u_x\,u_{w,x,x} - u_{w,x}\,u_{x,x} = 0, \\ u_{w,w,y} = u_{w,w,z} = u_{w,x,y} = u_{w,x,z} = 0, \\ (u\,u_{w,x} - u_w\,u_x)\,u_{y,z} - u_y\,u_z\,u_{w,x} = 0, & u_{w,y} = u_{w,z} = u_{x,y} = u_{x,z} = 0, \\ u \neq 0, & u_w \neq 0, & u_x \neq 0, & u_{w,x} \neq 0, & u\,u_{w,x} - u_w\,u_x \neq 0 \end{cases}$$
$$\{ u(w,x,y,z) = f_1(w) \cdot f_2(x) + f_3(y) \cdot f_4(z) \mid f_1' \neq 0,\ f_2' \neq 0,$$
$$u \text{ is not locally of the form } F_1(w) \cdot F_2(x) \text{ for analytic } F_1,\ F_2 \}$$
(3.11)

$$\begin{cases} u_y\,u_{y,y,z} - u_{y,y}\,u_{y,z} = 0, & u_z\,u_{y,z,z} - u_{y,z}\,u_{z,z} = 0, \\ u_{w,x} = u_{w,y} = u_{w,z} = 0, & u_x = 0, \\ u \neq 0, & u_y \neq 0, & u_z \neq 0, & u_{y,z} \neq 0, & u\,u_{y,z} - u_y\,u_z \neq 0 \end{cases}$$
$$\{ u(w,x,y,z) = f_1(w) + f_3(y) \cdot f_4(z) \mid f_3' \neq 0,\ f_4' \neq 0,$$
$$u \text{ is not locally of the form } F_3(y) \cdot F_4(z) \text{ for analytic } F_3,\ F_4 \}$$
(3.12)

$$\begin{cases} u_y\,u_{y,y,z} - u_{y,y}\,u_{y,z} = 0, & u_z\,u_{y,z,z} - u_{y,z}\,u_{z,z} = 0, \\ u_{x,y} = u_{x,z} = 0, & u_w = 0, \\ u \neq 0, & u_x \neq 0, & u_y \neq 0, & u_z \neq 0, & u_{y,z} \neq 0, & u\,u_{y,z} - u_y\,u_z \neq 0 \end{cases}$$
$$\{ f_2(x) + f_3(y) \cdot f_4(z) \mid f_2' \neq 0,\ f_3' \neq 0,\ f_4' \neq 0 \}$$
(3.13)

$$u\,u_{w,x} - u_w\,u_x = 0, \quad u_{w,y} = u_{w,z} = 0, \quad u_y = u_z = 0, \quad u \neq 0$$
$$\{ f_1(w) \cdot f_2(x) \mid f_1 \neq 0,\ f_2 \neq 0 \}$$
(3.14)

$$u\,u_{y,z} - u_y\,u_z = 0, \quad u_w = u_x = 0, \quad u \neq 0$$
$$\{ f_3(y) \cdot f_4(z) \mid f_3 \neq 0,\ f_4 \neq 0 \}$$
(3.15)

$$u_{w,x} = u_{w,y} = u_{w,z} = 0, \quad u_x = u_y = 0, \quad u \neq 0, \quad u_z \neq 0$$
$$\{ f_1(w) + f_4(z) \mid f_4' \neq 0 \}$$
(3.16)

$$u_{w,x} = u_{w,y} = u_{w,z} = 0, \quad u_x = u_z = 0, \quad u \neq 0$$
$$\{ f_1(w) + f_3(y) \mid f_1(w) + f_3(y) \neq 0 \}$$
(3.17)

$$u_{x,y} = u_{x,z} = 0, \quad u_w = u_z = 0, \quad u \neq 0, \quad u_x \neq 0$$
$$\{ f_2(x) + f_3(y) \mid f_2' \neq 0 \}$$
(3.18)

$$u_{x,y} = u_{x,z} = 0, \quad u_w = u_y = 0, \quad u \neq 0, \quad u_x \neq 0, \quad u_z \neq 0$$
$$\{ f_2(x) + f_4(z) \mid f_2' \neq 0,\ f_4' \neq 0 \}$$
(3.19)

$$u = 0$$
(3.20)

In the present case, the solution sets of systems (3.14) to (3.18) are not disjoint. The pairwise intersections can be described symbolically as follows.

$$\text{Sol}_\Omega(3.14) \cap \text{Sol}_\Omega(3.15) = \{ C \mid C \text{ a non-zero constant function} \},$$

$$\text{Sol}_\Omega(3.14) \cap \text{Sol}_\Omega(3.17) = \{ f_1(w) \mid f_1 \neq 0 \},$$

$$\text{Sol}_\Omega(3.14) \cap \text{Sol}_\Omega(3.18) = \{ f_2(x) \mid f_2' \neq 0 \},$$

$$\text{Sol}_\Omega(3.15) \cap \text{Sol}_\Omega(3.16) = \{ f_4(z) \mid f_4' \neq 0 \},$$

$$\text{Sol}_\Omega(3.15) \cap \text{Sol}_\Omega(3.17) = \{ f_3(y) \mid f_3 \neq 0 \}.$$

By a further splitting of (3.14) to (3.18) into simple systems, we may obtain a partition of the union of the solution sets of (3.11) to (3.20). (For instance, we may add $u_w \neq 0$, $u_x \neq 0$ to (3.14) and $u_y \neq 0$, $u_z \neq 0$ to (3.15), cf. also the next example.) The resulting systems form a Thomas decomposition of the set of analytic functions which is parametrized by

$$f_1(w) \cdot f_2(x) + f_3(y) \cdot f_4(z).$$

The context of the implicitization method performed above is explained in Remark 3.3.15, p. 203.

Example 3.1.41. Let $R = K\{a,b,u\}$ be the differential polynomial ring in the differential indeterminates a, b, u with commuting derivations ∂_w, ∂_x, ∂_y, ∂_z. We choose the block ranking $>$ with blocks $\{a,b\}$, $\{u\}$, where $a > b > u$, and which extends the degree-reverse lexicographical ordering on $\text{Mon}(\{\partial_w, \partial_x, \partial_y, \partial_z\})$ satisfying $\partial_z > \partial_y > \partial_x > \partial_w$.

Let us consider the following differential system S over R, which is an instance of an implicitization problem as described in Remark 3.3.17, p. 204:

$$u - a - b = 0, \quad a_y = a_z = 0, \quad \begin{vmatrix} a & a_w \\ a_x & a_{w,x} \end{vmatrix} = 0, \quad b_w = b_x = 0, \quad \begin{vmatrix} b & b_y \\ b_z & b_{y,z} \end{vmatrix} = 0.$$

We apply the Maple package `DifferentialThomas` (cf. Subsect. 2.2.6) to this system, in order to compute a Thomas decomposition of S with respect to $>$. The result consists of eight simple differential systems, whose intersections with $K\{u\}$ are given below. The left hand side of every inequation in these eight simple systems (defined over $K\{a,b,u\}$) is a differential polynomial in u. Hence, the intersections with $K\{u\}$ are simple differential systems over $K\{u\}$ (cf. also Rem. 3.1.33). Assuming that K is the field of meromorphic functions on a connected open subset Ω of $\mathbb{C}^4$ and that w, x, y, z are coordinates of $\mathbb{C}^4$, we also list the sets of analytic solutions of (3.21) to (3.28) in terms of arbitrary analytic functions f_1, f_2, f_3, f_4 of one argument.

$$
\begin{cases}
u_w\,\underline{u_{w,w,x}} - u_{w,w}\,u_{w,x} = 0, & u_x\,\underline{u_{w,x,x}} - u_{w,x}\,u_{x,x} = 0, \\
u_{w,w,y} = u_{w,w,z} = u_{w,x,y} = u_{w,x,z} = 0, \\
(u\,u_{w,x} - u_w\,u_x)\,u_{y,z} - u_y\,u_z\,u_{w,x} = 0, & u_{w,y} = u_{w,z} = u_{x,y} = u_{x,z} = 0, \\
u \neq 0, \quad u_w \neq 0, \quad u_x \neq 0, \quad u_{w,x} \neq 0, \quad u\,u_{w,x} - u_w\,u_x \neq 0
\end{cases}
$$
$$
\{u(w,x,y,z) = f_1(w)\cdot f_2(x) + f_3(y)\cdot f_4(z) \mid f_1' \neq 0,\, f_2' \neq 0,
$$
$$
u \text{ is not locally of the form } F_1(w)\cdot F_2(x) \text{ for analytic } F_1, F_2\}
$$
(3.21)

$$
\begin{cases}
u_y\,\underline{u_{y,y,z}} - u_{y,y}\,u_{y,z} = 0, & u_z\,\underline{u_{y,z,z}} - u_{y,z}\,u_{z,z} = 0, \\
u_{w,x} = u_{w,y} = u_{w,z} = 0, & u_x = 0, \quad u_y \neq 0, \quad u_z \neq 0, \quad u_{y,z} \neq 0
\end{cases}
$$
$$
\{f_1(w) + f_3(y)\cdot f_4(z) \mid f_3' \neq 0,\, f_4' \neq 0\}
$$
(3.22)

$$
\begin{cases}
u_y\,\underline{u_{y,y,z}} - u_{y,y}\,u_{y,z} = 0, & u_z\,\underline{u_{y,z,z}} - u_{y,z}\,u_{z,z} = 0, \\
u_{x,y} = u_{x,z} = 0, & u_w = 0, \quad u_x \neq 0, \quad u_y \neq 0, \quad u_z \neq 0, \quad u_{y,z} \neq 0
\end{cases}
$$
$$
\{f_2(x) + f_3(y)\cdot f_4(z) \mid f_2' \neq 0,\, f_3' \neq 0,\, f_4' \neq 0\}
$$
(3.23)

$$
\begin{cases}
u\,\underline{u_{w,w,x}} - u_x\,u_{w,w} = 0, & u_{w,w,y} = u_{w,w,z} = 0, \\
u\,\underline{u_{w,x}} - u_w\,u_x = 0, & u_{w,y} = u_{w,z} = u_{x,y} = u_{x,z} = 0, \\
u_y = u_z = 0, \quad u \neq 0, \quad u_w \neq 0, \quad u_x \neq 0
\end{cases}
$$
$$
\{f_1(w)\cdot f_2(x) \mid f_1' \neq 0,\, f_2' \neq 0\}
$$
(3.24)

$$
u_{w,x} = u_{w,y} = u_{w,z} = 0, \quad u_x = u_z = 0
$$
$$
\{f_1(w) + f_3(y)\}
$$
(3.25)

$$
u_{w,x} = u_{w,y} = u_{w,z} = 0, \quad u_x = u_y = 0, \quad u_z \neq 0
$$
$$
\{f_1(w) + f_4(z) \mid f_4' \neq 0\}
$$
(3.26)

$$
u_{x,y} = u_{x,z} = 0, \quad u_w = u_z = 0, \quad u_x \neq 0
$$
$$
\{f_2(x) + f_3(y) \mid f_2' \neq 0\}
$$
(3.27)

$$
u_{x,y} = u_{x,z} = 0, \quad u_w = u_y = 0, \quad u_x \neq 0, \quad u_z \neq 0
$$
$$
\{f_2(x) + f_4(z) \mid f_2' \neq 0,\, f_4' \neq 0\}
$$
(3.28)

In this case, the solution sets of the above eight simple systems are pairwise disjoint. Hence, they form a Thomas decomposition of the set of analytic functions which is parametrized by $f_1(w)\cdot f_2(x) + f_3(y)\cdot f_4(z)$. This illustrates an implicitization method introduced in Remark 3.3.17, p. 204. Note again that every polynomial in (3.21) to (3.28) can be written as a determinant (cf. also Ex. 3.3.6, p. 196). A comparison with the previous example yields:

$$(3.21) \iff (3.11) \quad \text{(the respective generic simple system)},$$

$$(3.22) \iff (3.12) \vee ((3.15) \wedge u_y \neq 0 \wedge u_z \neq 0),$$

$$(3.23) \iff (3.13),$$

$$(3.24) \iff (3.14) \wedge u_w \neq 0 \wedge u_x \neq 0,$$

$$(3.25) \iff (3.17) \vee (3.20),$$

$$(3.26) \iff (3.16),$$

$$(3.27) \iff (3.18),$$

$$(3.28) \iff (3.19).$$

3.1.5 Compatibility Conditions for Linear Systems

Returning to systems of linear functional equations as considered in the introduction to Sect. 2.1, we are going to study compatibility conditions for inhomogeneous linear systems.

An inhomogeneous system of linear functional equations, defined over an Ore algebra D, is written as

$$Ru = y, \tag{3.29}$$

where u is a vector of p unknown functions, $R \in D^{q \times p}$, and $y \in \mathscr{F}^{q \times 1}$ is a given vector, whose entries are elements of a left D-module $\mathscr{F}$. The module $\mathscr{F}$ is at the same time the set of candidates for the entries of u. Every matrix $R' \in D^{r \times q}$, $r \in \mathbb{N}$, satisfying $R'R = 0$ yields a necessary condition $R'y = 0$ for (3.29) to be solvable, i.e., a *compatibility condition* on the entries of y. Computation of a generating set for the left D-module of compatibility conditions is not only significant for the study of an inhomogeneous linear system (cf. Ex. 3.1.50), but is also a technique that is used to parametrize the solution set of a homogeneous linear system (cf. Rem. 3.1.55).

The matrix R defines a homomorphism

$$(.R) \colon D^{1 \times q} \longrightarrow D^{1 \times p}$$

of left D-modules, whose cokernel is discussed in the introduction to Sect. 2.1. Hence, we search for a generating set for the *syzygies* of the tuple $(e_1 R, \ldots, e_q R)$, where $e_1, \ldots, e_q$ are the standard basis vectors of $D^{1 \times q}$, and up to projective equivalence, we are interested in the syzygy module of $\mathrm{im}(.R)$ (cf. Subsect. A.1.3). Iterating the computation of syzygies, a free resolution of the cokernel of $(.R)$ is constructed.

In principle, the construction of all compatibility conditions for (3.29) (and also for nonlinear equations) is an elimination problem. The aim is to determine all consequences of (3.29) that are independent of u (which are then equations for the entries of y).

In what follows, we show that the computation of syzygies for the elements of a Janet basis J is a straightforward task. Each pair (p,v), where v is a non-multiplicative variable for the generator p in J, provides a syzygy. In fact, these syzygies form a Janet basis with respect to a suitable term ordering. The details below generalize the corresponding statement for commutative polynomial algebras over fields in [PR05, Thm. 19 4)]. For Gröbner bases they specialize to a well-known result by F.-O. Schreyer (cf. [Sch80] and also [Eis95, Sect. 15.5]).

Let D be an Ore algebra as described in the beginning of Subsect. 2.1.3 which satisfies Assumption 2.1.23 (p. 22), $q \in \mathbb{N}$, and $>$ an admissible term ordering on $\mathrm{Mon}(D^{1\times q})$ (cf. Assumption 2.1.29, p. 23). We denote by $e_1, \ldots, e_q$ the standard basis vectors of $D^{1\times q}$. A totally ordered set $X := \{x_1,\ldots,x_{n+l}\}$ of indeterminates and a bijection $\Xi\colon \mathrm{Mon}(D) \to \mathrm{Mon}(X)$ as in Remark 2.1.31 (p. 24) are fixed, with respect to which Janet decompositions of multiple-closed sets of monomials and their complements in $\mathrm{Mon}(D^{1\times q})$ are uniquely defined (cf. Def. 2.1.32, p. 24).

Definition 3.1.42. Let S be a subset of $\mathrm{Mon}(D^{1\times q})$ that is multiple-closed and let $T = \{(m_1,\mu_1),\ldots,(m_t,\mu_t)\}$ be a cone decomposition of S.

a) The *decomposition graph* of T is defined to be the labeled directed graph with vertices $m_1, \ldots, m_t$ whose set of edges is given as follows. There exists an edge from m_i to m_j labeled by v if and only if $v \in \overline{\mu_i}$ and (m_j,μ_j) is the unique element of T such that $\Xi(v) \cdot \Xi(m_i) \in \mathrm{Mon}(\Xi(\mu_j))\,\Xi(m_j)$.
b) If T is a Janet decomposition of S, then the decomposition graph of T is also called the *Janet graph* of T. If $J = \{(p_1,\mu_1),\ldots,(p_t,\mu_t)\}$ is a Janet basis for a submodule of $D^{1\times q}$, then we define the Janet graph of J to be the Janet graph of $\{(\mathrm{lm}(p_1),\mu_1),\ldots,(\mathrm{lm}(p_t),\mu_t)\}$.

Example 3.1.43. The Janet graph of the Janet decomposition

$$\{(x_1^3 x_2, \{x_1,x_2,x_3\}), (x_1^3 x_3, \{x_1,x_3\}), (x_1^2 x_2, \{x_2,x_3\}), (x_1 x_2, \{x_2,x_3\})\}$$

of the multiple-closed set in Example 2.1.7, p. 12, is given by

$$x_1^3 x_3 \xrightarrow{x_2} x_1^3 x_2 \xleftarrow{x_1} x_1^2 x_2 \xleftarrow{x_1} x_1 x_2.$$

Proposition 3.1.44. *Let S be a multiple-closed subset of $\mathrm{Mon}(D^{1\times q})$. The Janet graph of a Janet decomposition T of S has no cycles.*

Proof. Suppose $T = \{(m_1,\mu_1),\ldots,(m_t,\mu_t)\}$, where $m_1, \ldots, m_t \in \mathrm{Mon}(D^{1\times q})$ and $\mu_1, \ldots, \mu_t \subseteq \mathrm{Indet}(D)$. Since monomials m_i and m_j which are multiples of distinct standard basis vectors of $D^{1\times q}$ represent vertices belonging to different components of the Janet graph of T, we concentrate on the case $q = 1$ without of loss of generality. Moreover, for simplicity, we assume that the total ordering on X is given by $x_1 > x_2 > \ldots > x_{n+l}$. The definition of Janet division (cf. Def. 2.1.5, p. 10) suggests that we compare the monomials corresponding to vertices of the Janet graph with respect to the lexicographical ordering which extends this total ordering on X.

Let us consider an edge from m_1 to m_2 labeled by v in the Janet graph of T. Hence, v is a non-multiplicative variable for (m_1, μ_1), i.e., $v \notin \mu_1$. We are going to prove by contraposition that $\Xi(m_2)$ is greater than $\Xi(m_1)$ with respect to the lexicographical term ordering. Let $\Xi(m_1) = x^a$, $\Xi(m_2) = x^b$, where $a, b \in (\mathbb{Z}_{\geq 0})^{n+l}$, and $\Xi(v) = x_r$.

If there exists $j \in \{1, \ldots, r-1\}$ such that $a_j > b_j$, then there exists a minimal j with this property. For such an index j, the definition of Janet division implies that $\Xi^{-1}(x_j) \notin \mu_2$, which is a contradiction to $\Xi(v) \cdot \Xi(m_1) \in \mathrm{Mon}(\Xi(\mu_2)) \Xi(m_2)$.

Let us assume that $a_i = b_i$ for all $i = 1, \ldots, r-1$. Then, again by definition of Janet division, both assumptions $b_r = a_r$ and $b_r > a_r$ are contradictory to $v \notin \mu_1$ and $v \in \mu_2$.

Hence, $\Xi(m_2)$ is greater than $\Xi(m_1)$ with respect to the lexicographical term ordering, and a cycle in the Janet graph would yield a cycle in the lexicographical ordering, which is impossible. $\qquad \square$

Using the combinatorial information provided by the Janet graph, we are going to show that the non-multiplicative variables of a Janet basis J give rise to a Janet basis for the syzygies of J.

Let $J = \{(p_1, \mu_1), \ldots, (p_t, \mu_t)\}$ be a Janet basis for a submodule M of the free left D-module $D^{1 \times q}$ with respect to any admissible term ordering $>$ on $\mathrm{Mon}(D^{1 \times q})$. Let $f_1, \ldots, f_t$ be the standard basis vectors of $D^{1 \times t}$. We are going to determine a Janet basis for the kernel of the homomorphism of left D-modules

$$\varphi : D^{1 \times t} \longrightarrow D^{1 \times q} : f_i \longmapsto p_i, \qquad i = 1, \ldots, t, \tag{3.30}$$

i.e., for the syzygies of $(p_1, \ldots, p_t)$. To this end, we define

$$\Sigma(J) := \{(p, v) \mid (p, \mu) \in J, v \in \overline{\mu}\}.$$

Recall from Theorem 2.1.43 b), p. 30, that every $v \cdot p$, for $(p, v) \in \Sigma(J)$, has a unique representation $c_1 p_1 + \ldots + c_t p_t$, where each $c_i \in D$ is a left K-linear combination of elements of $\mathrm{Mon}(\mu_i)$, $i = 1, \ldots, t$. We define

$$\sigma_J(p, v) := v f_j - \sum_{i=1}^{t} c_i f_i, \qquad \text{if} \quad p = p_j, \quad (p, v) \in \Sigma(J). \tag{3.31}$$

Then we have $\sigma_J(p, v) \in \ker(\varphi)$ for all $(p, v) \in \Sigma(J)$.

The next step is to define an admissible term ordering on $D^{1 \times t}$ which traces comparisons of monomials in $D^{1 \times t}$ back to comparisons of the corresponding multiples of $p_1, \ldots, p_t$ via (3.30).

Let $>_J$ be a total ordering on $\{p_1, \ldots, p_t\}$ such that $p_i >_J p_j$ holds whenever there exists a directed path in the Janet graph of J from $\mathrm{lm}(p_i)$ to $\mathrm{lm}(p_j)$. Since the Janet graph has no cycles (cf. Prop. 3.1.44), such a total ordering exists (which can be obtained by topological sorting, cf. [Knu97, p. 263]). Finally, we define a total ordering $>_1$ on $\mathrm{Mon}(D^{1 \times t})$ by

$$m_1 f_i >_1 m_2 f_j \quad :\Longleftrightarrow \quad \begin{cases} \operatorname{lm}(m_1 p_i) > \operatorname{lm}(m_2 p_j) \quad \text{or} \\ (\ \operatorname{lm}(m_1 p_i) = \operatorname{lm}(m_2 p_j) \quad \text{and} \quad p_i >_J p_j\ ), \end{cases}$$

$m_1, m_2 \in \operatorname{Mon}(D)$. It follows from the properties of the admissible term ordering $>$ that $>_1$ is an admissible term ordering on $\operatorname{Mon}(D^{1 \times t})$. Using $>_1$ to define the leading monomial of non-zero elements of $D^{1 \times t}$, we have

$$\operatorname{lm}(\sigma_J(p,v)) = v f_j, \quad \text{if} \quad p = p_j, \quad (p,v) \in \Sigma(J). \tag{3.32}$$

In fact, Janet reduction (cf. Alg. 2.1.38, p. 27) constructs the unique representation $c_1 p_1 + \ldots + c_t p_t$ of $v \cdot p$ as above. Hence, there exists a unique $i \in \{1,\ldots,t\}$ such that $\operatorname{lm}(v \cdot p) = \operatorname{lm}(c_i \cdot p_i)$, and there exists an edge in the Janet graph of J from p to p_i, which ensures (3.32).

Note that we may attach the cones with generators $p_1, \ldots, p_t$ defined by J to the corresponding standard basis vectors $f_1, \ldots, f_t$, so that computation with the latter simulates to some extent computation with $p_1, \ldots, p_t$. We will not pursue this idea here but restrict our attention to the leading monomials of the syzygies $\sigma_J(p,v)$. Using the same total ordering of X as above, the distribution of multiplicative variables according to Janet division is achieved as follows. For every $v \in \operatorname{Indet}(D)$ define

$$v^{\le} := \Xi^{-1}(\{x \in X \mid x \le \Xi(v)\}), \tag{3.33}$$

where $\le$ is the reflexive closure of the total ordering on X chosen for constructing Janet decompositions.

Theorem 3.1.45. *Let* $J = \{(p_1, \mu_1), \ldots, (p_t, \mu_t)\}$ *be a Janet basis for a submodule* M *of* $D^{1 \times q}$ *with respect to any admissible term ordering* $>$ *on* $\operatorname{Mon}(D^{1 \times q})$. *For* $(p_j, v) \in \Sigma(J)$ *we define* $\mu_J(p_j, v) := \mu_j \cup v^{\le}$. *Then*

$$J' := \{(\sigma_J(p,v), \mu_J(p,v)) \mid (p,v) \in \Sigma(J)\}$$

is a Janet basis for the syzygies of $(p_1, \ldots, p_t)$ *with respect to* $>_1$.

Proof. As was noted already, we have $\sigma_J(p,v) \in \ker(\varphi)$ for every $(p,v) \in \Sigma(J)$. These syzygies form a generating set for the left D-module $\ker(\varphi)$ because any syzygy $s = (s_1, \ldots, s_t)$ of $(p_1, \ldots, p_t)$ can be written as left D-linear combination of these as follows. By Theorem 2.1.43 b), p. 30, the only representation of 0 as left D-linear combination of $p_1, \ldots, p_t$ in which no coefficient involves any non-multiplicative variable for the corresponding p_j is the one with zero coefficients. Hence, if s is non-zero, there exists $(p_j, v) \in \Sigma(J)$ for some $j \in \{1,\ldots,t\}$ such that v occurs in some monomial of s_j with non-zero coefficient. Subtraction of a suitable left multiple of $\sigma_J(p_j, v)$ from s (cf. (3.31)) yields a syzygy s' of $(p_1, \ldots, p_t)$ which is dealt with in the same way as s. If this reduction process handles the terms of s in decreasing order with respect to $>_1$, then an expression of s in terms of the syzygies $\sigma_J(p,v)$, $(p,v) \in \Sigma(J)$, is obtained in finitely many steps.

In order to prove the assertion of the theorem, we are going to show that J' is Janet complete and passive (cf. Def. 2.1.40, p. 28). Recall from (3.32) that the leading monomial of $\sigma_J(p_j, v)$ with respect to $>_1$ is $v f_j$, where $v \in \overline{\mu}_j$. For each $j = 1, \ldots, t$, the set $\overline{\mu}_j$ equals its Janet completion. In fact, the corresponding Janet decomposition of the multiple-closed subset of $\mathrm{Mon}(D)$ generated by $\overline{\mu}_j$ is given by $\{(w, \mu_j \cup w^\le) \mid w \in \overline{\mu}_j\}$, because Janet division assigns to the variable w in $\overline{\mu}_j$ the set of multiplicative variables $\mu_j \cup w^\le$ (cf. Def. 2.1.5, p. 10). Hence, J' is Janet complete. From the reduction process in the previous paragraph it is clear that

$$[\mathrm{Im}(\sigma_J(p, v)) \mid (p, v) \in \Sigma(J)] = \mathrm{Im}(\ker(\varphi)).$$

By Remark 2.1.41, p. 28, this is equivalent to the passivity of the Janet complete set J'. Thus, J' is a Janet basis for $\ker(\varphi)$ with respect to $>_1$. $\square$

Iterating the construction in Theorem 3.1.45 we obtain a free resolution of a given finitely presented module over an Ore algebra (cf. Subsect. A.1.3).

Corollary 3.1.46. *Let $J = \{(p_1, \mu_1), \ldots, (p_t, \mu_t)\}$ be a Janet basis for a submodule M of $D^{1 \times q}$ with respect to any admissible term ordering $>$ on $\mathrm{Mon}(D^{1 \times q})$. We define $r := \max\{|\overline{\mu}| \mid (p, \mu) \in J\}$ and $J^{(0)} := J$ and, using the notation of Theorem 3.1.45, for $i = 1, \ldots, r$,*

$$\Sigma^{(i)}(J) := \Sigma(J^{(i-1)}), \qquad J^{(i)} := (J^{(i-1)})'.$$

Then a free resolution of $D^{1 \times q}/M$ is given by

$$D^{1 \times q} \xleftarrow{d_0} D^{1 \times |J^{(0)}|} \xleftarrow{d_1} D^{1 \times |J^{(1)}|} \xleftarrow{d_2} \cdots \xleftarrow{d_r} D^{1 \times |J^{(r)}|} \longleftarrow 0, \qquad (3.34)$$

where for $i = 0, \ldots, r$, the homomorphism d_i of left D-modules maps the j-th standard basis vector of $D^{1 \times |J^{(i)}|}$ to the j-th element of $J^{(i)}$, $j = 1, \ldots, |J^{(i)}|$.

Proof. By Theorem 3.1.45, $J^{(i)}$ is a Janet basis for the syzygies of $J^{(i-1)}$ for every $i = 1, \ldots, r$. Note that we have $v \in \mu_J(p_j, v)$ for every $(p_j, v) \in \Sigma(J)$, and therefore $\mu_j \subsetneq \mu_J(p_j, v)$. Hence, the set of non-multiplicative variables is empty for every generator in $J^{(r)}$, which implies that $\Sigma(J^{(r)})$ is empty and d_r is a monomorphism. Hence, (3.34) is a chain complex of free left D-modules that is exact everywhere except at $D^{1 \times q}$, where the homology is isomorphic to $D^{1 \times q}/M$. $\square$

Corollary 3.1.47 (Hilbert's Syzygy Theorem). *Let K be a computable field (resp. $\mathbb{Z}$) and R the commutative polynomial algebra $K[x_1, \ldots, x_n]$ with positive grading. Then for every finitely generated graded R-module there exists a free resolution of length at most n (resp. $n + 1$) with finitely generated free R-modules and homomorphisms which map homogeneous elements to homogeneous elements.*

Proof. Let M be a graded submodule of $R^{1 \times q}$ for some $q \in \mathbb{N}$. Then (3.34) is a free resolution of $R^{1 \times q}/M$ of length $r + 1 \le n + 1$ with finitely generated free R-modules. If for $i = 0, \ldots, r$, the degree of each standard basis vector of $R^{1 \times |J^{(i)}|}$ is defined to be the degree of its image under d_i, then the homomorphisms d_i respect the gradings.

If K is a field, then the leading monomials of the elements of a minimal Janet basis for M form the minimal generating set for the multiple-closed set they generate (cf. Rem. 2.1.41, p. 28), i.e., no leading monomial divides (in the usual sense) any other of these leading monomials. Therefore, if x is the least element of X with respect to the chosen total ordering, then for all cones constructed by Algorithm 2.1.6 (*Decompose*), the indeterminate $\Xi^{-1}(x)$ is a multiplicative variable. Hence, for any minimal Janet basis J for M, we have $r \leq n-1$ in Corollary 3.1.46.   □

Example 3.1.48. Let D be the commutative polynomial algebra $\mathbb{Z}[x,y]$ with the degree-reverse lexicographical ordering extending $x > y$. Applying Corollary 3.1.46 to a Janet basis for the ideal I of D generated by $\{6, 2x, 3y, xy\}$, we construct a free resolution of D/I. The minimal Janet basis J for I (where x is given priority over y for Janet division) is

$$
\begin{aligned}
p_1 &:= 6, \quad \{ * , * \}, \\
p_2 &:= 3y, \quad \{ * , y \}, \\
p_3 &:= 2x, \quad \{ x , * \}, \\
p_4 &:= xy, \quad \{ x , y \}.
\end{aligned}
$$

Writing the four multiples of p_1, p_2, p_3 by their non-multiplicative variables as $\mathbb{Z}[x,y]$-linear combinations of $p_1, \ldots, p_4$ with coefficients involving only multiplicative variables yields

$$
\begin{aligned}
q_1 &:= \sigma_J(p_1,y) = (y, -2, 0, 0), \quad \{ * , y \}, \\
q_2 &:= \sigma_J(p_1,x) = (x, 0, -3, 0), \quad \{ x , y \}, \\
q_3 &:= \sigma_J(p_2,x) = (0, x, 0, -3), \quad \{ x , y \}, \\
q_4 &:= \sigma_J(p_3,y) = (0, 0, y, -2), \quad \{ x , y \},
\end{aligned}
$$

which is a Janet basis for the kernel of

$$
d_0 : D^{1\times 4} \longrightarrow D : (a_1, \ldots, a_4) \longmapsto \sum_{i=1}^{4} a_i \, p_i.
$$

Similarly, by normalizing $x \cdot \sigma_J(p_1, y)$ we compute the syzygy

$$
r := (x, -y, 2, -3), \quad \{ x , y \},
$$

which forms a Janet basis for the kernel of

$$
d_1 : D^{1\times 4} \longrightarrow D^{1\times 4} : (b_1, \ldots, b_4) \longmapsto \sum_{i=1}^{4} b_i \, q_i,
$$

and

$$
D \xleftarrow{\ d_0\ } D^{1\times 4} \xleftarrow{\ d_1\ } D^{1\times 4} \xleftarrow{\ d_2\ } D \longleftarrow 0
$$

is a free resolution of $\operatorname{coker}(d_0)$, where

$$
d_2 : D \longrightarrow D^{1\times 4} : c \longmapsto c \cdot r.
$$

Example 3.1.49. Note that the free resolution constructed along the lines of Corollary 3.1.46 is not minimal in general (in the sense that the ranks of the free modules are unnecessarily large). For instance, if I is an ideal of a commutative polynomial algebra D that is generated by a subset of the indeterminates of D, then the free resolution given by Corollary 3.1.46 is the Koszul complex defined by this regular sequence and is minimal (cf., e.g., [Eis95]). However, if I is the ideal of $\mathbb{Q}[x,y]$ which is generated by x^2 and y^2, we obtain the free resolution

$$D \xleftarrow{\;d_0\;} D^{1\times 3} \xleftarrow{\;d_1\;} D^{1\times 2} \longleftarrow 0, \tag{3.35}$$

where d_0 and d_1 are represented with respect to the standard bases by the matrices

$$\begin{pmatrix} y^2 \\ x^2 \\ xy^2 \end{pmatrix}, \qquad \begin{pmatrix} x & 0 & -1 \\ 0 & -y^2 & x \end{pmatrix},$$

respectively (i.e., row vectors are multiplied by these matrices from the right). The first row of the second matrix expresses the third element of the Janet basis for I in terms of the first one and indicates how (3.35) can be minimized.

The relevance of injective modules for the study of systems of linear functional equations and their compatibility conditions is demonstrated by the following example.

Example 3.1.50. Let K be either $\mathbb{R}$ or $\mathbb{C}$, and let $D = K[\partial_x, \partial_y]$ be the commutative polynomial algebra over K with indeterminates ∂_x, ∂_y, which we interpret as partial differential operators. Let $\mathscr{F}$ be a K-vector space of smooth functions defined on an open subset of K^2 with coordinates x, y. We assume that ∂_x, ∂_y act on $\mathscr{F}$ by partial differentiation with respect to x and y, respectively, which turns $\mathscr{F}$ into a D-module. Let us consider the inhomogeneous system

$$R_1 u = \begin{pmatrix} v_1 \\ v_2 \end{pmatrix}, \qquad \text{where} \quad R_1 := \begin{pmatrix} \partial_x \\ \partial_y \end{pmatrix} \in D^{2\times 1} \quad \text{and} \quad v_1, v_2 \in \mathscr{F}, \tag{3.36}$$

for one unknown function $u \in \mathscr{F}$. Obviously, we have the compatibility condition[4]

$$R_2 \begin{pmatrix} v_1 \\ v_2 \end{pmatrix} = 0, \qquad \text{where} \quad R_2 := (\partial_y \quad -\partial_x) \in D^{1\times 2}. \tag{3.37}$$

The matrices R_1, R_2 induce homomorphisms of D-modules

$$(.R_1): D^{1\times 2} \longrightarrow D, \qquad (.R_2): D \longrightarrow D^{1\times 2},$$

respectively, which give rise to the following exact sequence, where M is defined to be the cokernel of $(.R_1)$:

---

[4] Even though v_1 and v_2 are assumed to be given, the relevant module-theoretic constructions only involve the matrix R_1, which is defined over $\mathbb{Q}$. Hence, the computation of compatibility conditions (i.e., syzygies) can be performed effectively for the given system.

$$0 \longleftarrow M \longleftarrow D \xleftarrow{(.R_1)} D^{1\times 2} \xleftarrow{(.R_2)} D \longleftarrow 0. \tag{3.38}$$

In particular, $(.R_1)$ and $(.R_2)$ define a free resolution of M. By applying the left exact contravariant functor $\hom_D(-,\mathscr{F})$ to (3.38), we obtain a cochain complex of K-vector spaces[5]. Using the isomorphisms $\hom_D(D^{1\times k},\mathscr{F}) \cong \mathscr{F}^{k\times 1}$, the transposed maps of $(.R_1)$ and $(.R_2)$ are homomorphisms of K-vector spaces $(R_1).: \mathscr{F} \to \mathscr{F}^{2\times 1}$ and $(R_2).: \mathscr{F}^{2\times 1} \to \mathscr{F}$, respectively. Then the cochain complex may be written as

$$0 \longrightarrow \hom_D(M,\mathscr{F}) \longrightarrow \mathscr{F} \xrightarrow{(R_1).} \mathscr{F}^{2\times 1} \xrightarrow{(R_2).} \mathscr{F} \longrightarrow 0. \tag{3.39}$$

If $\mathscr{F}$ is an injective D-module, then this cochain complex is an exact sequence (cf. Rem. A.1.12, p. 237). In this case we recover the fact, mentioned in the introduction to Sect. 2.1, that the solution set $\{u \in \mathscr{F} \mid R_1 u = 0\}$ (being the kernel of the homomorphism $(R_1).$) is isomorphic to $\hom_D(M,\mathscr{F})$, i.e., Malgrange's isomorphism (2.2), p. 6. The exactness at $\mathscr{F}^{2\times 1}$ means that the inhomogeneous system (3.36) is solvable if and only if the compatibility condition (3.37) is fulfilled.

Often it is desirable not only that the solution sets inherit structural properties from the corresponding modules, which are defined by the systems of linear equations, but that, conversely, algebraic characteristics of the solution sets are also reflected by the modules. The notion of a cogenerator, which is defined next, implies a faithful duality in that sense. For more details, cf., e.g., [Lam99], [Obe90], [Rot09].

Definition 3.1.51. Let D be a ring with 1. A left D-module C is said to be a *cogenerator* for the category of left D-modules if for every left D-module M we have

$$\bigcap_{\varphi \in \hom_D(M,C)} \ker(\varphi) = \{0\}.$$

Remark 3.1.52. Let D be a ring with 1. A left D-module C is a cogenerator for the category of left D-modules if and only if, for every left D-module M and every $m \in M - \{0\}$, there exists $\varphi \in \hom_D(M,C)$ such that $\varphi(m) \neq 0$. In other words, C is a cogenerator for the category of left D-modules if and only if $\hom_D(M,C) = \{0\}$ implies $M = \{0\}$. Moreover, one can show that the argument used in Example 3.1.50, which proves that the exactness of (3.38) implies the exactness of (3.39), can be reversed if $\mathscr{F}$ is a cogenerator for the category of left D-modules: For the exactness of a chain complex $M_\bullet$ of left D-modules it is sufficient that $\hom_D(M_\bullet,\mathscr{F})$ is an exact sequence of abelian groups (cf. [Rob06, Prop. 4.4.3]).

Theorem 3.1.53 ([Rot09]). *Let D be a ring with 1. Then there exists a left D-module which is injective and a cogenerator for the category of left D-modules.*

The abstract construction proving this existence result is not useful for concrete computations. We list, however, examples of injective modules which are cogenerators and of crucial relevance for systems theory.

---

[5] More generally, if D is a non-commutative Ore algebra, (3.38) is an exact sequence of left D-modules, and (3.39) is a cochain complex of left K-vector spaces, because every element of K commutes with every element of D (cf. Rem. 2.1.15, p. 17).

Examples 3.1.54. a) Let $K \in \{\mathbb{R}, \mathbb{C}\}$ and D be the commutative polynomial algebra $K[s_1, \ldots, s_r]$ over K. Then the following D-modules are injective cogenerators for the category of D-modules (cf. [Mal62, Thm. 3.2], [Ehr70, Thm. 5.20,
Thm. 5.14], [Pal70, Thm. 3 in VII.8.2, Thm. 1 in VII.8.1], [Obe90, Thm. 2.54,
Sects. 3 and 4]):

  (i) the discrete signal space $K^{(\mathbb{Z}_{\geq 0})^r}$, whose elements can be represented by se-
      quences $a = (a_n)_{n \in (\mathbb{Z}_{\geq 0})^r}$, on which s_i acts by shifting the i-th index:

$$(s_i a)_n = a_{(n_1, \ldots, n_{i-1}, n_i+1, n_{i+1}, \ldots, n_r)}, \qquad n \in (\mathbb{Z}_{\geq 0})^r, \quad 1 \leq i \leq r.$$

  $K^{(\mathbb{Z}_{\geq 0})^r}$ can also be viewed as the set of formal power series in r variables
  with coefficients in K, in which case the action of D is given by left shifts of
  the coefficient sequences;
 (ii) the space of convergent power series in r variables with coefficients in K as a
      submodule of $K^{(\mathbb{Z}_{\geq 0})^r}$;
(iii) the space $C^\infty(\Omega, K)$ of K-valued smooth functions on an open and convex
      subset Ω of $\mathbb{R}^r$, where s_i acts by partial differentiation with respect to the i-th
      variable, $i = 1, \ldots, r$;
 (iv) the space $\mathscr{D}'(\Omega, K)$ of K-valued distributions on an open and convex subset
      Ω of $\mathbb{R}^r$, where D acts again by partial differentiation.

b) [Zer06] For the algebra of differential operators with rational function coeffi-
   cients $B_1(\mathbb{R})$ (cf. Ex. 2.1.18 b), p. 19) the $\mathbb{R}$-valued functions on $\mathbb{R}$ which are
   smooth except in finitely many points form an injective cogenerator for the cate-
   gory of left $B_1(\mathbb{R})$-modules.
c) [FO98] Let Ω be an open interval in the field of real numbers $\mathbb{R}$ and define
   the $\mathbb{C}$-algebra $\mathscr{R}_\Omega := \{ f/g \mid f, g \in \mathbb{C}[t], g(\lambda) \neq 0$ for all $\lambda \in \Omega \}$. Moreover, let
   $D := \mathscr{R}_\Omega[\partial]$ be the skew polynomial ring with commutation rule (cf. Def. 2.1.12,
   p. 16)

$$\partial a = a \partial + \frac{da}{dt}, \qquad a \in \mathscr{R}_\Omega.$$

Then the space of Sato's hyperfunctions on Ω is an injective cogenerator for the
category of left D-modules.

In the following remark we outline a formal algebraic approach to systems of
linear functional equations. Quite a few aspects have already been discussed above.
For more details, cf., e.g., [PQ99], [Qua99], [Zer00], [Pom01], [CQR05], [QR07],
[CQ08], [Qua10a], [Qua10b], [Rob14].

Remark 3.1.55. Given a system of (homogeneous) linear equations, a particularly
useful representation of the space of solutions is obtained if it can be recognized as
the image of a linear operator of the same kind as the one defining the equations. In
the case where the equations are defined over a field, Gaussian elimination achieves
this aim. If the system is underdetermined, then the corresponding Gauss-reduced
matrix singles out some variables of the system as parameters and specifies how all
other variables are expressed (linearly) in terms of the parameters. This procedure

can be viewed as identifying the kernel of the linear map induced by the system matrix as the image of another linear map. In particular, every tuple of values assigned to the parameters yields a solution, i.e., the parameters are not subject to any constraints.

More generally, if the equations are defined over a ring, which is not necessarily a (skew) field, the question arises whether it is still possible to construct an operator which is defined over the same ring and whose image equals the space of solutions. In general, the answer is negative. This question and the corresponding one for nonlinear differential equations, is known as *Monge's problem*; we refer to [Gou30, Zer32, Jan71] for historical details and [Car14, Hil12] for important contributions. Applications to control and systems theory can be found, e.g., in [PQ99, Qua99, Woo00, Zer00, Pom01, Sha01].

If solutions of a system of linear functional equations $Ry = 0$ defined over an Ore algebra D are to be found in $\mathscr{F}^{p \times 1}$, where $\mathscr{F}$ is a left D-module, then the duality defined by $\hom_D(-, \mathscr{F})$ allows, to a certain extent, to characterize structural information about the solution space in terms of properties of the left D-module $M := D^{1 \times p}/D^{1 \times q}R$, where $R \in D^{q \times p}$ (cf. also the introduction to Sect. 2.1). The duality is particularly precise if $\mathscr{F}$ is injective and a cogenerator for the category of left D-modules.

There is a one-to-one correspondence between the torsion elements of M and the left D-linear combinations of the entries of y which have a non-zero annihilator in D whenever y satisfies $Ry = 0$. Hence, non-trivial torsion elements give rise to constraints on the possible values of parameters for the solution space. In fact, if the left D-module $\mathscr{F}$ is chosen appropriately, parametrizability of the solution space is equivalent to the triviality of the torsion submodule $t(M)$ of M. If a parametrization exists, then an algorithm which computes $\mathrm{ext}_D^1(N, D) \cong t(M)$, where $N := D^{q \times 1}/RD^{p \times 1}$, constructs a parametrization at the same time. Methods from homological algebra allow to describe a hierarchy of parametrizability in terms of the vanishing of certain extension groups $\mathrm{ext}_D^i(N, D)$, $i \geq 1$. For instance, $\mathrm{ext}_D^1(N, D) \cong t(M)$ is trivial if and only if M is torsion-free, i.e., the solution space is parametrizable; $\mathrm{ext}_D^i(N, D) = \{0\}$ for $i \in \{1, 2\}$ holds if and only if M is reflexive, i.e., in addition, if a parametrization of the solution space is considered as a system of linear functional equations, then its solution space is again parametrizable, etc. The de Rham complex on a convex open subset of $\mathbb{R}^3$ is a well-known example which satisfies these conditions.

Whereas in the case of linear equations with coefficients in a field it is always possible to parametrize the solution set by an injective linear map, an injective parametrization may not exist in general. In control theoretic applications, linear and also nonlinear systems of differential equations whose solution sets have injective parametrizations in terms of arbitrary functions are nowadays said to be *flat* [FLMR95]. In the framework described above, a system $Ry = 0$ is flat if and only if M is a free left D-module. An algorithm which computes bases of free modules over the Weyl algebras, if the field of definition is of characteristic zero, has been obtained in [QR07] based on theorems of J. T. Stafford [Sta78] and constructive versions of them [HS01], [Ley04]. We also refer to [FQ07] for an implementation of the Quillen-Suslin theorem, which is relevant for computing bases of free modules over commutative polynomial algebras.

The methods developed in [CQR05] have been implemented in the Maple package `OreModules` [CQR07], which is available online together with a library of examples with origin in control theory and mathematical physics. Moreover, we refer to [QR05a, QR05b, Rob06, QR06, QR07, QR08, CQ08, QR10, Qua10a, Qua10b, QR13, Rob14] for recent work extending this approach. A Maple package `OreMorphisms` [CQ09], based on `OreModules`, implements methods developed in [CQ08] for factoring and decomposing linear functional systems.

In the rest of this subsection we will apply Janet's algorithm for the class of Ore algebras D which has been dealt with in Subsect. 2.1.3, in order to prove that $\mathscr{F} := \hom_K(D, K)$ is an injective left D-module. The proof uses Baer's criterion (cf. Lemma A.1.13, p. 238) and the fact that a Janet basis for a left ideal I of D provides a partition of $\mathrm{Mon}(D)$ into a set of monomials which are *parametric* for specifying a linear form $\lambda \in \hom_K(D, K)$ that is orthogonal to all elements of I with respect to the pairing

$$(\ , \ ) : D \times \mathscr{F} \longrightarrow K : (d, \lambda) \longmapsto \lambda(d)$$

(cf. (2.19) in Rem. 2.1.66, p. 50), and a set of monomials whose λ-values are uniquely determined by the choices of the values for the parametric ones. Note that Janet's algorithm provides both generalized Hilbert series enumerating the respective sets of monomials (cf. Subsect. 2.1.5).

Theorem 3.1.56 ([Rob06], Cor. 4.3.7). *If D satisfies Assumption 2.1.23, then the left D-module $\mathscr{F} := \hom_K(D, K)$ is injective.*

Proof. Let I be a left ideal of D and $\varphi : I \to \mathscr{F}$ a homomorphism of left D-modules. Then φ defines a system of linear equations with coefficients in D for an unknown $\lambda \in \mathscr{F}$:

$$d \cdot \lambda = \varphi(d), \quad d \in I. \tag{3.40}$$

It is easily seen from (2.21), p. 50, and (2.22) that this system is equivalent to a system of linear equations with coefficients in K for the unknown values (m, λ), $m \in \mathrm{Mon}(D)$, defining λ.

Since D is left Noetherian, the left ideal I is finitely generated. Let J be a Janet basis for I with respect to an admissible term ordering (cf. Def. 2.1.40 and Assumption 2.1.29). Then

$$\biguplus_{(p, \mu) \in J} \mathrm{Mon}(\mu) p \tag{3.41}$$

is a basis for the left K-vector space I (cf. Thm. 2.1.43 b)). For every choice of values for those unknowns (m, λ) with $m \in \mathrm{Mon}(D) - \mathrm{lm}(I)$, a solution of (3.40) is uniquely determined. More precisely, for every map

$$\mathrm{Mon}(D) - [\mathrm{lm}(p) \mid (p, \mu) \in J] \longrightarrow K$$

the values for the unknowns in $\{ (\mathrm{lm}(p), \lambda) \mid (p, \mu) \in J \}$ are uniquely determined by the equations with left hand sides given by (3.41). Then all values $(m, \lambda) \in K$

for $m \in \mathrm{Mon}(D)$ are specified, which uniquely determines a solution λ of (3.40). Moreover, the homomorphism $\psi \colon D \to \mathscr{F}$ defined by $\psi(1) := \lambda$ restricts to φ on I. By Baer's criterion (cf. Lemma A.1.13), $\mathscr{F}$ is injective. $\square$

Remark 3.1.57. In fact, the injective left D-module $\mathscr{F} = \mathrm{hom}_K(D, K)$ is a cogenerator for the category of left D-modules (cf. [Rob06, Thm. 4.4.7]; cf. also [Bou07, § 1.8, Prop. 11, Prop. 13], [Obe90, Cor. 3.12, Rem. 3.13]).

The following remark revisits systems of linear functional equations as introduced in Sect. 2.1. It will be drawn upon in Subsect. 3.2.1. (The set of solutions is also referred to as a behavior in the systems-theoretic context, but we are not going to employ this notion here.)

Remark 3.1.58. Let two systems S_1 and S_2 of linear functional equations for unknown functions $y_1, \ldots, y_p$ be given by $R_i y = 0$, where $R_i \in D^{q_i \times p}$, $i = 1, 2$, and where D is a (not necessarily commutative) algebra over a field K. Then

$$0 \longleftarrow D^{1 \times p}/(D^{1 \times q_1} R_1 + D^{1 \times q_2} R_2)$$

$$\overset{\alpha}{\longleftarrow} (D^{1 \times p}/D^{1 \times q_1} R_1) \oplus (D^{1 \times p}/D^{1 \times q_2} R_2) \qquad (3.42)$$

$$\overset{\beta}{\longleftarrow} D^{1 \times p}/(D^{1 \times q_1} R_1 \cap D^{1 \times q_2} R_2) \longleftarrow 0$$

is a short exact sequence of left D-modules, where β maps a residue class to the pair of its images under the canonical projections and α takes the difference of the images (cf., e.g., [Bou98a, II, § 1, no. 7, Prop. 10]). Let $\mathscr{F}$ be a left D-module. Then

$$0 \longrightarrow \{y \in \mathscr{F}^{p \times 1} \mid R_1 y = R_2 y = 0\}$$

$$\overset{\iota}{\longrightarrow} \{y \in \mathscr{F}^{p \times 1} \mid R_1 y = 0\} \oplus \{y \in \mathscr{F}^{p \times 1} \mid R_2 y = 0\} \qquad (3.43)$$

$$\overset{\pi}{\longrightarrow} \{y \in \mathscr{F}^{p \times 1} \mid R_1 y = 0\} + \{y \in \mathscr{F}^{p \times 1} \mid R_2 y = 0\} \longrightarrow 0,$$

where

$$\iota \colon y \longmapsto (y, -y), \qquad \pi \colon (y_1, y_2) \longmapsto y_1 + y_2,$$

is a short exact sequence of K-vector spaces. By applying the left exact contravariant functor $\mathrm{hom}_D(-, \mathscr{F})$ to (3.42), we obtain the exact sequence of K-vector spaces

$$0 \longrightarrow \mathrm{hom}_D(D^{1 \times p}/(D^{1 \times q_1} R_1 + D^{1 \times q_2} R_2), \mathscr{F})$$

$$\longrightarrow \mathrm{hom}_D((D^{1 \times p}/D^{1 \times q_1} R_1) \oplus (D^{1 \times p}/D^{1 \times q_2} R_2), \mathscr{F}) \qquad (3.44)$$

$$\longrightarrow \mathrm{hom}_D(D^{1 \times p}/(D^{1 \times q_1} R_1 \cap D^{1 \times q_2} R_2), \mathscr{F}),$$

the last homomorphism being not necessarily surjective. Using Malgrange's isomorphism (2.2), p. 6, the last three K-vector spaces may be identified with the solution set of the combined system $S_1 + S_2$ given by $R_1 y = 0$, $R_2 y = 0$, with the direct sum of the solution sets of S_1 and S_2, and with the solution set of the system $S_1 \cap S_2$ whose equations are the common linear consequences of both systems, respectively. Since the first two terms (ignoring the leading 0) in (3.43) and (3.44) coincide up to K-vector space isomorphism, non-trivial cohomology at the last term in (3.44) means that not every solution of $S_1 \cap S_2$ can be written as a sum of a solution of S_1 and a solution of S_2.

Malgrange's isomorphism is functorial in the following sense. For $i = 1$, 2, let $A_i \in D^{q_i \times p_i}$ and assume that a homomorphism

$$D^{1 \times p_1} / D^{1 \times q_1} A_1 \longrightarrow D^{1 \times p_2} / D^{1 \times q_2} A_2$$

of left D-modules is defined by a matrix $T \in D^{p_1 \times p_2}$, i.e., there exists $U \in D^{q_1 \times q_2}$ such that $A_1 T = U A_2$. Then the following diagram is commutative, where the vertical maps are induced by T:

$$\hom_D(D^{1 \times p_1} / D^{1 \times q_1} A_1, \mathscr{F}) \longrightarrow \{y_1 \in \mathscr{F}^{p_1 \times 1} \mid A_1 y_1 = 0\}$$
$$\uparrow \qquad\qquad\qquad\qquad\qquad\qquad \uparrow$$
$$\hom_D(D^{1 \times p_2} / D^{1 \times q_2} A_2, \mathscr{F}) \longrightarrow \{y_2 \in \mathscr{F}^{p_2 \times 1} \mid A_2 y_2 = 0\}.$$

In fact, the identification of the K-vector spaces in (3.43) and (3.44) respects the given homomorphisms in this sense.

If the left D-module $\mathscr{F}$ is injective, then the last homomorphism in (3.44) is surjective, and we conclude that the third terms (ignoring the leading 0) in (3.43) and (3.44) are isomorphic K-vector spaces (cf. also [Sha01] for the cases where D is a commutative polynomial algebra in n partial differential operators and $\mathscr{F}$ is, e.g., the space of (tempered) distributions on $\mathbb{R}^n$ or the space of complex-valued smooth functions on $\mathbb{R}^n$).

In Subsect. 3.2.1 (more precisely, Thm. 3.2.10, p. 164) the above duality is used in the context of linear partial differential equations. Then D is the skew polynomial ring $K \langle \partial_1, \ldots, \partial_n \rangle$ of differential operators with coefficients in a field K of meromorphic functions. In this case the center of expansion of a power series solution needs to be chosen outside the variety of smaller dimension defined by the denominators and leading coefficients of the equations under consideration. If this requirement is taken into account, then a Janet basis for the system of linear PDEs allows to determine the analytic solutions in the same way as in the case of constant coefficients (cf. Rem. 2.1.70, p. 53). Using moreover the fact that, in this situation, the duality between equations and solutions is also faithful (cf. Rem. 3.1.57), we conclude that the sum and the intersection of two left D-modules, which are generated by the left hand sides of two systems of linear PDEs, correspond faithfully to the intersection and the sum of their solution spaces, respectively.

3.2 Linear Differential Elimination

Let $\Omega \subseteq \mathbb{C}^n$ be open and connected, and let $z_1, \ldots, z_n$ be coordinates of $\mathbb{C}^n$. We abbreviate the tuple of coordinates by z.

The objective of this section is to develop formal methods to deal with a given family of (complex) analytic functions $u\colon \Omega \to \mathbb{C}$ of the form

$$f_1(\alpha_1(z))\,g_1(z) + \ldots + f_k(\alpha_k(z))\,g_k(z), \tag{3.45}$$

where $k \in \mathbb{N}$ is fixed,

$$g_i\colon \Omega \longrightarrow \mathbb{C}, \quad \alpha_i\colon \Omega \longrightarrow \mathbb{C}^{\nu(i)}, \qquad i = 1,\ldots,k,$$

are given analytic functions, and

$$f_i\colon \alpha_i(\Omega) \longrightarrow \mathbb{C}, \qquad i = 1,\ldots,k,$$

are arbitrary functions such that each $f_i \circ \alpha_i$ is analytic. (If $\nu(i) = 0$, then f_i is understood as an arbitrary constant.)

For a given analytic function $u\colon \Omega \to \mathbb{C}$, the task is to decide whether u is a member of the above family, and if so, to find parameters $f_1, \ldots, f_k$ such that u can be written as in (3.45).

In order to attack the first problem, we use differential elimination to compute an implicit representation of the given parametrized family of analytic functions. The result is supposed to be a system of partial differential equations whose set of analytic solutions coincides with the parametrized family (with appropriate conditions on the domain of definition of the analytic functions under consideration). The membership test for the parametrized family is then reduced to a substitution of the given function into the system of PDEs.

Problems similar to the above ones come up in the context of computing conservation laws for partial differential equations [WBM99]. When it is possible to give the general solution to a system of PDEs in closed form, it will in general depend in a redundant way on arbitrary constants and arbitrary functions of certain of the independent variables. Ways of identifying this redundancy are discussed in [WBM99, Sect. 4.1]. Linear expressions as (3.45), in which, however, unknown functions depend on certain of the independent variables and not on given functions, are treated there using different methods. A collection of powerful routines for solving such problems is the software package CRACK, developed in the computer algebra system REDUCE [Hea99] since the 1990s, cf., e.g., [Wol04] for a short description.

The methods presented in this section build on Janet's algorithm, which defines normal forms for systems of linear partial differential equations (cf. Sect. 2.1). The examples were dealt with using the Maple package Janet, cf. Subsect. 2.1.6.

Some classes of nonlinear parametrizations of families of analytic functions are treated in Sect. 3.3. As expected, the present section provides more efficient methods to solve problems emerging from linear parametrizations.

Subsection 3.2.1 solves the implicitization problem for linear parametrizations; the system of PDEs is obtained as Janet basis for the intersection of the left ideals of linear differential operators which annihilate the respective summands in (3.45). Since computation of this intersection (e.g., by using Zassenhaus' trick, Alg. 3.1.31, p. 138) can be a hard task in practice, alternative methods producing generators for this intersection are developed in Subsect. 3.2.2. The second problem of finding parameters realizing an analytic function of the family is dealt with in Subsect. 3.2.3. In Subsect. 3.2.4 we study classes of linear parametrizations whose annihilating left ideals have generating sets of differential operators with additional properties (e.g., having constant coefficients). Some applications are presented in Subsect. 3.2.5.

The contents of this section are based on [PR10].

3.2.1 Implicitization Problem for Linear Parametrizations

In what follows, let K be the field of meromorphic functions on the open and connected subset Ω of $\mathbb{C}^n$, and let $\partial_1, \ldots, \partial_n$ be the partial derivative operators[6] with respect to the coordinates $z_1, \ldots, z_n$ of $\mathbb{C}^n$. We denote by D the skew polynomial ring $K\langle \partial_1, \ldots, \partial_n \rangle$ of differential operators with coefficients in K. The monomial in $\partial_1, \ldots, \partial_n$ with exponent vector $I \in (\mathbb{Z}_{\geq 0})^n$ is also written briefly as ∂^I.

Let $k \in \mathbb{N}$, and let $g_i \colon \Omega \to \mathbb{C}$, $\alpha_i \colon \Omega \to \mathbb{C}^{v(i)}$, $i = 1, \ldots, k$, be analytic functions. We assume that no g_i is the zero function. Hence, division by g_i is defined in the field of meromorphic functions.

The analytic functions in the family under consideration are of the form

$$f_1(\alpha_1(z)) g_1(z) + \ldots + f_k(\alpha_k(z)) g_k(z), \tag{3.46}$$

where $f_1, \ldots, f_k$ are arbitrary functions such that each $f_i \circ \alpha_i$ is analytic. Therefore, we may assume without loss of generality that the components of each function $\alpha_i \colon \Omega \to \mathbb{C}^{v(i)}$ are functionally independent. In fact, we make the stronger

Assumption 3.2.1. The Jacobian matrix

$$\left(\frac{\partial \alpha_i}{\partial z} \right) = \left(\frac{\partial \alpha_{i,j}}{\partial z_r} \right)_{1 \leq j \leq v(i), 1 \leq r \leq n}$$

of each α_i has rank $v(i)$ at every point of Ω.

---

[6] In examples we frequently denote coordinates also by x, y, z, in which case ∂_x, ∂_y, ∂_z are the corresponding partial derivative operators.

Lemma 3.2.2. *Let* $F\colon \Omega \to \mathbb{C}$ *be an analytic function which factors over* $\alpha_i(\Omega)$ *for some* $i \in \{1,\ldots,k\}$. *Then the function* $f_i\colon \alpha_i(\Omega) \to \mathbb{C}$ *satisfying* $F(z) = f_i(\alpha_i(z))$ *for all* $z \in \Omega$ *is uniquely determined and analytic.*

Proof. The case $v(i) = 0$, i.e., α_i is constant, is trivial. If α_i is not constant, then the open mapping theorem implies that $\alpha_i(\Omega)$ is open and connected in $\mathbb{C}^{v(i)}$. Obviously, $f_i\colon \alpha_i(\Omega) \to \mathbb{C}$ is uniquely determined. In order to prove that f_i is analytic, let $w \in \Omega$ be arbitrary. Since the Jacobian matrix of α_i has rank $v(i)$ at w, among the coordinates $z_1, \ldots, z_n$ there exist $\beta_{v(i)+1}, \ldots, \beta_n$ and a neighborhood Ω' of w in Ω such that

$$z \longmapsto (\alpha_i(z), \beta) := (\alpha_{i,1}(z), \ldots, \alpha_{i,v(i)}(z), \beta_{v(i)+1}, \ldots, \beta_n)$$

defines an analytic diffeomorphism $\kappa\colon \Omega' \to \Omega'' \subseteq \mathbb{C}^n$. Denote by $\pi\colon \mathbb{C}^n \to \mathbb{C}^{v(i)}$ the projection onto the first $v(i)$ components and by ι an analytic right inverse of the restriction of π to $\Omega'' = \mathrm{im}(\kappa)$. Then the restriction of f_i to Ω'' equals $F \circ \kappa^{-1} \circ \iota$ and is therefore analytic. The uniqueness of f_i implies its analyticity as a function $\alpha_i(\Omega) \to \mathbb{C}$. $\square$

The following example shows that a reasonable definition of the family of functions parametrized by (3.46) requires more care regarding the domain of definition of the parameters $f_1, \ldots, f_k$.

Example 3.2.3. Let us specialize (3.46) to $f(z^2)$ on an open and connected subset Ω of $\mathbb{C}$ with coordinate $z = z_1$, i.e., $k = 1$, $g_1(z) = 1$, $\alpha_1(z) = z^2$, $f = f_1$. Assumption 3.2.1 requires Ω to be a subset of $\mathbb{C} - \{0\}$. We consider the analytic function $u(z) = z$. Restricted to any simply connected neighborhood of any point w in Ω, the function u is of the form $z \mapsto f(z^2)$. However, since a square root function on any open annulus with center 0 in $\mathbb{C}$ would be multi-valued, u cannot be represented as $z \mapsto f(z^2)$ on such open subsets.

Removable singularities constitute another reason why membership of an analytic function to a parametrized family may be justified in a neighborhood of almost every, but not every point in the domain of definition.

Example 3.2.4. Let Ω be an open and connected subset of $\mathbb{C}$ with coordinate $z = z_1$ and consider the family of analytic functions on Ω of the form

$$f(z) \cdot \sin(z).$$

Even if $0 \in \mathbb{C}$ belongs to Ω, the function $\sin(z)/z$ is analytic on Ω, but $1/z$ is not allowed as a coefficient function f. If we use a definition which asks for representations as $f(z) \cdot \sin(z)$ only around generic points of Ω, then the above family coincides with the family of all analytic functions on Ω.

Our aim is to characterize analytic functions of the form (3.46) in terms of differential equations. The next definition offers enough flexibility regarding the domain of definition of these functions.

Definition 3.2.5. We define the family $S(\alpha_1, g_1; \alpha_2, g_2; \ldots; \alpha_k, g_k)$ of analytic functions on Ω *parametrized by* (3.46) as the $\mathbb{C}$-vector space

$$\{u\colon \Omega \longrightarrow \mathbb{C} \mid u \text{ is complex analytic; for almost every } w \in \Omega$$

$$\text{there exists an open neighborhood } \Omega' \subseteq \Omega \text{ of } w \text{ and}$$

$$\text{analytic functions } f_i\colon \alpha_i(\Omega') \to \mathbb{C}, i = 1, \ldots, k, \text{ such that}$$

$$u(z) = f_1(\alpha_1(z))\, g_1(z) + \ldots + f_k(\alpha_k(z))\, g_k(z) \text{ for all } z \in \Omega'\}.$$

(The meaning of "almost every $w \in \Omega$" is that the set of $w \in \Omega$ for which the condition given in the previous lines is fulfilled is supposed to be dense in Ω.)

We refer to $f_1, \ldots, f_k$ as the *parameters* in (3.46).

For $i = 1, \ldots, k$, the family $S(\alpha_i, g_i)$ is parametrized by the i-th summand in (3.46). We have

$$S(\alpha_1, g_1; \alpha_2, g_2; \ldots; \alpha_k, g_k) = S(\alpha_1, g_1) + S(\alpha_2, g_2) + \ldots + S(\alpha_k, g_k).$$

An implicit description of each $S(\alpha_i, g_i)$ in terms of linear partial differential equations of order one is provided by the following lemma. The left hand sides of these equations are obtained as compositions of certain multiplication operators with directional derivatives. Let $i \in \{1, \ldots, k\}$.

Lemma 3.2.6. *Let $b_1, \ldots, b_{n-\nu(i)} \in K^{n \times 1}$ be a basis for the kernel of the Jacobian matrix $\left(\frac{\partial \alpha_i}{\partial z}\right)$ of α_i. (If $\nu(i) = 0$, then let $b_1, \ldots, b_n$ be the standard basis vectors of $K^{n \times 1}$.) Then for any analytic function $u\colon \Omega \to \mathbb{C}$ we have $u \in S(\alpha_i, g_i)$ if and only if u is annihilated by*

$$\left(\sum_{r=1}^{n} (b_j)_r \partial_r\right) \circ \frac{1}{g_i}, \qquad j = 1, \ldots, n - \nu(i). \tag{3.47}$$

Proof. In order to identify the annihilating operators, we divide functions in $S(\alpha_i, g_i)$ by g_i and apply a coordinate transformation in the same way as in the proof of Lemma 3.2.2. Division by the non-zero analytic function g_i is well-defined in the field of meromorphic functions. For any point of Ω there exist a neighborhood Ω' in Ω and elements $\beta_{\nu(i)+1}, \ldots, \beta_n$ of $\{z_1, \ldots, z_n\}$ such that

$$z \longmapsto (\alpha_i(z), \beta) := (\alpha_{i,1}(z), \ldots, \alpha_{i,\nu(i)}(z), \beta_{\nu(i)+1}, \ldots, \beta_n)$$

defines an analytic diffeomorphism $\kappa\colon \Omega' \to \Omega'' \subseteq \mathbb{C}^n$. Using the components of κ as new coordinates, the function u/g_i transforms to one which does not depend on $\beta_{\nu(i)+1}, \ldots, \beta_n$. Hence, its partial derivatives with respect to $\beta_{\nu(i)+1}, \ldots, \beta_n$ are zero. Conversely, every analytic function satisfying the latter condition defines a function in $S(\alpha_i, g_i)$ after writing it with respect to the coordinates $z_1, \ldots, z_n$ and multiplying it by g_i. Therefore, the assertion follows by noticing that the partial differential operators with respect to the last $n - \nu(i)$ new coordinates are transformed by the inverse coordinate change to the directional derivatives in (3.47). $\qquad\square$

Definition 3.2.7. We denote by $\mathscr{I}(\alpha_i, g_i)$ the left ideal of $D = K\langle \partial_1, \ldots, \partial_n \rangle$ which is generated by the differential operators in (3.47) and call it the *annihilator* of the family $S(\alpha_i, g_i)$.

Remark 3.2.8. Lemma 3.2.6 can be restated as

$$S(\alpha_i, g_i) = \mathrm{Sol}_\Omega(\mathscr{I}(\alpha_i, g_i)).$$

In the special case $\nu(i) = n$ no condition is imposed on the coefficient function f_i by the choice of its arguments (for instance, if α_i is a restriction of the identity map $\mathbb{C}^n \to \mathbb{C}^n$). Then we have $\mathscr{I}(\alpha_i, g_i) = \{0\}$, and any analytic function u on Ω is representable as $f_i \cdot g_i$, where $f_i := u/g_i$. Dealing with a family parametrized by (3.46) with $\nu(1) = \ldots = \nu(k) = n$ reduces to linear algebra over the field K of meromorphic functions on Ω.

Example 3.2.9. Let x, y be coordinates of $\mathbb{C}^2$ and consider the analytic functions $\alpha_1(x, y) := x$, $g_1(x, y) := (\cos(x+y))^2$. Then $(0, 1)^\top$ forms a basis for the kernel of the Jacobian matrix of α_1, so that the annihilator $\mathscr{I}(\alpha_1, g_1)$ of $S(\alpha_1, g_1)$ is generated by the differential operator

$$\partial_y \circ \frac{1}{(\cos(x+y))^2} = \frac{1}{(\cos(x+y))^2} \partial_y + \frac{2\sin(x+y)}{(\cos(x+y))^3}.$$

The implicitization problem for families of analytic functions parametrized by (3.46) is solved by the following theorem.

Theorem 3.2.10. *We have*

$$S(\alpha_1, g_1; \alpha_2, g_2; \ldots; \alpha_k, g_k) = \mathrm{Sol}_\Omega(\mathscr{I}(\alpha, g)),$$

where the annihilating left ideal

$$\mathscr{I}(\alpha, g) := \bigcap_{i=1}^{k} \mathscr{I}(\alpha_i, g_i)$$

of D is the intersection of the annihilating left ideals of $S(\alpha_1, g_1)$, ..., $S(\alpha_k, g_k)$ given in Lemma 3.2.6.

Proof. By the duality between systems of homogeneous linear partial differential equations and their spaces of analytic solutions, sums and intersections of left modules of linear PDEs correspond respectively to intersections and sums of the solution spaces (cf. also Rem. 3.1.58, p. 158). $\qquad\square$

Remark 3.2.11. A Janet basis for the intersection of left ideals in Theorem 3.2.10 can be computed, e.g., by using Zassenhaus' trick (cf. Alg. 3.1.31, p. 138). In practice, such a computation can be a hard task. Therefore, alternative methods to construct generators for this intersection are discussed in the following subsections.

3.2.2 Differential Elimination using Linear Algebra

In this subsection we present a method to compute implicit equations for a family of analytic functions using linear algebra techniques. A given linear parametrization and the partial derivatives of this expression up to some order are encoded as a matrix. The linear relations which are satisfied by the rows of this matrix yield linear partial differential equations which are satisfied by all analytic functions of the family. A generalization of this construction to polynomially nonlinear parametrizations will be given in Subsect. 3.3.3.

As before the implicit description of a given family $S(\alpha_1, g_1; \alpha_2, g_2; \ldots; \alpha_k, g_k)$ of analytic functions will be given by the left ideal $\mathscr{I}(\alpha, g)$ of the skew polynomial ring $D = K\langle \partial_1, \ldots, \partial_n \rangle$ (cf. the previous subsection). Note that $\mathscr{I}(\alpha, g)$ is finitely generated. Hence, by taking into account more and more partial derivatives of functions of the family, a generating set for $\mathscr{I}(\alpha, g)$ will eventually be obtained from the matrix constructed below. Clearly, as this left ideal is not known beforehand, a termination criterion is needed.

Recall that the family $S(\alpha_1, g_1; \alpha_2, g_2; \ldots; \alpha_k, g_k)$ consists of analytic functions u of the form

$$u(z) = f_1(\alpha_1(z)) g_1(z) + \ldots + f_k(\alpha_k(z)) g_k(z) \tag{3.48}$$

(cf. Def. 3.2.5 for details). Since every function that appears is analytic, partial derivatives of the equation (3.48) exist of any order. In what follows, we read the linear equation (3.48) as a definition of u in terms of a certain linear combination of $f_1 \circ \alpha_1, \ldots, f_k \circ \alpha_k$ (as opposed to the understanding of $S(\alpha_1, g_1; \alpha_2, g_2; \ldots; \alpha_k, g_k)$ as the set of linear combinations of $g_1, \ldots, g_k$ with analytic coefficients of prescribed type) and we deal with its derivatives in a similar way. Linearity of the right hand side of (3.48) in $f_1 \circ \alpha_1, \ldots, f_k \circ \alpha_k$ implies that the dependence of any finite collection of derivatives $\partial^I u$ on the derivatives of $f_i \circ \alpha_i$ can be written as a matrix equation, which we are going to introduce next.

In order to apply techniques from linear algebra, we use jet variables in the symbol U to refer to the left hand sides of (3.48) and its partial derivatives, i.e., U_I, where $I \in (\mathbb{Z}_{\geq 0})^n$, are (algebraically independent) indeterminates representing $\partial^I u$. Analogously, we introduce the jet variables in $F_1, \ldots, F_k$, more precisely, $(F_l)_L$, where $l \in \{1, \ldots, k\}$, $L \in (\mathbb{Z}_{\geq 0})^{v(l)}$, as names for $(\partial^L f_l) \circ \alpha_l$. The jet variable U is a synonym for $U_{(0,\ldots,0)}$, and similar statements hold for $F_1, \ldots, F_k$. Note that if $v(i) = 0$, i.e., f_i is an arbitrary constant, then exactly one jet variable in F_i is introduced, namely the one of order zero.

We assume that total orderings $>$ have been chosen on the set of jet variables in U and on the set of jet variables in $F_1, \ldots, F_k$ which respect the differentiation order and for which the variable U and one of $F_1, \ldots, F_k$ is the least element, respectively. For $\omega \in \mathbb{Z}_{\geq 0}$, the number of jet variables in U up to order ω and the number of jet variables in $F_1, \ldots, F_k$ up to order ω are

$$\zeta(\omega) := \binom{n+\omega}{\omega}, \qquad \sigma(\omega) := \sum_{i=1}^{k} \binom{v(i)+\omega}{\omega}.$$

We define $I_1, \ldots, I_m \in (\mathbb{Z}_{\geq 0})^n$, where $m := \zeta(\omega)$, such that

$$U_{I_m} > \ldots > U_{I_2} > U_{I_1} = U$$

are the $\zeta(\omega)$ lowest jet variables in U with respect to the chosen ordering. Moreover, we define $l_j \in \{1, \ldots, k\}$ and $L_j \in (\mathbb{Z}_{\geq 0})^{v(l_j)}$, $j = 1, \ldots, \sigma(\omega)$, such that

$$(F_{l_j})_{L_j}, \quad j = 1, \ldots, \sigma(\omega),$$

is the sequence of $\sigma(\omega)$ lowest jet variables in $F_1, \ldots, F_k$, increasingly sorted with respect to the chosen ordering.

Definition 3.2.12. For $i = 1, \ldots, k$, let $g_i \colon \Omega \to \mathbb{C}$ be non-zero analytic functions and $\alpha_i \colon \Omega \to \mathbb{C}^{v(i)}$ analytic functions with Jacobian matrix of rank $v(i)$ at every point of Ω. For $j = 1, \ldots, m$, define p_j to be the linear polynomial in the jets of $F_1, \ldots, F_k$ with coefficients in K such that substitution of $(\partial^L f_l) \circ \alpha_l$ for $(F_l)_L$ yields $\partial^{l_j} u$, as obtained by formally differentiating (3.48). Let $\omega \in \mathbb{Z}_{\geq 0}$. Then the *derivative matrix* of order ω

$$\Delta_\omega(\alpha, g) \in K^{\zeta(\omega) \times \sigma(\omega)}$$

is defined to be the $\zeta(\omega) \times \sigma(\omega)$ matrix whose entry at position (i, j) is the coefficient of $(F_{l_j})_{L_j}$ in p_i.

Remark 3.2.13. The polynomials p_j in the previous definition are uniquely defined because the expressions $(\partial^L f_l) \circ \alpha_l$ are treated as functionally independent when (3.48) is formally differentiated. Since the differential order of the jets $U_{I_1}, \ldots, U_{I_m}$ is less than or equal to ω and since the sequence of jets $(F_{l_j})_{L_j}$ for $j = 1, \ldots, \sigma(\omega)$ consists of all jets of $F_1, \ldots, F_k$ up to differentiation order ω, every derivative $\partial^{l_j} u$, $j = 1, \ldots, m$, is representable as a row in $\Delta_\omega(\alpha, g)$.

Example 3.2.14. Let x, y be coordinates for $\mathbb{C}^2$ and define the analytic functions $g_1(x,y) := x$, $g_2(x,y) := y$, $\alpha_1(x,y) := y + x^2$, and $\alpha_2(x,y) := x + y^2$. Hence, we deal with analytic functions of the form

$$f_1(y+x^2) \cdot x + f_2(x+y^2) \cdot y.$$

For $\omega = 2$ we get

$$\Delta_2(\alpha, g) = \left(\begin{array}{cc|cc|cc}
x & y & 0 & 0 & 0 & 0 \\
0 & 1 & x & 2y^2 & 0 & 0 \\
1 & 0 & 2x^2 & y & 0 & 0 \\
\hline
0 & 0 & 0 & 6y & x & 4y^3 \\
0 & 0 & 1 & 1 & 2x^2 & 2y^2 \\
0 & 0 & 6x & 0 & 4x^3 & y
\end{array}\right),$$

where the rows are indexed by $U, U_{(0,1)}, U_{(1,0)}, U_{(0,2)}, U_{(1,1)}, U_{(2,0)}$ and the columns are indexed by $F_1, F_2, (F_1)_{(1)}, (F_2)_{(1)}, (F_1)_{(2)}, (F_2)_{(2)}$.

Example 3.2.15. Let x, y be coordinates for $\mathbb{C}^2$ and define the analytic functions $g_1(x,y) := y$, $g_2(x,y) := 1$, $\alpha_1(x,y) := x$, and $\alpha_2(x,y) := 1$. Hence, we deal with analytic functions of the form

$$f_1(x) \cdot y + f_2.$$

For $\omega = 2$ we get

$$\Delta_2(\alpha, g) = \begin{pmatrix} \begin{array}{c|c|c|c} y & 1 & 0 & 0 \\ \hline 1 & 0 & 0 & 0 \\ \hline 0 & 0 & y & 0 \\ \hline 0 & 0 & 0 & 0 \\ \hline 0 & 0 & 1 & 0 \\ \hline 0 & 0 & 0 & y \end{array} \end{pmatrix},$$

where the rows are indexed by $U, U_{(0,1)}, U_{(1,0)}, U_{(0,2)}, U_{(1,1)}, U_{(2,0)}$ and the columns are indexed by $F_1, F_2, (F_1)_{(1)}, (F_1)_{(2)}$.

Remark 3.2.16. The derivative matrix has a lower block triangular structure due to the sorting of the jet variables with respect to differentiation order. The block at position (i, j) (corresponding to order i for the jets of U and order j for the jets of $F_1, \ldots, F_k$) is obtained from the two blocks at position $(i-1, j)$ and $(i-1, j-1)$ according to the rules of differentiation, if $i > 0$, $j > 0$, which allows for an efficient way of extending $\Delta_\omega(\alpha, g)$ to $\Delta_{\omega+1}(\alpha, g)$.

Remark 3.2.17. Recall that the skew polynomial ring $D = K\langle \partial_1, \ldots, \partial_n \rangle$ has an increasing filtration given by the total degree of polynomials (or total order of differential operators in this context; cf. Rem. 2.1.61, p. 44). The Hilbert series of D with respect to this filtration is the generating function whose i-th coefficient equals the dimension of the left K-vector space

$$D_{\leq i} := \{p \in D \mid p = 0 \text{ or } \deg(p) \leq i\}, \qquad i \in \mathbb{Z}_{\geq 0}.$$

Let I be a left ideal of D. Then I inherits an increasing filtration $(I_{\leq i})_{i \in \mathbb{Z}_{\geq 0}}$ given by $I_{\leq i} := D_{\leq i} \cap I$ from D, and we endow the left D-module D/I with the increasing filtration $\Phi := ((D/I)_{\leq i})_{i \in \mathbb{Z}_{\geq 0}}$, where $(D/I)_{\leq i}$ is defined as the image of $D_{\leq i}$ under the canonical projection $D \to D/I$. Moreover, we denote by Γ the grading of the associated graded module $\mathrm{gr}(D/I)$ over the associated graded ring $\mathrm{gr}(D)$.

A Janet basis J for I with respect to any degree-compatible term ordering yields via the generalized Hilbert series $H_{\mathrm{Mon}(D)-\mathrm{lm}(I)}(\partial_1, \ldots, \partial_n)$ of D/I the Hilbert series of D/I with respect to Φ (cf. Def. 2.1.57, p. 41, Rem. 2.1.59, and Rem. 2.1.61):

$$H_{D/I, \Phi}(\lambda) = H_{\mathrm{gr}(D/I), \Gamma}(\lambda) \cdot \frac{1}{1-\lambda}, \qquad H_{\mathrm{gr}(D/I), \Gamma}(\lambda) = H_{\mathrm{Mon}(D)-\mathrm{lm}(I)}(\lambda, \ldots, \lambda).$$

The generalized Hilbert series of D/I enumerates the parametric derivatives of the system of linear PDEs corresponding to I, i.e., it enumerates exactly those monomials $\partial_1^{i_1} \cdot \ldots \cdot \partial_n^{i_n}$, $(i_1, \ldots, i_n) \in (\mathbb{Z}_{\geq 0})^n$, corresponding to Taylor coefficients in a power

series solution whose value can be chosen arbitrarily according to the selection of
the highest term in each equation (cf. Rem. 2.1.67, p. 50).

In this situation we denote by $\Delta_{\omega,J}(\alpha,g)$ the submatrix of $\Delta_\omega(\alpha,g)$ which is
formed by the rows corresponding to the $U_{(i_1,\ldots,i_n)}$ for the parametric derivatives
$\partial_1^{i_1} \cdot \ldots \cdot \partial_n^{i_n}$ determined by J, and we also use $\Delta_{\omega,J}$ as an abbreviation of $\Delta_{\omega,J}(\alpha,g)$.
Furthermore, we let $\Delta_{\omega,J}^{(i,j)}$ be the submatrix of $\Delta_{\omega,J}$ which is defined by restricting
further to the (i,j)-block referred to in Remark 3.2.16, and we define

$$\Delta_{\omega,J}^{(i,<j)} := \begin{pmatrix} \Delta_{\omega,J}^{(i,0)} & \cdots & \Delta_{\omega,J}^{(i,j-1)} \end{pmatrix}, \qquad i,j \in \mathbb{Z}_{\geq 0}.$$

Example 3.2.18. Resuming Example 3.2.15, the analytic functions of the family
$S(\alpha_1,g_1;\alpha_2,g_2)$ under consideration satisfy the linear partial differential equations

$$u_{y,y} = 0, \qquad y \cdot u_{x,y} - u_x = 0,$$

which form a Janet basis J with respect to the degree-reverse lexicographical or-
dering which extends $\partial_x > \partial_y$ (the sets of multiplicative variables being $\{\partial_y\}$ and
$\{\partial_x,\partial_y\}$, respectively). The parametric derivatives are enumerated by the generalized Hilbert series

$$1 + \partial_y + \frac{\partial_x}{1-\partial_x}.$$

Hence, for $\omega = 2$ we have

$$\Delta_{2,J}(\alpha,g) = \left(\begin{array}{cc|c|c} y & 1 & 0 & 0 \\ 1 & 0 & 0 & 0 \\ 0 & 0 & y & 0 \\ 0 & 0 & 0 & y \end{array} \right).$$

Now we establish an important relationship between the rank of the derivative
matrix of order ω and the dimension of the K-vector space $\mathscr{I}(\alpha,g)_{\leq\omega}$.

Remark 3.2.19. The derivative matrix $\Delta_\omega(\alpha,g) \in K^{\zeta(\omega)\times\sigma(\omega)}$ has the following
property, which is essential for the elimination method to be described below:

$$\zeta(\omega) - \mathrm{rank}(\Delta_\omega(\alpha,g)) = \dim_K \mathscr{I}(\alpha,g)_{\leq\omega}.$$

This number is the K-dimension of the kernel of the linear map which is induced by
$\Delta_\omega(\alpha,g)$ on row vectors. In fact,

$$\{v \in K^{1\times\zeta(\omega)} \mid v \cdot \Delta_\omega(\alpha,g) = 0\} \longrightarrow \mathscr{I}(\alpha,g)_{\leq\omega} : v \longmapsto \sum_{i=1}^{\zeta(\omega)} v_i \cdot \partial^{l_i}$$

is an isomorphism of K-vector spaces by construction of the derivative matrix.

In the previous remark the linear relations among the rows of the derivative matrix are interpreted as partial differential equations that are satisfied by all func-

tions in the family $S(\alpha_1, g_1; \ldots; \alpha_k, g_k)$. Before investigating linear dependency of the columns, we use the previous remark in the following algorithm which computes a Janet basis for the left ideal $_D\langle \mathscr{I}(\alpha,g)_{\leq\omega}\rangle$ of $D = K\langle \partial_1, \ldots, \partial_n \rangle$ that is generated by annihilating differential operators up to order ω. In order to obtain a Janet basis for $\mathscr{I}(\alpha,g)$, the value of ω has to be chosen large enough, a problem which we are going to treat afterwards.

Algorithm 3.2.20 (Implicitization using linear algebra).

Input: Analytic functions $g_i \colon \Omega \to \mathbb{C}$, $g_i \neq 0$, and $\alpha_i \colon \Omega \to \mathbb{C}^{\nu(i)}$ with Jacobian matrix of rank $\nu(i)$ at every point of Ω, $i = 1, \ldots, k$, a term ordering $>$ on $\mathrm{Mon}(D)$ which is compatible with the differentiation order, and $\omega \in \mathbb{Z}_{\geq 0}$

Output: A Janet basis J for $_D\langle \mathscr{I}(\alpha,g)_{\leq\omega}\rangle$ with respect to $>$

Algorithm:

1: $J \leftarrow \{0\}$; $\Delta' \leftarrow \Delta_0(\alpha,g)$; $U' \leftarrow U_{(0,\ldots,0)}$
2: **for** $i = 1, \ldots, \omega$ **do**
3:      let $U^{(1)}, \ldots, U^{(m)}$ be the jet variables corresponding to the rows of $\Delta_{i,J}^{(i,i)}$
4:      compute a K-basis $\{b_1, \ldots, b_r\}$ of $\{v \in K^{1\times m} \mid v \cdot \Delta_{i,J}^{(i,i)} = 0\}$
5:      apply Gaussian elimination of rows with priority given to the first entries to

$$\left(\begin{array}{c|c} \Delta' & U' \\ b_1 \cdot \Delta_{i,J}^{(i,<i)} & b_1 \cdot (U^{(1)}, \ldots, U^{(m)})^\top \\ \vdots & \vdots \\ b_r \cdot \Delta_{i,J}^{(i,<i)} & b_r \cdot (U^{(1)}, \ldots, U^{(m)})^\top \end{array} \right)$$

resulting in a block matrix of the form

$$\left(\begin{array}{c|c} \Delta'' & U'' \\ \hline 0 & W \end{array} \right),$$

where Δ'' is a matrix whose rows are non-zero and W is a column vector whose entries are K-linear combinations of jet variables in U
6:      compute a Janet basis with respect to $>$ for the left ideal of D which is generated by J and the differential operators corresponding to the entries of W, and replace J with this result

7:      $\Delta' \leftarrow \left(\begin{array}{c|c} \Delta'' & 0 \\ \Delta_{i,J}^{(i,<i)} & \Delta_{i,J}^{(i,i)} \end{array} \right)$; $U' \leftarrow \left(\begin{array}{c} U'' \\ U^{(1)} \\ \vdots \\ U^{(m)} \end{array} \right)$

8: **end for**
9: **return** J

Proof. We take advantage of the lower block triangular structure of $\Delta_i(\alpha, g)$ (cf. Rem. 3.2.16) by first determining a basis for the K-vector space of relations satisfied by the rows of $\Delta_{i,J}^{(i,i)}$ in step 4. Then Gaussian elimination in step 5 only deals with the corresponding linear combinations of the order i rows of $\Delta_{i,J}(\alpha, g)$ whose entries in order i columns are all zero. In the beginning of every round i of the loop, the rows of Δ' generate the row space of $\Delta_{i-1,J}(\alpha, g)$. Therefore, step 5 computes, in particular, a generating set for the K-vector space of relations satisfied by the rows of $\Delta_{i,J}(\alpha, g)$, the generators being written as K-linear combinations of jet variables in U. Rows of $\Delta_\omega(\alpha, g)$ which correspond to leading monomials of such relations and their derivatives will not be considered again in later rounds because the parametric derivatives defined by J are used as indices of relevant rows. By Remark 3.2.19, after step 6, J is a Janet basis for the linear relations satisfied by the partial derivatives up to order i of analytic functions $f_1(\alpha_1(z)) g_1(z) + \ldots + f_k(\alpha_k(z)) g_k(z)$. $\qquad\qquad\square$

Example 3.2.21. We demonstrate Algorithm 3.2.20 on the family of analytic functions

$$u(x, y, z) = f_1(x) + f_2(y) + f_3(x + y, z)$$

on $\mathbb{C}^3$ with coordinates x, y, z. A Janet basis will be determined for $_D\langle \mathscr{I}(\alpha, g)_{\leq \omega} \rangle$, where $\omega := 3$. We have

$$\Delta_2(\alpha, g) = \left(\begin{array}{ccc|cccc|ccccc}
1 & 1 & 1 & 0 & 0 & 0 & 0 & 0 & 0 & 0 & 0 & 0 \\
\hline
0 & 0 & 0 & 0 & 0 & 1 & 0 & 0 & 0 & 0 & 0 & 0 \\
0 & 0 & 0 & 0 & 1 & 0 & 1 & 0 & 0 & 0 & 0 & 0 \\
0 & 0 & 0 & 1 & 0 & 0 & 1 & 0 & 0 & 0 & 0 & 0 \\
\hline
0 & 0 & 0 & 0 & 0 & 0 & 0 & 0 & 0 & 1 & 0 & 0 \\
0 & 0 & 0 & 0 & 0 & 0 & 0 & 0 & 0 & 0 & 1 & 0 \\
0 & 0 & 0 & 0 & 0 & 0 & 0 & 0 & 0 & 0 & 1 & 0 \\
0 & 0 & 0 & 0 & 0 & 0 & 0 & 0 & 1 & 0 & 0 & 1 \\
0 & 0 & 0 & 0 & 0 & 0 & 0 & 0 & 0 & 0 & 0 & 1 \\
0 & 0 & 0 & 0 & 0 & 0 & 0 & 1 & 0 & 0 & 0 & 1 \\
\end{array}\right),$$

where the rows are indexed by U, $U_{(0,0,1)}$, $U_{(0,1,0)}$, $U_{(1,0,0)}$, $U_{(0,0,2)}$, $U_{(0,1,1)}$, $U_{(1,0,1)}$, $U_{(0,2,0)}$, $U_{(1,1,0)}$, $U_{(2,0,0)}$ and the columns are indexed by F_1, F_2, F_3, $(F_1)_{(1)}$, $(F_2)_{(1)}$, $(F_3)_{(0,1)}$, $(F_3)_{(1,0)}$, $(F_1)_{(2)}$, $(F_2)_{(2)}$, $(F_3)_{(0,2)}$, $(F_3)_{(1,1)}$, $(F_3)_{(2,0)}$. Therefore, in the round of the loop for $i = 2$, a K-basis of $\{v \in K^{1 \times 6} \mid v \cdot \Delta_{2,\{0\}}^{(2,2)} = 0\}$ is given by $\{(0, 1, -1, 0, 0, 0)\}$. The matrix

$$\left(\begin{array}{ccc|cccc|c}
1 & 1 & 1 & 0 & 0 & 0 & 0 & U \\
\hline
0 & 0 & 0 & 0 & 0 & 1 & 0 & U_{(0,0,1)} \\
0 & 0 & 0 & 0 & 1 & 0 & 1 & U_{(0,1,0)} \\
0 & 0 & 0 & 1 & 0 & 0 & 1 & U_{(1,0,0)} \\
\hline
0 & 0 & 0 & 0 & 0 & 0 & 0 & U_{(0,1,1)} - U_{(1,0,1)} \\
\end{array}\right)$$

considered in step 5 is row-reduced. Hence, a Janet basis J for $_D\langle \mathscr{I}(\alpha, g)_{\leq 2} \rangle$ in step 6 is $\{U_{(1,0,1)} - U_{(0,1,1)}\}$, which singles out the parametric derivatives $U_{(2,0,0)}$,

$U_{(1,1,0)}$, $U_{(0,2,0)}$, $U_{(0,1,1)}$, $U_{(0,0,2)}$ of order 2 and $U_{(3,0,0)}$, $U_{(2,1,0)}$, $U_{(1,2,0)}$, $U_{(0,3,0)}$, $U_{(0,2,1)}$, $U_{(0,1,2)}$, $U_{(0,0,3)}$ of order 3. In the next round of the loop, the linear relations satisfied by the rows of $\Delta_{3,j}^{(3,3)}$ are generated by $(0, 1, -1, 0, 0, 0, 0)$, and after an additional Gaussian elimination, step 6 finishes with the Janet basis

$$\begin{cases} \underline{U_{(1,0,1)}} - U_{(0,1,1)}, \; \{ \, * \,, \partial_2, \partial_3 \, \}, \\[2mm] \underline{U_{(2,0,1)}} - U_{(0,2,1)}, \; \{ \partial_1, \, * \,, \partial_3 \, \}, \\[2mm] \underline{U_{(2,1,0)}} - U_{(1,2,0)}, \; \{ \partial_1, \partial_2, \partial_3 \, \}. \end{cases} \tag{3.49}$$

In order to solve the implicitization problem using Algorithm 3.2.20, a criterion is needed which allows to decide whether the bound ω on the differentiation order is large enough. To this end we examine the linear relations which are satisfied by the columns of the derivative matrix $\Delta_\omega(\alpha, g)$.

Remark 3.2.22. The columns of the derivative matrix $\Delta_\omega(\alpha, g)$ are indexed by the jet variables $(F_{l_j})_{L_j}$, $j = 1, \ldots, \sigma(\omega)$. Every representation of the zero function $\Omega \to \mathbb{C}$ as $u_1 + \ldots + u_k$, where $u_i \in S(\alpha_i, g_i)$, $i = 1, \ldots, k$ (cf. Def. 3.2.5), yields a column vector $v \in K^{\sigma(\omega) \times 1}$ satisfying $\Delta_\omega(\alpha, g) \cdot v = 0$ as follows.

For $j = 1, \ldots, k$, let $v_j = u_j / g_j$. Assume that v_j for $j = 1, \ldots, \sigma(\omega - 1)$ are already defined. Let $i \in \{1, \ldots, k\}$. First recall that all jets of F_i up to differentiation order ω are included in the sequence $(F_{l_j})_{L_j}$, $j = 1, \ldots, \sigma(\omega)$. Let $F^{(1)}, \ldots, F^{(r)}$ be the jets of F_i of order ω, which correspond to the entries of v to be defined now, $r := \binom{v(i) + \omega - 1}{\omega}$. By forming symbolically all partial derivatives of order ω of the equation $u_i(z) / g_i(z) = f_i(\alpha_i(z))$ and substituting v_j for the expression $(\partial^{L_j} f_{l_j}) \circ \alpha_{l_j}$, $j = 1, \ldots, \sigma(\omega - 1)$, we obtain an inhomogeneous system of linear equations over K for the v_j corresponding to $F^{(1)}, \ldots, F^{(r)}$.

Using the same technique as in the proofs of Lemmas 3.2.2 and 3.2.6, we may assume that coordinates have been chosen in such a way that at most the r partial derivatives of u_i / g_i of order ω with respect to the first $v(i)$ coordinates are non-zero. Then the matrix associated with the linear system is of shape $r \times r$, and its determinant is a certain power of the functional determinant (cf. Rem. A.2.7, p. 246, with $\dim_\mathbb{C} F = 1$) of the restriction of α_i to the first $v(i)$ coordinates, which is non-zero by Assumption 3.2.1. Then the entries v_j which correspond to $F^{(1)}, \ldots, F^{(r)}$ are uniquely determined as solution of this inhomogeneous linear system.

We encode the ambiguity of a representation of an analytic function in the family $S(\alpha_1, g_1; \ldots; \alpha_k, g_k)$ by the $\mathbb{C}$-vector space

$$R := \{ (u_1, \ldots, u_k) \mid u_i \in S(\alpha_i, g_i), \, i = 1, \ldots, k, \, u_1 + \ldots + u_k = 0 \}.$$

Then the above construction defines a $\mathbb{C}$-linear map

$$\gamma_\omega : R \longrightarrow \{ v \in K^{\sigma(\omega) \times 1} \mid \Delta_\omega(\alpha, g) \cdot v = 0 \}.$$

Conversely, every vector $v \in K^{\sigma(\omega) \times 1}$ satisfying $\Delta_\omega(\alpha, g) \cdot v = 0$ gives rise to the system of equations

$$(\partial^{L_j} f_{l_j}) \circ \alpha_{l_j} = v_j, \qquad j = 1, \ldots, \sigma(\omega), \tag{3.50}$$

which, if solvable, furnishes a representation of the zero function. The candidate for a solution of (3.50) is given by $(f_1 \circ \alpha_1, \ldots, f_k \circ \alpha_k) = (v_1, \ldots, v_k)$. Moreover, a necessary condition for solvability of (3.50) is that for each j, the function v_j factors over $\alpha_{l_j}(\Omega)$. In general, there may exist vectors v as above for which (3.50) is not solvable (cf. Ex. 3.2.24 below).

Example 3.2.23. Let us consider the family of analytic functions on $\mathbb{C}^2$ with coordinates x, y of the form $f_1(x) + f_2(x+y) \cdot e^{x-y}$. The derivative matrix of order 1 is given by

$$\Delta_1(\alpha, g) = \left(\begin{array}{cc|cc} 1 & e^{x-y} & 0 & 0 \\ 0 & -e^{x-y} & 0 & e^{x-y} \\ 0 & e^{x-y} & 1 & e^{x-y} \end{array} \right),$$

whose rows and columns are indexed by $U, U_{(0,1)}, U_{(1,0)}$ and $F_1, F_2, (F_1)_{(1)}, (F_2)_{(1)}$, respectively. Then the K-vector space $\{v \in K^{4 \times 1} \mid \Delta_1(\alpha, g) \cdot v = 0\}$ is generated by

$$(e^{2x}, -e^{x+y}, 2e^{2x}, -e^{x+y})^\top.$$

This vector gives rise to the representation of the zero function with coefficients $f_1(x) = e^{2x}$, $f_2(x+y) = -e^{x+y}$. Note that

$$(e^{x-y}, -1, 2e^{x-y}, -1)^\top$$

is another generator, whose first entry, however, is not of the form $f_1(x)$. A solution of (3.50) can be determined as K-multiple of any generator v by applying the annihilating operators ∂_y and $\partial_x - \partial_y$ of $f_1(x)$ and $f_2(x+y)$, respectively, to $\eta(x,y) \cdot v_1$ and $\eta(x,y) \cdot v_2$. In fact, the intersection of $S(\alpha_1, g_1)$ and $S(\alpha_2, g_2)$ is the $\mathbb{C}$-vector space generated by e^{2x}, which allows to determine all representations of the zero function as an element of $S(\alpha_1, g_1; \alpha_2, g_2)$.

Example 3.2.24. Let x, y, z be coordinates of $\mathbb{C}^3$ and consider the family of analytic functions on $\mathbb{C}^3$ which are locally of the form

$$f_1(x+y, x+z) + f_2(x^2+y, z).$$

The derivative matrix of order 1 is given by

$$\Delta_1(\alpha, g) = \left(\begin{array}{cc|cccc} 1 & 1 & 0 & 0 & 0 & 0 \\ 0 & 0 & 1 & 1 & 0 & 0 \\ 0 & 0 & 0 & 0 & 1 & 1 \\ 0 & 0 & 1 & 0 & 1 & 2x \end{array} \right),$$

whose rows and columns are indexed by U, $U_{(0,0,1)}$, $U_{(0,1,0)}$, $U_{(1,0,0)}$ and F_1, F_2, $(F_1)_{(0,1)}$, $(F_2)_{(0,1)}$, $(F_1)_{(1,0)}$, $(F_2)_{(1,0)}$, respectively. Then the two vectors

$$(1, -1, 0, 0, 0, 0)^\top, \qquad (0, 0, 2x-1, -(2x-1), 1, -1)^\top \qquad (3.51)$$

form a K-basis for $\{v \in K^{6 \times 1} \mid \Delta_1(\alpha, g) \cdot v = 0\}$. Clearly, the first basis vector originates from the fact that the intersection of $S(\alpha_1, g_1)$ and $S(\alpha_2, g_2)$ contains all constant functions. However, there exists no non-constant function in K that is annihilated by both $\partial_x - \partial_y - \partial_z$ and $\partial_x - 2x\partial_y$, which are the annihilating operators of $S(\alpha_1, g_1)$ and $S(\alpha_2, g_2)$, respectively. In fact, $\{\partial_x, \partial_y, \partial_z\}$ is a Janet basis for the left ideal of $D = K\langle \partial_x, \partial_y, \partial_z \rangle$ which is generated by these two operators. Therefore, the second basis vector (or a non-trivial K-linear combination of both basis vectors) does not account for any additional non-zero solution of (3.50).

The second basis vector in (3.51) originates from the relation

$$\alpha_{1,1} - \alpha_{1,2} - \alpha_{2,1} + \alpha_{2,2} + (\alpha_{1,2} - \alpha_{2,2})^2 = 0, \qquad (3.52)$$

which is satisfied by the components of the functions $\alpha_1(x,y,z) = (x+y, x+z)$ and $\alpha_2(x,y,z) = (x^2+y, z)$. Applying any derivation $\partial \in \{\partial_x, \partial_y, \partial_z\}$ to (3.52) yields

$$0 = \partial\alpha_{1,1} - \partial\alpha_{1,2} - \partial\alpha_{2,1} + \partial\alpha_{2,2} + 2(\alpha_{1,2} - \alpha_{2,2})(\partial\alpha_{1,2} - \partial\alpha_{2,2})$$
$$= (2(\alpha_{1,2} - \alpha_{2,2}) - 1)\partial\alpha_{1,2} - (2(\alpha_{1,2} - \alpha_{2,2}) - 1)\partial\alpha_{2,2} + \partial\alpha_{1,1} - \partial\alpha_{2,1}.$$

We denote by $\Delta_\omega^{(i,j)}(\alpha, g)$ the block at position (i, j) in $\Delta_\omega(\alpha, g)$ referred to in Remark 3.2.16. By the definition of the derivative matrix, the coefficients of the above linear combination of $\partial\alpha_{1,2}$, $\partial\alpha_{2,2}$, $\partial\alpha_{1,1}$, $\partial\alpha_{2,1}$ form a vector $c \in K^{4 \times 1}$ satisfying $\Delta_1^{(1,1)}(\alpha, g) \cdot c = 0$. In fact, c gives rise to the second vector in (3.51). However, (3.52) is not a representation of the zero function of the form $f_1 \circ \alpha_1 + f_2 \circ \alpha_2$ because of the product of $\alpha_{1,2}$ and $\alpha_{2,2}$ occurring in the last term.

Moreover, the derivative matrix of order 2 is given by

$$\Delta_2(\alpha, g) = \begin{pmatrix}
1 & 1 & 0 & 0 & 0 & 0 & 0 & 0 & 0 & 0 & 0 & 0 \\
0 & 0 & 1 & 1 & 0 & 0 & 0 & 0 & 0 & 0 & 0 & 0 \\
0 & 0 & 0 & 0 & 1 & 1 & 0 & 0 & 0 & 0 & 0 & 0 \\
0 & 0 & 1 & 0 & 1 & 2x & 0 & 0 & 0 & 0 & 0 & 0 \\
0 & 0 & 0 & 0 & 0 & 0 & 1 & 1 & 0 & 0 & 0 & 0 \\
0 & 0 & 0 & 0 & 0 & 0 & 0 & 0 & 1 & 1 & 0 & 0 \\
0 & 0 & 0 & 0 & 0 & 0 & 1 & 0 & 1 & 2x & 0 & 0 \\
0 & 0 & 0 & 0 & 0 & 0 & 0 & 0 & 0 & 0 & 1 & 1 \\
0 & 0 & 0 & 0 & 0 & 0 & 0 & 0 & 1 & 0 & 1 & 2x \\
0 & 0 & 0 & 0 & 0 & 2 & 1 & 0 & 2 & 0 & 1 & 4x^2
\end{pmatrix},$$

the additional rows and columns being indexed by $U_{(0,0,2)}$, $U_{(0,1,1)}$, $U_{(1,0,1)}$, $U_{(0,2,0)}$, $U_{(1,1,0)}$, $U_{(2,0,0)}$ and $(F_1)_{(0,2)}$, $(F_2)_{(0,2)}$, $(F_1)_{(1,1)}$, $(F_2)_{(1,1)}$, $(F_1)_{(2,0)}$, $(F_2)_{(2,0)}$, respectively. We note that the submatrix of $\Delta_2(\alpha, g)$ which is formed by the columns indexed by $(F_1)_{(0,1)}$, $(F_2)_{(0,1)}$, $(F_1)_{(1,0)}$, $(F_2)_{(1,0)}$ has full column rank. Hence, in a solution of

$$\begin{pmatrix} \Delta_2^{(1,1)}(\alpha,g) & 0 \\ \Delta_2^{(2,1)}(\alpha,g) & \Delta_2^{(2,2)}(\alpha,g) \end{pmatrix} \begin{pmatrix} w_1 \\ w_2 \end{pmatrix} = 0$$

the component $w_1 \in K^{4\times 1}$ is uniquely determined by the component $w_2 \in K^{6\times 1}$. The compatibility conditions on $-\Delta_2^{(2,2)}(\alpha,g)\,w_2$ for the solvability of the inhomogeneous system of linear equations

$$\Delta_2^{(2,1)}(\alpha,g)\,w_1 = -\Delta_2^{(2,2)}(\alpha,g)\,w_2$$

imply $\Delta_2^{(2,2)}(\alpha,g)\,w_2 = 0$, and thus $w_1 = 0$. In particular, we conclude that the second vector in (3.51) cannot be extended to a vector in $\{v \in K^{12\times 1} \mid \Delta_2(\alpha,g)\cdot v = 0\}$. A K-basis for this kernel is given by

$$(1,-1,0,\ldots,0)^{\top},\ (0,\ldots,0,(2x-1)^2,-(2x-1)^2,2x-1,-(2x-1),1,-1)^{\top}.$$

Again, the second basis vector originates from (3.52). In fact, by multiplying the relation $c_1\,\partial\alpha_{1,2} + c_2\,\partial\alpha_{2,2} + c_3\,\partial\alpha_{1,1} + c_4\,\partial\alpha_{2,1} = 0$ derived on the previous page by $c_1\,\partial'\alpha_{1,2} - c_2\,\partial'\alpha_{2,2} + c_3\,\partial'\alpha_{1,1} - c_4\,\partial'\alpha_{2,1}$ for $\partial,\partial' \in \{\partial_x,\partial_y,\partial_z\}$, by exchanging the roles played by ∂ and ∂', and by averaging the two results, we obtain

$$c_1^2\,\partial'\alpha_{1,2}\,\partial\alpha_{1,2} - c_2^2\,\partial'\alpha_{2,2}\,\partial\alpha_{2,2} + c_1\,c_3\,(\partial'\alpha_{1,1}\,\partial\alpha_{1,2} + \partial'\alpha_{1,2}\,\partial\alpha_{1,1}) -$$
$$c_2\,c_4\,(\partial'\alpha_{2,1}\,\partial\alpha_{2,2} + \partial'\alpha_{2,2}\,\partial\alpha_{2,1}) + c_3^2\,\partial'\alpha_{1,1}\,\partial\alpha_{1,1} - c_4^2\,\partial'\alpha_{2,1}\,\partial\alpha_{2,1} = 0.$$

The coefficients of this linear combination give rise to the second basis vector.

As the previous example has demonstrated, investigating the representability of the zero function as an element of $S(\alpha_1,g_1;\ldots;\alpha_k,g_k)$ only yields a lower bound for the dimension of $\{v \in K^{\sigma(\omega)\times 1} \mid \Delta_\omega(\alpha,g)\cdot v = 0\}$, or an upper bound for the column rank of $\Delta_\omega(\alpha,g)$. Confining ourselves to methods of linear algebra in this subsection, we are going to formulate only a termination criterion for the implicitization problem based on this upper bound. However, the polynomial relations which are satisfied by the components of $\alpha_1,\ldots,\alpha_k$, such as (3.52) above, can be computed by the techniques discussed in Subsect. 3.1.1, if these components are given by polynomial or rational functions in $z_1,\ldots,z_n$.

Next we examine the image of the $\mathbb{C}$-linear map

$$\gamma_\omega: R \longrightarrow \{v \in K^{\sigma(\omega)\times 1} \mid \Delta_\omega(\alpha,g)\cdot v = 0\}$$

defined in Remark 3.2.22, where

$$R := \{(u_1,\ldots,u_k) \mid u_i \in S(\alpha_i,g_i),\, i = 1,\ldots,k,\, u_1 + \ldots + u_k = 0\}.$$

Let $>$ be the total ordering on the jet variables in U which is used to define the derivative matrix (cf. Def. 3.2.12).

Lemma 3.2.25. *Let $u^{(1)}, \ldots, u^{(r)} \in R$. We fix an arbitrary point p in Ω. Moreover, for $j = 1, \ldots, r$ and $i = 1, \ldots, k$, we define $c_i^{(j)} \in \mathbb{C}^{\zeta(\omega) \times 1}$ to be the vector of coefficients in the Taylor expansion of $u_i^{(j)}$ around p up to order ω, the ordering of the entries being determined according to $>$. Let $c^{(j)}$ be the vector whose entries are successively given by $c_1^{(j)}, \ldots, c_k^{(j)}$. If $c^{(1)}, \ldots, c^{(r)}$ are linearly independent over $\mathbb{C}$, then $\gamma_\omega(u^{(1)}), \ldots, \gamma_\omega(u^{(r)})$ are linearly independent over K.*

Proof. Let $\eta_1, \ldots, \eta_r \in K$ be such that

$$\sum_{j=1}^{r} \eta_j \cdot \gamma_\omega(u^{(j)}) = 0. \tag{3.53}$$

Suppose that not all of $\eta_1, \ldots, \eta_r$ are zero. After possibly replacing $(\eta_1, \ldots, \eta_r)$ with a suitable K-multiple, we may assume that p is not a pole of any η_j and that the Taylor coefficient of order zero of at least one of them is non-zero. Let $i \in \{1, \ldots, k\}$ and $I \in (\mathbb{Z}_{\geq 0})^n$, $|I| \leq \omega$, be arbitrary. In (3.53) we neglect all entries of the vectors $\gamma_\omega(u^{(j)})$ except those which express $\partial^I u_i^{(j)}$ when the restriction of $\Delta_\omega(\alpha, g)$ to the corresponding columns and the relevant row is multiplied from the left. Using this multiplication, we deduce from (3.53) that we have

$$\sum_{j=1}^{r} \eta_j \cdot \partial^I u_i^{(j)} = 0. \tag{3.54}$$

By expanding the functions in (3.54) into Taylor series around p, substituting p for z, and assembling the results for all $i \in \{1, \ldots, k\}$ and $I \in (\mathbb{Z}_{\geq 0})^n$, $|I| \leq \omega$, in the ordering determined by $>$, we obtain a linear dependency of $c^{(1)}, \ldots, c^{(r)}$ given by the Taylor coefficients of $\eta_1, \ldots, \eta_r$ of order zero, which is a contradiction. $\square$

Remark 3.2.26. The elements of the free left module $D^{1 \times k}$ over the skew polynomial ring $D = K\langle \partial_1, \ldots, \partial_n \rangle$ represent linear differential polynomials in k differential indeterminates in terms of k-tuples of differential operators. We choose a term-over-position ordering $>$ on $\mathrm{Mon}(D^{1 \times k})$ which for each of the k components restricts to the total ordering on the jets of U used in the definition of the derivative matrix. In particular, $>$ respects the differentiation order.

Define M to be the submodule of $D^{1 \times k} = \bigoplus_{i=1}^{k} D$ generated by $\bigoplus_{i=1}^{k} \mathscr{I}(\alpha_i, g_i)$ and the sum of the standard basis vectors of $D^{1 \times k}$. The increasing filtration of D by the total differentiation order is extended to $D^{1 \times k}$ by taking the maximum of the total differentiation orders of the non-zero components of a non-zero element of $D^{1 \times k}$. It defines an increasing filtration Φ of $D^{1 \times k}/M$ (cf. Rem. 3.2.17). We denote by Γ the grading of the associated graded module $\mathrm{gr}(D^{1 \times k}/M)$ over the associated graded ring $\mathrm{gr}(D)$.

Using finite generating sets for the annihilating left ideals $\mathscr{I}(\alpha_i, g_i)$ (e.g., as provided by Lemma 3.2.6), we compute the Janet basis J for M with respect to $>$, and we obtain the Hilbert series $H_{D^{1\times k}/M,\Phi}$ of $D^{1\times k}/M$ with respect to Φ from the generalized Hilbert series of $D^{1\times k}/M$:

$$H_{D^{1\times k}/M,\Phi}(\lambda) = H_{D^{1\times k}/M,\Gamma}(\lambda) \cdot \frac{1}{1-\lambda} = \left(\sum_{j=1}^{k} H_{\overline{S},j}(\lambda,\ldots,\lambda) \right) \cdot \frac{1}{1-\lambda}, \quad (3.55)$$

where $\overline{S} := \mathrm{Mon}(D^{1\times k}) - \mathrm{lm}(M)$. The following proposition gives an upper bound for the column rank of $\Delta_\omega(\alpha,g)$ in terms of the coefficients of this Hilbert series.

Proposition 3.2.27. *For $\omega \in \mathbb{Z}_{\geq 0}$ let $\tau(\omega)$ be the coefficient of λ^ω in the Taylor expansion of the Hilbert series $H_{D^{1\times k}/M,\Phi}(\lambda)$ of the previous remark. Then, for every $\omega \in \mathbb{Z}_{\geq 0}$, we have*

$$\mathrm{rank}(\Delta_\omega(\alpha,g)) \leq \sigma(\omega) - \tau(\omega).$$

Proof. The analytic solutions on Ω of the system of linear partial differential equations defined by the Janet basis J of the previous remark form the $\mathbb{C}$-vector space R defined in Remark 3.2.22. The parametric derivatives up to order ω allow to define $u^{(1)}, \ldots, u^{(r)} \in R$, where $r := \tau(\omega)$, whose Taylor coefficients up to order ω form $\mathbb{C}$-linearly independent vectors. By Lemma 3.2.25, the images of $u^{(1)}, \ldots, u^{(r)}$ under the $\mathbb{C}$-linear map

$$\gamma_\omega : R \longrightarrow \{v \in K^{\sigma(\omega)\times 1} \mid \Delta_\omega(\alpha,g) \cdot v = 0\}$$

are linearly independent over K. Therefore, $\sigma(\omega) - \tau(\omega)$ is an upper bound for the column rank of $\Delta_\omega(\alpha,g)$. $\qquad\square$

Proposition 3.2.28. *For $i \in \mathbb{Z}_{\geq 0}$, we denote by $h_i^{\leq \omega}$ the coefficient of λ^i in the Taylor expansion of the Hilbert series of $D/_D\langle \mathscr{I}(\alpha,g)_{\leq \omega}\rangle$ with respect to the filtration defined by the total differentiation order. Then we have:*

a) *If $h_i^{\leq \omega} > \sigma(i) - \tau(i)$ for some $i > \omega$, then $_D\langle \mathscr{I}(\alpha,g)_{\leq \omega}\rangle \neq \mathscr{I}(\alpha,g)$.*
b) *If $h_i^{\leq \omega} = \sigma(i) - \tau(i)$ for all $i > \omega$, then $_D\langle \mathscr{I}(\alpha,g)_{\leq \omega}\rangle = \mathscr{I}(\alpha,g)$.*

Proof. For $i \in \mathbb{Z}_{\geq 0}$, we denote by h_i the coefficient of λ^i in the Taylor expansion of the Hilbert series of $D/\mathscr{I}(\alpha,g)$ with respect to the filtration defined by the differentiation order. Note that for every non-negative integer j, the coefficient h_j is a lower bound for $h_j^{\leq \omega}$. Moreover, we have $h_j^{\leq \omega+r} \geq h_j^{\leq \omega+r+1}$ for all $r \in \mathbb{Z}_{\geq 0}$, and equality holds if ω is large enough. By Proposition 3.2.27 and Remark 3.2.19,

$$\sigma(j) - \tau(j) \geq \mathrm{rank}(\Delta_j(\alpha,g))$$

$$= \zeta(j) - \dim_K \mathscr{I}(\alpha,g)_{\leq j} = \dim_K(D_{\leq j}/_D\langle \mathscr{I}(\alpha,g)_{\leq j}\rangle) = h_j.$$

Hence, the hypothesis of a) implies $h_j^{\leq \omega} > h_j$ for all $j \geq i$, whereas the hypothesis of b) implies $h_j^{\leq \omega} = h_j$ for all $j \in \mathbb{Z}_{\geq 0}$. $\qquad\square$

Example 3.2.29. The family of analytic functions on $\mathbb{C}^3$ which is parametrized by $u(x,y,z) = f_1(x) + f_2(y) + f_3(x+y,z)$ was dealt with in Example 3.2.21. Using Algorithm 3.2.20, the Janet basis $\{ U_{(1,0,1)} - U_{(0,1,1)} \}$ for $_D\langle \mathscr{I}(\alpha,g)_{\le 2} \rangle$ was determined. The generalized Hilbert series

$$\frac{1}{(1-\partial_y)(1-\partial_z)} + \frac{\partial_x}{(1-\partial_x)(1-\partial_y)}$$

of the associated graded residue class module yields via (3.55) the Hilbert series

$$\sum_{\omega \ge 0} h_{\bar{\omega}}^{\le 2} \lambda^{\omega} = \frac{1+\lambda}{(1-\lambda)^3}.$$

Similarly, a Janet basis for $_D\langle \mathscr{I}(\alpha,g)_{\le 3} \rangle$ is given by (3.49). The corresponding parametric derivatives of U, which are enumerated by

$$\frac{1}{(1-\partial_y)(1-\partial_z)} + \frac{\partial_x}{1-\partial_y} + \frac{\partial_x^2}{1-\partial_x},$$

are counted up to a certain differentiation order by the Hilbert series

$$\sum_{\omega \ge 0} h_{\bar{\omega}}^{\le 3} \lambda^{\omega} = \frac{1+\lambda - \lambda^3}{(1-\lambda)^3}.$$

On the other hand, following Remark 3.2.26, we compute a Janet basis for the left $K\langle \partial_x, \partial_y, \partial_z \rangle$-module which is generated by the left hand sides of

$$\partial_y u_1 = 0, \ \partial_z u_1 = 0, \ \partial_x u_2 = 0, \ \partial_z u_2 = 0, \ \partial_x u_3 - \partial_y u_3 = 0, \ u_1 + u_2 + u_3 = 0.$$

Choosing $u_1 > u_2 > u_3$, the parametric derivatives are then given by u_2, u_3, $\partial_y u_3$, and therefore

$$\sum_{\omega \ge 0} \tau(\omega) \lambda^{\omega} = 2 + \frac{3\lambda}{1-\lambda}.$$

The following table summarizes some parts of the information we have obtained:

ω	0	1	2	3	4	5
$\zeta(\omega) = \binom{\omega+3}{\omega}$	1	4	10	20	35	56
$\sigma(\omega) = 2\binom{\omega+1}{\omega} + \binom{\omega+2}{\omega}$	3	7	12	18	25	33
$\tau(\omega)$	2	3	3	3	3	3
$\sigma(\omega) - \tau(\omega)$	1	4	9	15	22	30
$h_{\bar{\omega}}^{\le 2}$	1	4	9	16	25	36
$h_{\bar{\omega}}^{\le 3}$	1	4	9	15	22	30

In fact, we have

$$\sum_{\omega \geq 0} (\sigma(\omega) - \tau(\omega)) \lambda^\omega = \frac{2\lambda - 3}{(1-\lambda)^3} - \left(2 + \frac{3\lambda}{1-\lambda}\right) = \frac{1+\lambda-\lambda^3}{(1-\lambda)^3} = \sum_{\omega \geq 0} h_{\bar{\omega}}^{\leq 3} \lambda^\omega.$$

Hence, part b) of Proposition 3.2.28 implies $_D\langle \mathscr{I}(\alpha,g)_{\leq 3}\rangle = \mathscr{I}(\alpha,g)$.

3.2.3 Constructing a Representation

We continue to use the notation of the previous subsection.

Remark 3.2.30. Let $\Gamma_0 \in K^{k \times 1}$ be a right inverse of $\Delta_0(\alpha,g)$. Since the analytic functions g_i are assumed to be non-zero (and therefore invertible in K), such a matrix exists. Assume that $X_1, \ldots, X_{k-1} \in K^{k \times 1}$ form a basis of the kernel of the linear map induced by $\Delta_0(\alpha,g)$. Then

$$u(z) = f_1(\alpha_1(z)) g_1(z) + \ldots + f_k(\alpha_k(z)) g_k(z)$$

is equivalent to the existence of meromorphic functions $a_1, \ldots, a_{k-1}$ on Ω such that

$$\begin{pmatrix} f_1(\alpha_1(z)) \\ \vdots \\ f_k(\alpha_k(z)) \end{pmatrix} = u(z)\Gamma_0 + a_1(z)X_1 + \ldots + a_{k-1}(z)X_{k-1} \tag{3.56}$$

holds almost everywhere on Ω. Interpreting $a_1, \ldots, a_{k-1}$ as unknown functions and applying the annihilators given by Lemma 3.2.6 to each entry of the vector in (3.56) yields linear partial differential equations for $a_1, \ldots, a_{k-1}$ if a given analytic function on Ω is substituted for $u(z)$.

Example 3.2.31. Let x, y, z be coordinates of $\mathbb{C}^3$. We consider the family of analytic functions on $\mathbb{C}^3$ which are locally of the form

$$u(x,y,z) = f_1(x+y^2) \cdot x + f_2(x,y+z) \cdot y + f_3(y,x-z), \tag{3.57}$$

and we would like to determine all possible ways of representing the function $u(x,y,z) = z$ in the above form. The families of analytic functions parametrized respectively by $f_1(x+y^2)$, $f_2(x,y+z)$, $f_3(y,x-z)$ have the following implicit descriptions in terms of linear partial differential equations:

$$\{U_z = 0,\, 2yU_x - U_y = 0\}, \quad \{U_y - U_z = 0\}, \quad \{U_x + U_z = 0\}. \tag{3.58}$$

A right inverse Γ_0 of the linear map induced by $\Delta_0(\alpha,g)$ and a basis $\{X_1, X_2\}$ of its kernel are given by

$$\Gamma_0 := \begin{pmatrix} 0 \\ 0 \\ 1 \end{pmatrix}, \qquad X_1 := \begin{pmatrix} 1 \\ 0 \\ -x \end{pmatrix}, \qquad X_2 := \begin{pmatrix} y \\ -x \\ 0 \end{pmatrix}.$$

Hence, the right hand side in (3.56) equals

$$
\begin{pmatrix}
a_1(x,y,z) + y \cdot a_2(x,y,z) \\
-x \cdot a_2(x,y,z) \\
u(x,y,z) - x \cdot a_1(x,y,z)
\end{pmatrix}.
\tag{3.59}
$$

Substitution of the i-th entry of the previous vector for $U(x,y,z)$ in the i-th system of PDEs in (3.58), $i = 1$, 2, 3, yields

$$
\begin{cases}
\dfrac{\partial a_1}{\partial z} + y\,\dfrac{\partial a_2}{\partial z} = 0, \quad 2y\,\dfrac{\partial a_1}{\partial x} - \dfrac{\partial a_1}{\partial y} + 2y^2\,\dfrac{\partial a_2}{\partial x} - y\,\dfrac{\partial a_2}{\partial y} - a_2 = 0, \\[2mm]
x\,\dfrac{\partial a_2}{\partial y} - x\,\dfrac{\partial a_2}{\partial z} = 0, \\[2mm]
\dfrac{\partial u}{\partial x} + \dfrac{\partial u}{\partial z} - x\,\dfrac{\partial a_1}{\partial x} - x\,\dfrac{\partial a_1}{\partial z} - a_1 = 0.
\end{cases}
\tag{3.60}
$$

In order to find all analytic coefficient functions f_1, f_2, f_3 such that (3.57) represents the analytic function $u(x,y,z) = z$, we substitute z for $u(x,y,z)$ in (3.60). The analytic solutions of this system of linear PDEs are given by

$$
a_1(x,y,z) = -\frac{y}{x} \cdot h(x-y-z) + 1, \quad a_2(x,y,z) = \frac{1}{x} \cdot h(x-y-z),
$$

where h is an arbitrary analytic function of one argument. Although points in $\mathbb{C}^3$ whose x-coordinate is zero have to be excluded from the domain of definition of a_1 and a_2, substitution into (3.59) specializes (3.56) to

$$
\begin{pmatrix}
f_1(x+y^2) \\
f_2(x,y+z) \\
f_3(y,x-z)
\end{pmatrix}
=
\begin{pmatrix}
1 \\
-h(x-y-z) \\
z-x+y \cdot h(x-y-z)
\end{pmatrix},
$$

whose right hand side is defined on $\mathbb{C}^3$ if h is defined on $\mathbb{C}$. Note that this equation is solvable for f_1, f_2, f_3 for any admissible function h.

Remark 3.2.32. If $u(z)$ is not substituted with a concrete function in the system of linear partial differential equations constructed in Remark 3.2.30, then elimination of $a_1, \ldots, a_{k-1}$ and their partial derivatives from these equations results in a generating set for $\mathscr{I}(\alpha, g)$. In other words, computing a Janet basis J for the above PDE system with respect to a term ordering that ranks every partial derivative of any a_i higher than any partial derivative of u is another method to determine $\mathscr{I}(\alpha, g)$. If the term ordering is chosen appropriately, determining the subset of J consisting of the elements which only involve u is equivalent to computing the intersection in Theorem 3.2.10 or to applying Algorithm 3.2.20 for ω large enough.

Example 3.2.33. Resuming Example 3.2.31, the annihilating left ideal $\mathscr{I}(\alpha, g)$ may be obtained from a Janet basis for (3.60) with respect to a block ordering with blocks $\{a_1, a_2\}$, $\{u\}$ such that $a_1 > u$ and $a_2 > u$. The result of such a computation is described in Example 3.1.28, p. 136.

Remark 3.2.34. If an implicit description in terms of homogeneous linear PDEs for the sum of certain but not all $f_i(\alpha_i(z))g_i(z)$ is known, then, by construction, the result of substituting the sum of all $f_i(\alpha_i(z))g_i(z)$ into these equations involves only those coefficient functions f_i which are not part of the implicitly described family. Applying the differential operators of the left hand sides of the homogeneous system of linear PDEs to both sides of

$$u(z) = f_1(\alpha_1(z))g_1(z) + \ldots + f_k(\alpha_k(z))g_k(z) \tag{3.61}$$

for a concrete analytic function u then yields an inhomogeneous system of linear PDEs for these remaining coefficient functions in a representation of u as in (3.61). In this way, coefficient functions in such a representation can be determined separately as solutions to differential equations that we expect to be easier to handle than the PDE system for $a_1, \ldots, a_{k-1}$ in Remark 3.2.30 (cf. also (3.60)).

Let us suppose that $f_i(\alpha_i(z))g_i(z)$ for $i \in \{1, \ldots, r\}$ are the summands yet to be treated. If all components of $\alpha_1, \ldots, \alpha_r$ are functionally independent (which is in particular the case if $r = 1$), or, more precisely, if the Jacobian matrix of $(\alpha_1, \ldots, \alpha_r)$ has rank $m := v(1) + \ldots + v(r)$ at $w \in \Omega$ (so that, in particular, $m \leq n$), then among the coordinates $z_1, \ldots, z_n$ there exist $\beta_{m+1}, \ldots, \beta_n$ and a neighborhood Ω' of w in Ω such that $(\alpha_1, \ldots, \alpha_r, \beta_{m+1}, \ldots, \beta_n)$ defines coordinates on Ω' (cf. also the technique used in the proof of Lemma 3.2.6). Inversion of the diffeomorphism defined by $(\alpha_1, \ldots, \alpha_r, \beta_{m+1}, \ldots, \beta_n)$ allows to write the differential equations for $f_1, \ldots, f_r$ in the new coordinates. In cases in which this inversion can be carried out effectively, the resulting equations involve $\beta_{m+1}, \ldots, \beta_n$ as parameters, which may be specialized to suitable values. Taking derivatives with respect to these parameters also produces new equations. By combining sufficiently many of such equations and performing elimination of unknown derivatives, computation of the coefficient functions $f_1, \ldots, f_r$ often becomes an easy task.

Example 3.2.35. We demonstrate the strategy discussed in the previous remark on the family of analytic functions on $\mathbb{C}^3$ with coordinates x, y, z which are locally of the form

$$u(x, y, z) = f_1(x + y^2) \cdot x + f_2(x, y + z) \cdot y + f_3(y, x - z). \tag{3.62}$$

This family has been dealt with already in the previous two examples. In this case, the annihilating left ideal for the family $S(\alpha_2, g_2; \alpha_3, g_3)$ is generated by the composition

$$(\partial_x + \partial_z)(y(\partial_y - \partial_z) - 1)$$

of the two commuting differential operators annihilating $S(\alpha_2, g_2)$ and $S(\alpha_3, g_3)$, respectively. Applying this operator to (3.62) for $u(x, y, z) = z$ results in

$$-1 = 2xy^2 \cdot f_1''(x + y^2) + (2y^2 - x) \cdot f_1'(x + y^2) - f_1(x + y^2).$$

Changing coordinates to $(X, y, z) := (x + y^2, y, z)$ yields

$$-1 = 2(X - y^2)y^2 \cdot f_1''(X) + (3y^2 - X) \cdot f_1'(X) - f_1(X),$$

which is an equation that holds for any value of y. By substituting $y = 0$ and $y = 1$ respectively, we obtain the ordinary differential equations

$$\begin{cases} X \cdot f_1'(X) + f_1(X) = 1, \\ 2(1 - X) \cdot f_1''(X) + (X - 3) \cdot f_1'(X) + f_1(X) = 1, \end{cases}$$

whose only common solution is $f_1(X) = 1$, cf. also the end of Example 3.2.31.

Remark 3.2.36. Instead of applying appropriate differential operators to a concrete function u, an analogous computation with symbols representing an analytic function u and its partial derivatives may result in formulas for the coefficient functions $f_1, \ldots, f_k$ in terms of u, usually involving derivatives. Clearly, formulas yielding the f_i via substitution (and possibly integration) are valid only if the given function u is in $S(\alpha_1, g_1; \ldots; \alpha_k, g_k)$.

Example 3.2.37. Let x, y, z be coordinates of $\mathbb{C}^3$ and consider the family of analytic functions on $\mathbb{C}^3$ of the form $f_1(x + y, z) + f_2(x, y + z)$. The families $S(\alpha_1, g_1)$ and $S(\alpha_2, g_2)$ which are defined by the summands (where g_1 and g_2 are the constant function 1) are the sets of analytic solutions of the linear PDEs

$$\{U_x - U_y = 0\}, \quad \{U_y - U_z = 0\},$$

respectively. Suppose that u is an analytic function on $\mathbb{C}^3$ of the above form. Then substitution of $f_2(x, y + z) - u(x, y, z)$ for $U(x, y, z)$ in the first system yields zero. Equivalently, the partial derivatives $(f_2)_i$ of f_2 with respect to its i-th argument, $i = 1, 2$, satisfy

$$(f_2)_1 - (f_2)_2 = \frac{\partial u}{\partial x} - \frac{\partial u}{\partial y}. \tag{3.63}$$

Similarly, substitution of $f_1(x + y, z) - u(x, y, z)$ for $U(x, y, z)$ in the second system results in the formula

$$(f_1)_1 - (f_1)_2 = \frac{\partial u}{\partial y} - \frac{\partial u}{\partial z}. \tag{3.64}$$

In order to find, if possible, analytic functions $f_1(x + y, z)$, $f_2(x, y + z)$ which represent $u(x, y, z) = xy + y^2 + yz$, we solve the PDEs (3.63) and (3.64) with given right hand sides. We obtain

$$f_1(x + y, z) = (x + y)^2 + (x + y)z + g(x + y + z),$$

$$f_2(x, y + z) = -x^2 - x(y + z) + h(x + y + z),$$

where g and h are arbitrary analytic functions of one argument.

Example 3.2.38. We consider the family of analytic functions on $\mathbb{C}^2$ with coordinates x, y which are locally of the form $f_1(x) + f_2(x + y) \cdot e^{x - y^2}$. The derivative matrix of order 2

$$\Delta_2(\alpha,g) = \begin{pmatrix} 1 & e^{x-y^2} & 0 & 0 & 0 & 0 \\ 0 & -2y\,e^{x-y^2} & 0 & e^{x-y^2} & 0 & 0 \\ 0 & e^{x-y^2} & 1 & e^{x-y^2} & 0 & 0 \\ 0 & 2\,(2y^2-1)\,e^{x-y^2} & 0 & -4y\,e^{x-y^2} & 0 & e^{x-y^2} \\ 0 & -2y\,e^{x-y^2} & 0 & -(2y-1)\,e^{x-y^2} & 0 & e^{x-y^2} \\ 0 & e^{x-y^2} & 0 & 2\,e^{x-y^2} & 1 & e^{x-y^2} \end{pmatrix}$$

is invertible, which is, apart from the circumstance that $\Delta_2(\alpha,g)$ is of square shape, due to the fact that representations of analytic functions of the above form are unique. If u is an analytic function that is locally of the above form, then $\Delta_2(\alpha,g)^{-1}$ allows to express the functions $\partial^{L_j}(f_{l_j}) \circ \alpha_{l_j}$ which correspond to the entries of the vector $(F_1, F_2, (F_1)_{(1)}, (F_2)_{(1)}, (F_1)_{(2)}, (F_2)_{(2)})^\top$ as K-linear combinations of the derivatives of the function u that correspond to the entries of the vector $(U, U_{(0,1)}, U_{(1,0)}, U_{(0,2)}, U_{(1,1)}, U_{(2,0)})^\top$. For the present case we obtain:

$$f_1(x) = u(x,y) + \left(y+\frac{1}{2}\right)\frac{\partial u}{\partial y}(x,y) + \frac{1}{2}\frac{\partial^2 u}{\partial y^2}(x,y) - \frac{1}{2}\frac{\partial^2 u}{\partial x \partial y}(x,y),$$

$$f_2(x+y) = -\frac{1}{2}e^{y^2-x}\left((2y+1)\frac{\partial u}{\partial y}(x,y) + \frac{\partial^2 u}{\partial y^2}(x,y) - \frac{\partial^2 u}{\partial x \partial y}(x,y)\right).$$

For instance, substitution of $u(x,y) = \exp(-y(1+y))$ into these formulas yields $f_1(x) = 0$, $f_2(x+y) = \exp(-(x+y))$, whereas substitution of $u(x,y) = x \cdot y$ is meaningless because $x \cdot y$ is not of the above form. An implicit description of the above family is given by the system of linear PDEs

$$\begin{cases} \dfrac{\partial^3 u}{\partial x \partial y^2} - \dfrac{\partial^3 u}{\partial y^3} - (2y+1)\dfrac{\partial^2 u}{\partial y^2} - 4\dfrac{\partial u}{\partial y} = 0, \\[2mm] \dfrac{\partial^3 u}{\partial x^2 \partial y} - \dfrac{\partial^3 u}{\partial y^3} - 2\,(2y+1)\dfrac{\partial^2 u}{\partial x \partial y} + (4y^2+4y-5)\dfrac{\partial u}{\partial y} = 0. \end{cases}$$

It may be obtained by computing the intersection $\mathscr{I}(\alpha_1,g_1) \cap \mathscr{I}(\alpha_2,g_2)$ or by collecting linear relations that are satisfied by the rows of the derivative matrix $\Delta_\omega(\alpha,g)$, where $\omega = 3$ is large enough.

3.2.4 Annihilators for Linear Parametrizations

As before, K denotes the field of meromorphic functions on the open and connected subset Ω of $\mathbb{C}^n$ with coordinates $z = (z_1,\ldots,z_n)$, and $D = K\langle\partial_1,\ldots,\partial_n\rangle$ is the skew polynomial ring of differential operators with coefficients in K. Let $k \in \mathbb{N}$, and let $g_i\colon \Omega \to \mathbb{C}$, $\alpha_i\colon \Omega \to \mathbb{C}^{v(i)}$, $i = 1, \ldots, k$, be analytic functions as described in Subsect. 3.2.1.

In this subsection we relate properties of the given analytic functions g_i and α_i to the existence of generating sets for the annihilating left ideal $\mathscr{I}(\alpha, g)$ consisting of differential operators of a particular kind. These results may be used to recognize the type of analytic solutions of a given system of partial differential equations (cf. Subsect. 3.3.5).

Remark 3.2.39. Implicitization of a family $S := S(\alpha_1, g_1; \ldots; \alpha_k, g_k)$ of analytic functions produces a system of linear differential equations whose set of analytic solutions equals S. The fact that S is a $\mathbb{C}$-vector space corresponds to the fact that the left hand sides of the equations are homogeneous differential polynomials of degree one. In particular, the family S satisfies that $c \cdot u \in S$ for every $u \in S$ and every $c \in \mathbb{C}$, a property which is reminiscent of projective varieties, cf. also Lemma A.3.14, p. 253. For a generalization to nonlinear differential elimination, cf. Remark 3.3.19, p. 205.

Proposition 3.2.40. *Let $g_1, \ldots, g_k \colon \Omega \to \mathbb{C}$ be non-zero constant or exponential functions composed with polynomials in z of degree one, and let $\alpha_i \colon \Omega \to \mathbb{C}^{v(i)}$, $i = 1, \ldots, k$, be analytic functions, all of whose components are polynomials in z of degree one. Then $\mathscr{I}(\alpha, g)$ has a generating set consisting of differential operators with constant coefficients.*

Proof. If all g_i are constant, then, for any $\omega \in \mathbb{Z}_{\geq 0}$, all entries of the derivative matrix $\Delta_\omega(\alpha, g)$ are constant. In this case, the linear dependencies of the rows of $\Delta_\omega(\alpha, g)$ are generated by vectors with constant entries, and since these form a generating set for $\mathscr{I}(\alpha, g)$ if ω is chosen large enough, the claim follows. For every $j = 1, \ldots, k$ such that g_j is an exponential function as in the hypothesis, the columns of $\Delta_\omega(\alpha, g)$ which are indexed by jets of F_j are constant vectors multiplied by g_j. The linear dependencies of the rows are determined independently of the factor g_j, which shows that they are again generated by constant vectors. $\qquad\square$

We illustrate Proposition 3.2.40 and its proof on the following example.

Example 3.2.41. Let us consider the family of analytic functions on $\mathbb{C}^2$ with coordinates x, y of the form $f_1(x) + f_2(x+y) \cdot e^{x-y}$, which has been considered already in Example 3.2.23. The derivative matrix of order 2 is given by

$$\Delta_2(\alpha, g) = \left(\begin{array}{ccc|ccc} 1 & e^{x-y} & 0 & 0 & 0 & 0 \\ 0 & -e^{x-y} & 0 & e^{x-y} & 0 & 0 \\ 0 & e^{x-y} & 1 & e^{x-y} & 0 & 0 \\ 0 & e^{x-y} & 0 & -2e^{x-y} & 0 & e^{x-y} \\ 0 & -e^{x-y} & 0 & 0 & 0 & e^{x-y} \\ 0 & e^{x-y} & 0 & 2e^{x-y} & 1 & e^{x-y} \end{array} \right),$$

whose rows and columns are indexed by U, $U_{(0,1)}$, $U_{(1,0)}$, $U_{(0,2)}$, $U_{(1,1)}$, $U_{(2,0)}$ and F_1, F_2, $(F_1)_{(1)}$, $(F_2)_{(1)}$, $(F_1)_{(2)}$, $(F_2)_{(2)}$, respectively. The linear dependencies of the rows of $\Delta_2(\alpha, g)$ are generated by $(0, -2, 0, -1, 1, 0)$, which corresponds to the annihilating operator $\partial_x \partial_y - \partial_y^2 - 2\partial_y$. In fact, Lemma 3.2.6 and Theorem 3.2.10 show that the annihilating left ideal $\mathscr{I}(\alpha, g) = \mathscr{I}(\alpha_1, g_1) \cap \mathscr{I}(\alpha_2, g_2)$ of $K\langle \partial_x, \partial_y \rangle$ is generated by $\partial_y(\partial_x - \partial_y - 2)$.

Proposition 3.2.42. *Let I be an ideal of the polynomial algebra $\mathbb{C}[\partial_1,\dots,\partial_n]$. Then the following two statements are equivalent.*

a) *There exist $k \in \mathbb{N}$, $v(1)$, ..., $v(k) \in \{0,\dots,n-1\}$, and non-zero analytic functions g_1, ..., $g_k \colon \Omega \to \mathbb{C}$ which are constant or exponential functions composed with polynomials in z of degree one, and for every $i = 1$, ..., k, there exists $\alpha_i \colon \Omega \to \mathbb{C}^{v(i)}$, all of whose components are polynomials in z of degree one, with Jacobian matrix of rank $v(i)$ at every point of Ω, such that*

$$D \cdot I = \mathscr{I}(\alpha,g).$$

b) *The ideal I is radical and each of its prime components is generated by polynomials of degree one.*

Proof. Let us assume that b) holds. Let a prime component of I be generated by the polynomials $p_1, \dots, p_m \in \mathbb{C}[\partial_1,\dots,\partial_n]$ of degree one, and define q_i as the homogeneous component of p_i of degree one, $i = 1, \dots, m$. We define

$$h := \exp(a_1 z_1 + \dots + a_n z_n)$$

for any zero $(a_1,\dots,a_n) \in \mathbb{C}^n$ of the chosen prime component of I, i.e., we have $p_i(a_1,\dots,a_n) = 0$ for all $i = 1, \dots, m$. (If each p_i is homogeneous, then the constant 1 may be chosen for h.) If $m < n$, then we choose a basis $(\beta_1,\dots,\beta_{n-m})$ of the vector space of linear homogeneous polynomials in $\mathbb{C}[z_1,\dots,z_n]$ which are annihilated by $q_1, \dots, q_m$ (acting by partial differentiation).

Now, if I_1, ..., I_k are the prime components of I, then we define g_i to be the corresponding analytic function h from above, $v(i)$ to be the difference $n - m$, and in case $v(i) > 0$, the analytic function α_i to be given by $(\beta_1,\dots,\beta_{n-m})$. Lemma 3.2.6 and Theorem 3.2.10 imply a).

In order to prove the implication a) $\Rightarrow$ b), we note that, for every $i \in \{1,\dots,k\}$, by Lemma 3.2.6, $\mathscr{I}(\alpha_i,g_i)$ has a generating set consisting of first order differential operators in D. By interpreting these annihilating differential operators as linear dependencies of the rows of the corresponding derivative matrix and using the same argument as in the proof of Proposition 3.2.40, we conclude that these generators may be chosen to have constant coefficients. The ideal I_i of $\mathbb{C}[\partial_1,\dots,\partial_n]$ which they generate is prime because the generators are polynomials in ∂_1, ..., ∂_n of degree one. Since $\mathscr{I}(\alpha,g)$ is the intersection of the left ideals $\mathscr{I}(\alpha_1,g_1)$, ..., $\mathscr{I}(\alpha_k,g_k)$ by Theorem 3.2.10 and since we have $D \cdot I = \mathscr{I}(\alpha,g)$, the ideal I is the intersection of $I_1, \dots, I_k$ and is therefore radical. $\qquad \square$

An application of the previous proposition to symbolic solving of partial differential equations will be given in Subsect. 3.3.5 (cf. Ex. 3.3.40, p. 221).

Proposition 3.2.43. *Let g_1, ..., $g_k \colon \Omega \to \mathbb{C}$ be non-zero products of (complex) rational functions in z and exponential functions whose arguments are rational functions in z, and let $\alpha_i \colon \Omega \to \mathbb{C}^{v(i)}$, $i = 1$, ..., k, be analytic functions, all of whose components are rational functions in z. Then $\mathscr{I}(\alpha,g)$ has a generating set consisting of differential operators with polynomial coefficients.*

Proof. The proof is similar to the one of Proposition 3.2.40. The assumptions imply that, for any $\omega \in \mathbb{Z}_{\geq 0}$, each column of $\Delta_\omega(\alpha, g)$ is a vector of rational functions in z multiplied by an exponential function whose argument is a rational function in z. The linear dependencies of the rows of $\Delta_\omega(\alpha, g)$ are therefore generated by vectors with polynomial entries. $\qquad\square$

Example 3.2.44. Let x, y be coordinates of $\mathbb{C}^2$. We consider the family of analytic functions which are locally of the form $f_1(\frac{x}{1-y}) \cdot \frac{1}{y} + f_2(\frac{y}{1-x}) \cdot \frac{1}{x}$. The derivative matrix of order 2 is given by

$$
\Delta_2(\alpha, g) = \left(
\begin{array}{cc|cc|cc}
\frac{1}{y} & \frac{1}{x} & 0 & 0 & 0 & 0 \\[4pt]
-\frac{1}{y^2} & 0 & \frac{x}{y(1-y)^2} & \frac{1}{x(1-x)} & 0 & 0 \\[4pt]
0 & -\frac{1}{x^2} & \frac{1}{y(1-y)} & \frac{y}{x(1-x)^2} & 0 & 0 \\[4pt]
\frac{2}{y^3} & 0 & \frac{2x(2y-1)}{y^2(1-y)^3} & 0 & \frac{x^2}{y(1-y)^4} & \frac{1}{x(1-x)^2} \\[4pt]
0 & 0 & \frac{2y-1}{y^2(1-y)^2} & \frac{2x-1}{x^2(1-x)^2} & \frac{x}{y(1-y)^3} & \frac{y}{x(1-x)^3} \\[4pt]
0 & \frac{2}{x^3} & 0 & \frac{-2y(2x-1)}{x^2(1-x)^3} & \frac{1}{y(1-y)^2} & \frac{y^2}{x(1-x)^4}
\end{array}
\right),
$$

whose rows and columns are indexed by U, $U_{(0,1)}$, $U_{(1,0)}$, $U_{(0,2)}$, $U_{(1,1)}$, $U_{(2,0)}$ and F_1, F_2, $(F_1)_{(1)}$, $(F_2)_{(1)}$, $(F_1)_{(2)}$, $(F_2)_{(2)}$, respectively. The linear dependencies of the rows of $\Delta_2(\alpha, g)$ are generated by the vector

$$
\big(-(2xy-x-y+1),\, -y(4xy-3x-y+1),\, -x(4xy-x-3y+1),
$$
$$
xy^2(1-y),\, -xy(2xy-x-y+1),\, x^2y(1-x)\big),
$$

which corresponds to the annihilating operator

$$
d := x^2 y(1-x)\partial_x^2 - xy(2xy-x-y+1)\partial_x\partial_y + xy^2(1-y)\partial_y^2 -
$$
$$
x(4xy-x-3y+1)\partial_x - y(4xy-3x-y+1)\partial_y - (2xy-x-y+1).
$$

In fact, the intersection $\mathscr{I}(\alpha, g)$ of $\mathscr{I}(\alpha_1, g_1) = {}_D\langle xy\partial_x - y(1-y)\partial_y - (1-y)\rangle$ and $\mathscr{I}(\alpha_2, g_2) = {}_D\langle x(1-x)\partial_x - xy\partial_y + (1-x)\rangle$ is generated by d.

The following proposition discusses the case of constant functions $g_1, \ldots, g_k$.

Proposition 3.2.45. Let $g_1, \ldots, g_k : \Omega \to \mathbb{C}$ be non-zero constant functions and let $\alpha_i : \Omega \to \mathbb{C}^{\nu(i)}$ be analytic functions with Jacobian matrix of rank $\nu(i)$ at every point of Ω, $i = 1, \ldots, k$. Then we have:

a) No element of the left ideal $\mathscr{I}(\alpha, g)$ of D has a term of differentiation order zero, i.e., every element of $\mathscr{I}(\alpha, g)$ is a left K-linear combination of ∂^I, $I \in (\mathbb{Z}_{\geq 0})^n$, not involving $\partial^{(0,\ldots,0)}$.

b) If all components of each α_i are polynomials in z of degree one, then $\mathscr{I}(\alpha, g)$ has a generating set with the property that each of its elements P has constant coefficients and the differentiation order is the same for each term in P.

c) If all components of each α_i are homogeneous polynomials in z (not necessarily of the same degree), $0 \in \Omega$, and

$$X(\Omega) := \{ c \in \mathbb{C} \mid c \cdot z \in \Omega \text{ for all } z \in \Omega \}$$

is an infinite set, then $\mathscr{I}(\alpha, g)$ has a generating set with the property that for each of its elements P, the linear differential polynomial $P(u)$ is isobaric around 0 (cf. Def. A.3.17, p. 254), where u is a differential indeterminate.

Proof. a) Let $\Delta_\omega(\alpha, g)$ be the derivative matrix of order ω for $\omega \in \mathbb{Z}_{\geq 0}$ sufficiently large. The first k columns of $\Delta_\omega(\alpha, g)$ correspond to the zero order jet variables $F_1, \ldots, F_k$. It follows from the assumption that all non-zero entries of these columns are in the row corresponding to the zero order jet variable U. Therefore, no non-trivial linear relation among the rows of $\Delta_\omega(\alpha, g)$ is a relation involving the row corresponding to U.

b) By Proposition 3.2.40, the generating set may be chosen to consist of differential operators with constant coefficients. Moreover, for any $\omega \in \mathbb{Z}_{\geq 0}$, the derivative matrix $\Delta_\omega(\alpha, g)$ has a block diagonal structure, because the off-diagonal blocks in the structure described in Remark 3.2.16 have non-zero entries only if the product rule of differentiation is applied in a non-trivial way. Hence, there exists a generating set for the linear relations satisfied by the rows of $\Delta_\omega(\alpha, g)$ whose elements are relations involving only rows corresponding to jet variables of the same differentiation order.

c) The assumption on the functions $\alpha_1, \ldots, \alpha_k$ implies that for any analytic function $u \in S := S(\alpha_1, g_1; \ldots; \alpha_k, g_k)$ we have $u(c \cdot -) \in S$ for all $c \in X(\Omega)$, where $u(c \cdot -)$ denotes the analytic function on Ω defined by

$$u(c \cdot -)(z_1, \ldots, z_n) := u(c \cdot z_1, \ldots, c \cdot z_n), \qquad (z_1, \ldots, z_n) \in \Omega.$$

Now the claim follows from Lemma A.3.20, p. 255.   □

3.2.5 Applications

In this subsection we present a few applications which use constructive solutions of the implicitization problem. Implicit descriptions of certain families of analytic functions, which are given by linear parametrizations, allow, e.g., to decide whether a given function may be written as sum of functions which depend only on specified arguments (cf. Examples 3.2.46 and 3.2.47). We exemplify a method for constructing matrix representations of certain Lie groups (cf. Example 3.2.48). In the context of solving partial differential equations, implicit descriptions of families as above may allow to examine a set of solutions to a PDE system with regard to completeness (cf. Example 3.2.49) and may give rise to algebraic simplifications in the search for solutions of a particular form (e.g., solutions to PDE systems in terms of special functions; cf. Example 3.2.51). Both applications to PDEs profit from the use of Thomas' algorithm and improve symbolic solving of PDEs.

Example 3.2.46. Let the spherical coordinates r, φ, θ of $\mathbb{R}^3 - \{0\}$ be defined in terms of the Cartesian coordinates x, y, z by

$$x = r\cos(\varphi)\cos(\theta), \quad y = r\sin(\varphi)\cos(\theta), \quad z = r\sin(\theta).$$

Given an analytic function u on some connected open subset of $\mathbb{R}^3 - \{0\}$ in spherical coordinates, deciding whether it is of the form

$$f_1(r) + f_2(\varphi) + f_3(\theta) \tag{3.65}$$

for some analytic functions f_1, f_2, f_3, or of the form

$$f_{1,2}(r, \varphi) + f_{1,3}(r, \theta) + f_{2,3}(\varphi, \theta) \tag{3.66}$$

for some analytic functions $f_{1,2}$, $f_{1,3}$, $f_{2,3}$, boils down to checking whether certain differential operators annihilate u. We are going to express the same conditions for analytic functions given in Cartesian coordinates.

Let κ be the coordinate transformation from Cartesian coordinates to spherical coordinates defined by

$$r = \sqrt{x^2 + y^2 + z^2}, \quad \varphi = \arctan\left(\frac{y}{x}\right), \quad \theta = \arctan\left(\frac{z}{\sqrt{x^2 + y^2}}\right).$$

By considering the partial derivatives with respect to r, φ, θ as vector fields on $\mathbb{R}^3 - \{0\}$, we express these operators in terms of $\frac{\partial}{\partial x}, \frac{\partial}{\partial y}, \frac{\partial}{\partial z}$ using the inverse of the Jacobian matrix of κ. Appropriate scaling yields three pairwise commuting differential operators defined by

$$R := \frac{1}{\sqrt{x^2 + y^2 + z^2}} \left(x\frac{\partial}{\partial x} + y\frac{\partial}{\partial y} + z\frac{\partial}{\partial z} \right),$$

$$\Phi := -y\frac{\partial}{\partial x} + x\frac{\partial}{\partial y},$$

$$\Theta := \frac{-z}{\sqrt{x^2 + y^2}} \left(x\frac{\partial}{\partial x} + y\frac{\partial}{\partial y} \right) + \sqrt{x^2 + y^2}\,\frac{\partial}{\partial z}.$$

The intersection of the annihilators of the summands in (3.65) is generated by

$$R \cdot \Phi, \quad R \cdot \Theta, \quad \Phi \cdot \Theta,$$

whose simultaneous vanishing characterizes analytic functions of the form (3.65). Similarly, for (3.66) we obtain the annihilating differential operator $R \cdot \Phi \cdot \Theta$.

The following example replaces the spherical coordinates by the elementary symmetric functions in the Cartesian coordinates. The annihilating operators allow to decide whether a given analytic function is of a particular form and invariant under the symmetric group, which permutes the Cartesian coordinates.

Example 3.2.47. Let

$$e_1 := x+y+z, \quad e_2 := yz+xz+xy, \quad e_3 := xyz$$

be the elementary symmetric functions in the Cartesian coordinates x, y, z of $\mathbb{C}^3$. Using the same technique as in the previous example, suitably scaled differential operators ε_1, ε_2, ε_3 which pairwise commute and satisfy $\varepsilon_i(e_j) = \delta_{i,j}$, where $\delta_{i,j}$ is the Kronecker symbol, are given by

$$\varepsilon_1 := \frac{1}{\Delta}(x^2\,(y-z)\partial_x + y^2\,(z-x)\partial_y + z^2\,(x-y)\,\partial_z),$$

$$\varepsilon_2 := -\frac{1}{\Delta}(x\,(y-z)\partial_x + y\,(z-x)\partial_y + z\,(x-y)\,\partial_z),$$

$$\varepsilon_3 := \frac{1}{\Delta}((y-z)\,\partial_x + (z-x)\,\partial_y + (x-y)\,\partial_z)$$

with $\Delta := (x-y)\,(x-z)\,(y-z)$. The analytic functions of x, y, z which are annihilated by $\varepsilon_1 \cdot \varepsilon_2 \cdot \varepsilon_3$ are those of the form $f_1(e_2, e_3) + f_2(e_3, e_1) + f_3(e_1, e_2)$.

Example 3.2.48. For the Lie group $(\mathbb{R}, +)$ we would like to find representations $\rho \colon \mathbb{R} \to \mathrm{GL}(n, \mathbb{C})$ (i.e., analytic group homomorphisms) for some $n \in \mathbb{N}$, where the entries of the first row of the matrix $\rho(x)$ are given (complex-valued) analytic functions on $\mathbb{R}$. We denote the entry at position (i, j) of $\rho(x)$ by $\rho_{i,j}(x)$.

The first row of $\rho(0)$ is necessarily equal to $(1, 0, 0, \ldots, 0)$. Moreover, for any $x, y \in \mathbb{R}$ we have $\rho(x+y) = \rho(x)\rho(y)$. If the first row of $\rho(x)$ is chosen to be an analytic function $\gamma \colon \mathbb{R} \to \mathbb{C}^{1 \times n}$ (with at least two non-zero components), then we have

$$\gamma_j(x+y) = \gamma_1(x)\,\gamma_j(y) + \gamma_2(x)\rho_{2,j}(y) + \ldots + \gamma_n(x)\rho_{n,j}(y), \qquad j = 1, \ldots, n,$$

i.e., the function $\gamma_j(x+y) - \gamma_1(x)\gamma_j(y)$ belongs to $S(\alpha_1, g_1; \ldots; \alpha_k, g_k)$, where

$$\alpha_j(x, y) := y, \quad g_j(x, y) := \gamma_{i_j}(x), \qquad j = 1, \ldots, k,$$

and $\gamma_{i_1}, \ldots, \gamma_{i_k}$ are the non-zero entries in $(\gamma_2, \ldots, \gamma_n)$. If we choose, for instance, in case $n = 3$,

$$\gamma(x) = (1 + a_1\,x + a_2\,x^2, b\,x, c\,x),$$

where a_1, a_2, b, $c \in \mathbb{C}$, and we assume $b \neq 0$ for this particular example, then the analytic functions

$$1 + a_1(x+y) + a_2(x+y)^2 - (1 + a_1 x + a_2 x^2)(1 + a_1 y + a_2 y^2),$$
$$b(x+y) - (1 + a_1 x + a_2 x^2)by,$$
$$c(x+y) - (1 + a_1 x + a_2 x^2)cy$$

are necessarily of the form $bx \cdot \rho_{2,i}(y) + cx \cdot \rho_{3,i}(y)$ for $i = 1, 2, 3$, respectively. Analytic functions of this form are the analytic solutions of

$$x\frac{\partial u}{\partial x} - u = 0. \tag{3.67}$$

By substitution of the previous three expressions for u in (3.67) we conclude that, if γ is the first row of a matrix representation $\rho \colon \mathbb{R} \to \mathrm{GL}(3, \mathbb{C})$, then $a_2 = 0$. In that case, using the right inverse

$$\Gamma_0 := \begin{pmatrix} 1/(bx) \\ 0 \end{pmatrix}$$

of $\Delta_0(\alpha, g) = (bx, cx)$, we obtain (cf. also Rem. 3.2.30)

$$\rho_{2,1}(x) = -\frac{a_1^2 x}{b} + c\,\xi_1(x), \qquad \rho_{3,1}(x) = -b\,\xi_1(x),$$

$$\rho_{2,2}(x) = 1 - a_1 x + c\,\xi_2(x), \qquad \rho_{3,2}(x) = -b\,\xi_2(x),$$

$$\rho_{2,3}(x) = \frac{c}{b}(1 - a_1 x) + c\,\xi_3(x), \qquad \rho_{3,3}(x) = -b\,\xi_3(x),$$

where ξ_1, ξ_2, ξ_3 are arbitrary complex-valued analytic functions on $\mathbb{R}$ satisfying $\xi_2(0) = 0$ and $\xi_3(0) = -b^{-1}$. Further conditions on ξ_1, ξ_2, ξ_3 are imposed by $\rho(x+y) = \rho(x)\rho(y)$. We do not go into these details here, but only mention that for $\xi_1(x) = \xi_2(x) = 0$, $\xi_3(x) = -b^{-1}$ we obtain the solution

$$\rho(x) = \begin{pmatrix} 1 + a_1 x & bx & cx \\ -\frac{a_1^2}{b}x & 1 - a_1 x & -\frac{a_1 c}{b}x \\ 0 & 0 & 1 \end{pmatrix}.$$

The methods discussed in the previous subsections provide techniques to improve symbolic solution procedures for linear and nonlinear partial differential equations. In many cases they allow to detect that a given family of analytic solutions misses some solutions (cf. Ex. 3.2.49). Moreover, if analytic solutions of a special form are to be found, an implicit description of this specification in terms of differential equations may be added to the given PDE system, and computing the algebraic consequences of this combined system (e.g., using Thomas' algorithm) may result in PDEs that are easier to solve (cf. Ex. 3.2.51 for the Laplace equation, Ex. 3.3.49, p. 228, for the Korteweg-de Vries equation, and Ex. 3.3.50, p. 230, for the Navier-Stokes equations).

A prominent example of software that computes symbolic solutions to systems of partial differential equations is the `pdsolve` command in Maple [CTvB95]. Both

deterministic and heuristic methods (e.g., the method of characteristics and separation of variables) are applied by `pdsolve` and it tries to detect patterns in a given
PDE. Due to the enormous difficulty of the problem of solving PDEs explicitly, if
possible at all, `pdsolve` may often fail to produce an answer or returns an incomplete answer.

Example 3.2.49. Let us consider the nonlinear PDE

$$\left(\frac{\partial^2 u}{\partial x^2}\right)^2 + \frac{\partial^2 u}{\partial y^2} + \frac{\partial u}{\partial x} + \frac{\partial u}{\partial y} = 0 \tag{3.68}$$

for one unknown real analytic function u of x, y. In the present case, the command
`pdsolve` in Maple 13 tries the method of separation of variables and returns

$$u(x,y) = F_1(x) + F_2(y), \qquad (F_1'')^2 = c - F_1', \quad F_2'' = -c - F_2', \qquad c = \text{const}.$$

In order to judge the quality of this answer, we first note that the analytic functions
of the form $F_1(x) + F_2(y)$ are the analytic solutions of the linear PDE

$$\frac{\partial^2 v}{\partial x \partial y} = 0. \tag{3.69}$$

By applying Thomas' algorithm (cf. Sect. 2.2) to (3.68) we obtain the following
Thomas decomposition of the set of analytic solutions of (3.68):

$u_{x,x}{}^2 + u_{y,y} + u_x + u_y = 0 \ \{\partial_x, \partial_y\}$
$u_{y,y} + u_x + u_y \neq 0$

$u_{x,y,y} + u_{x,y} = 0 \ \{\partial_x, \partial_y\}$
$u_{x,x} = 0 \ \{\, *, \partial_y\}$
$u_{y,y} + u_x + u_y = 0 \ \{\, *, \partial_y\}$

Note that the second system consists of linear equations only. Its analytic solutions
are given by

$$u(x,y) = (c_4(x+y+1) - c_3)e^{-y} + c_2(x-y) + c_1, \qquad c_1, \dots, c_4 \in \mathbb{R}. \tag{3.70}$$

Every solution of (3.68) which is defined by (3.70) with $c_4 \neq 0$ is not of the form
$F_1(x) + F_2(y)$ for analytic functions F_1, F_2, as can be checked by substitution into
(3.69). On the other hand, further polynomial solutions of (3.68) can be determined
easily. For instance,

$$u(x,y) = (x-y)^2 - 3(x+y) + 1$$

is a solution of the first simple system in the above Thomas decomposition and is
not a solution of (3.69).

Remark 3.2.50. Note that some or all of the analytic functions $\alpha_1, \dots, \alpha_k$ in the
parametrization $u(z) = f_1(\alpha_1(z))g_1(z) + \dots + f_k(\alpha_k(z))g_k(z)$ may be chosen to be
constant, i.e., the corresponding coefficient functions among $f_1, \dots, f_k$ are constant.
(Due to Assumption 3.2.1 the corresponding $v(i)$ are zero.)

Let α_i be constant. If one of the equivalent statements in Proposition 3.2.42 holds, then we may consider the annihilating left ideal $\mathscr{I}(\alpha, g)$ as an ideal of a commutative polynomial algebra $\mathbb{C}[\partial_1, \ldots, \partial_n]$, and the annihilating ideal of $S(\alpha_i, g_i)$ is a zero-dimensional ideal of $\mathbb{C}[\partial_1, \ldots, \partial_n]$.

Functions g_i such that $S(\alpha_i, g_i)$, where α_i is constant, is the solution set of a system of linear differential equations with polynomial (or rational function) coefficients are also referred to as *holonomic functions* (cf., e.g., [Zei90], [Chy98, CS98], [Cou95, Chap. 20], [SST00], and the references therein).

In the following example we take the intersection of two vector spaces of functions rather than their sum. In this context, forming this intersection amounts to restricting an unknown set of solutions of a given PDE system to solutions of a special form. We recover the well-known solutions of the Laplace equation in cylindrical coordinates in terms of Bessel functions.

Example 3.2.51. Let r, φ, z be the cylindrical coordinates of $\mathbb{R}^3 - \{0\} \times \{0\} \times \mathbb{R}$ defined by

$$x = r\cos(\varphi), \quad y = r\sin(\varphi), \quad z = z$$

in terms of Cartesian coordinates x, y, z. In cylindrical coordinates the Laplace equation is given by

$$\frac{\partial^2 u}{\partial r^2} + \frac{1}{r}\frac{\partial u}{\partial r} + \frac{1}{r^2}\frac{\partial^2 u}{\partial \varphi^2} + \frac{\partial^2 u}{\partial z^2} = 0. \tag{3.71}$$

We would like to solve this partial differential equation in terms of Bessel functions $J_\nu(r)$, which are defined to be the non-trivial solutions of

$$r^2 \frac{d^2 F}{dr^2} + r\frac{dF}{dr} + (r^2 - \nu^2)F = 0,$$

where ν is a real number. If a solution u of (3.71) is a Bessel function, then it satisfies the partial differential equation

$$r^2 \frac{\partial^2 u}{\partial r^2} + r\frac{\partial u}{\partial r} + (r^2 - \nu^2)u = 0. \tag{3.72}$$

We combine (3.71) with (3.72). Computing a Janet basis for this combination achieves a separation of variables. In fact, we obtain the consequences

$$\frac{\partial^2 u}{\partial z^2} = u, \qquad \frac{\partial^2 u}{\partial \varphi^2} = -\nu^2 u.$$

The parametric derivatives are u, u_z, u_φ, u_r, $u_{\varphi,z}$, $u_{r,z}$, $u_{r,\varphi}$, $u_{r,\varphi,z}$, and the solutions of the combination of (3.71) and (3.72) are

$$e^{z+\nu\varphi i} J_{\pm\nu}(r), \quad e^{-z+\nu\varphi i} J_{\pm\nu}(r), \quad e^{z-\nu\varphi i} J_{\pm\nu}(r), \quad e^{-z-\nu\varphi i} J_{\pm\nu}(r),$$

and all their $\mathbb{R}$-linear combinations, where $i^2 = -1$.

3.3 Nonlinear Differential Elimination

Let Ω be an open and connected subset of $\mathbb{C}^n$ with coordinates $z = (z_1, \ldots, z_n)$. We denote by $\mathcal{O}$ the ring of analytic functions on Ω (which is an integral domain).

Instead of linear combinations of analytic functions, as dealt with in Sect. 3.2, we consider in this section a family of (complex) analytic functions of the form

$$p(f_1(\alpha_1(z)), \ldots, f_k(\alpha_k(z))), \tag{3.73}$$

later referred to as $S(p; \alpha_1, \ldots, \alpha_k)$, where $p \in \mathcal{O}[F_1, \ldots, F_k]$ is a given multilinear polynomial, i.e., each F_i occurs at most to the first power in p. Here $k \in \mathbb{N}$ is fixed,

$$\alpha_i \colon \Omega \longrightarrow \mathbb{C}^{\nu(i)}, \qquad i = 1, \ldots, k,$$

are given analytic functions, and

$$f_i \colon \alpha_i(\Omega) \longrightarrow \mathbb{C}, \qquad i = 1, \ldots, k,$$

are arbitrary functions such that each $f_i \circ \alpha_i$ is analytic. (If $\nu(i) = 0$, then f_i is understood as an arbitrary constant.)

This section develops methods which compute polynomial partial differential equations for an implicit description of a family of analytic functions as above. Similarly to Sect. 3.2, this task can be formulated as an elimination problem. The nonlinear nature of the problems, however, requires more general techniques.

The problem of characterizing functions (with sufficient regularity) which are representable in a specified form in terms of functional parameters has a long history. As expected for questions of linear dependence, it turns out that (generalized) Wronskian determinants often provide a concise answer to such a problem. Our source for the following historical remarks is the book [RŠ95].

Hilbert's 13th problem asks whether there exist continuous functions of several variables which cannot be expressed as superpositions of continuous functions of two variables. This problem was solved by V. I. Arnold in 1957 [Arn63] using previous work of his supervisor A. N. Kolmogorov [Kol63]. He proved that every continuous function of several variables can be represented as a composition of finitely many continuous functions of two variables.

In 1747, J. d'Alembert [d'A47] proved that a sufficiently smooth scalar function u of the form $u(x, y) = f(x) \cdot g(y)$ satisfies the partial differential equation

$$\frac{\partial^2 \log(u)}{\partial x \partial y} = 0.$$

This observation has been extended and generalized in several directions up to the present day.

At the third International Congress of Mathematicians in 1904, C. Stéphanos announced a result stating that u is of the form

$$\sum_{i=1}^{n} f_i(x) \cdot g_i(y) \tag{3.74}$$

if and only if u solves the partial differential equation

$$\begin{vmatrix} u & u_x & \cdots & u_{x^n} \\ u_y & u_{x,y} & \cdots & u_{x^n,y} \\ \vdots & \vdots & \ddots & \vdots \\ u_{y^n} & u_{x,y^n} & \cdots & u_{x^n,y^n} \end{vmatrix} = 0, \tag{3.75}$$

where indices denote partial differentiations. He probably assumed analyticity of the functions involved, but made no remark about regularity and did not give a proof of the claim.

The above statement was proved for functions of class C^n by the Czech mathematician F. Neuman in 1980 [Neu82]. The corrected assertion requires the non-vanishing of the corresponding Wronskian determinant of next smaller size[7] in order to conclude that u is of the form (3.74), but ensures then that the tuples $(f_1, \ldots, f_n)$ and $(g_1, \ldots, g_n)$ may be chosen to be linearly independent. In 1984, T. M. Rassias gave a counterexample showing that Stéphanos' claim does not hold in general for functions with less regularity [Ras86] (cf. also [GR89, Prop. 3.1] for a counterexample of class C^∞). For more work on decomposing functions in the above sense, we refer to, e.g., [Neu90], [NR91], [GR89], [ČŠ90], [ČŠ91].

Some implicitization problems for differential equations in analogy to the classical problem in algebraic geometry were dealt with in [Gao03] and [RS10], by using the characteristic set method by J. F. Ritt and Wen-tsün Wu on the one hand and by applying differential resultants to systems of linear ordinary differential equations on the other hand. In both approaches, however, the arguments of the parametric functions are not restricted, as is the case in (3.73).

Fundamental remarks about the implicitization problem for families of analytic functions parametrized by (3.73) are given in Subsect. 3.3.1. Subsection 3.3.2 solves the implicitization problem for certain families of products of analytic functions and presents elimination methods for the general problem. Thomas' algorithm (cf. Sect. 2.2) is vitally important for most of the implicitization techniques discussed below. The concept of derivative matrix, used for the linear implicitization problem in Subsect. 3.2.2, is generalized to multilinear parametrizations in Subsect. 3.3.3. In Subsect. 3.3.4 the problem of finding a representation of a given analytic function as member of a parametrized family is addressed. Subsection 3.3.5 presents some applications to systems of PDEs.

---

[7] When dealing with analytic functions on a simply connected open set, no assumption on Wronskian determinants of smaller size is necessary due to the fact that functions of this kind are uniquely determined once their restriction to a non-empty open subset is known.

3.3.1 Implicitization Problem for Multilinear Parametrizations

Let $\mathcal{O}$ be the ring of analytic functions on $\Omega \subseteq \mathbb{C}^n$. We denote by K its field of fractions and let $\partial_1, \ldots, \partial_n$ be the partial derivative operators with respect to the coordinates $z_1, \ldots, z_n$ of $\mathbb{C}^n$. In what follows, all differential algebras over K will be endowed with commuting derivations $\partial_1, \ldots, \partial_n$ extending those of K. In particular, we denote by $K\{U\}$ the differential polynomial ring in one differential indeterminate U with commuting derivations $\partial_1, \ldots, \partial_n$. (For definitions of these notions of differential algebra, cf. Sect. A.3.)

Let $k \in \mathbb{N}$, and let $\alpha_i \colon \Omega \to \mathbb{C}^{\nu(i)}$, $i = 1, \ldots, k$, be analytic functions. We consider a family of analytic functions of the form

$$p(f_1(\alpha_1(z)), \ldots, f_k(\alpha_k(z))), \tag{3.76}$$

where $p \in \mathcal{O}[F_1, \ldots, F_k]$ is a multilinear polynomial with analytic coefficients and $f_1, \ldots, f_k$ are arbitrary functions such that each $f_i \circ \alpha_i$ is analytic. Therefore, we may assume without loss of generality that the components of each $\alpha_i \colon \Omega \to \mathbb{C}^{\nu(i)}$ are functionally independent. In fact, keeping the setting of Subsect. 3.2.1, we again make Assumption 3.2.1, p. 161, i.e., we assume that the Jacobian matrix

$$\left(\frac{\partial \alpha_i}{\partial z} \right) = \left(\frac{\partial \alpha_{i,j}}{\partial z_r} \right)_{1 \leq j \leq \nu(i), 1 \leq r \leq n}$$

of each α_i has rank $\nu(i)$ at every point of Ω. Then Lemma 3.2.2 applies verbatim to the present situation, which implies that each function f_i is necessarily analytic.

More precisely, the family of analytic functions referred to above is defined as follows.

Definition 3.3.1. We define the family $S(p; \alpha_1, \ldots, \alpha_k)$ of analytic functions on Ω *parametrized* by (3.76) as the set

$$\{ u \colon \Omega \longrightarrow \mathbb{C} \mid u \text{ is complex analytic; for almost every } w \in \Omega$$

$$\text{there exists an open neighborhood } \Omega' \subseteq \Omega \text{ of } w \text{ and}$$

$$\text{analytic functions } f_i \colon \alpha_i(\Omega') \to \mathbb{C}, \, i = 1, \ldots, k, \text{ such that}$$

$$u(z) = p(f_1(\alpha_1(z)), \ldots, f_k(\alpha_k(z))) \text{ for all } z \in \Omega' \}.$$

(The meaning of "almost every $w \in \Omega$" is that the set of $w \in \Omega$ for which the condition given in the previous lines is fulfilled is supposed to be dense in Ω.)

We refer to $f_1, \ldots, f_k$ as the *parameters* in (3.76). When the focus is on p and not so much on $\alpha_1, \ldots, \alpha_k$, we also say that $S(p; \alpha_1, \ldots, \alpha_k)$ *is parametrized by p.*

Our first step is to study the ideal of differential polynomials which vanish on the given family S of analytic functions.

Remark 3.3.2. If $S \subseteq K$ is an algebra over the subfield of constants of K, then a necessary and sufficient condition that the zero polynomial be the only differential polynomial in $K\{U\}$ vanishing on S is that there exist $u_1, \ldots, u_n \in S$ such that

$$\det(\partial_i u_j)_{1 \leq i, j \leq n} \neq 0$$

(cf. [Kol73, Cor. to Thm. 3 in Sect. II.6]). Note that most of the families S considered in this subsection are not algebras, e.g., because most of them are not closed under addition (cf., e.g., Example 3.3.6). As we will see, for instance in Corollary 3.3.30, for every family S of analytic functions of the form (3.76) with $v(i) < n$ for all $i = 1, \ldots, k$, there exists a non-zero polynomial partial differential equation which is solved by all $u \in S$. If the Jacobian matrix of some function α_i has rank n, then we have $S(p; \alpha_1, \ldots, \alpha_k) = \mathcal{O}$, because p is a multilinear polynomial. Of course, the above "Jacobian criterion" applies to the latter case and, e.g., to cases where S equals the set of all analytic functions on Ω depending on a certain proper subset of $\{z_1, \ldots, z_n\}$.

Remark 3.3.3. Let a family S of analytic functions be parametrized by a polynomial $p \in \mathcal{O}[F_1, \ldots, F_k]$. We consider $\mathcal{O}[F_1, \ldots, F_k]$ as a subring of the differential polynomial ring $K\{F_1, \ldots, F_k\}$ over K, the latter being endowed with commuting derivations $\partial_1, \ldots, \partial_n$ to be identified later with those of $K\{U\}$. In what follows, we will work with a residue class differential ring of $K\{F_1, \ldots, F_k\}$, which allows to use the residue class of F_i as a name for the composed function $f_i(\alpha_i(z))$ which satisfies the same differential relations.

For each $i = 1, \ldots, k$, the family of analytic functions of the form $f_i(\alpha_i(z))$ equals the set of analytic solutions of a differential ideal Z_i of $K\{F_i\}$ which is generated by certain linear differential polynomials of order one, cf. Lemma 3.2.6, p. 163 (viz. the special case $g_i(z) = 1$). We interpret each Z_i as a subset of $K\{F_1, \ldots, F_k\}$ and define the differential ideal Z of $K\{F_1, \ldots, F_k\}$ which is generated by $Z_1, \ldots, Z_k$. Since it is generated by linear differential polynomials, Z is prime.

Now $U \longmapsto p + Z$ defines a homomorphism

$$\Phi \colon K\{U\} \longrightarrow K\{F_1, \ldots, F_k\}/Z$$

of differential algebras over K. We adjoin U as a differential indeterminate to $K\{F_1, \ldots, F_k\}/Z$, and we consider $K\{U\}$ as a subring of the resulting differential algebra $(K\{F_1, \ldots, F_k\}/Z)\{U\}$ over K. Moreover, let I be the differential ideal of $(K\{F_1, \ldots, F_k\}/Z)\{U\}$ which is generated by $U - (p + Z)$. Then we have

$$\mathcal{I}_{K\{U\}}(S) = I \cap K\{U\} = \ker(\Phi).$$

Since $K\{F_1, \ldots, F_k\}/Z$ is an integral domain, the image of Φ is also an integral domain, and thus $\mathcal{I}_{K\{U\}}(S)$ is a prime ideal by the homomorphism theorem.

Proposition 3.3.4. *The ideal $\mathcal{I}_{K\{U\}}(S)$ of $K\{U\}$ of all differential polynomials which vanish under substitution of $p(f_1(\alpha_1(z)), \ldots, f_k(\alpha_k(z)))$ for U is prime and finitely generated as a radical differential ideal.*

Proof. The first statement is clear from Remark 3.3.3. Finite generation follows from the Basis Theorem of Ritt-Raudenbush for radical differential ideals (cf. Thm. A.3.22, p. 256, or the references given in Subsect. A.3.4).   □

Remark 3.3.5. The computation of the intersection $I \cap K\{U\}$ in Remark 3.3.3 is an elimination problem of the type discussed in Subsect. 3.1.4. Effective ways to solve this particular elimination problem will be presented in Subsect. 3.3.2. Of course, the special case where the parametrization p is a linear polynomial in $\mathscr{O}[F_1, \ldots, F_k]$ is dealt with in Sect. 3.2 using Janet's algorithm for systems of linear partial differential equations, which is a method to solve elimination problems as given in Subsect. 3.1.3.

In general, the parametrized family $S := S(p; \alpha_1, \ldots, \alpha_k)$ of analytic functions is a proper subset of the set of analytic solutions of the differential ideal $\mathscr{I}_{K\{U\}}(S)$.

Example 3.3.6. Let K be the field of meromorphic functions on an open and connected subset Ω of $\mathbb{C}^4$ with coordinates w, x, y, z, and let S denote the family of analytic functions on Ω of the form

$$u(w, x, y, z) = f_1(w) \cdot f_2(x) + f_3(y) \cdot f_4(z).$$

The ideal $\mathscr{I}_{K\{U\}}(S)$ in $K\{U\}$ of differential polynomials which vanish under substitution of functions of this family for U is prime by Proposition 3.3.4. It contains, e.g., the four linear differential polynomials

$$U_{w,y}, \quad U_{w,z}, \quad U_{x,y}, \quad U_{x,z} \tag{3.77}$$

and the (generalized) Wronskian determinants

$$\begin{vmatrix} U_{w^i, x^j} & U_{w^i, x^k} \\ U_{w^l, x^j} & U_{w^l, x^k} \end{vmatrix}, \quad \begin{vmatrix} U_{y^i, z^j} & U_{y^i, z^k} \\ U_{y^l, z^j} & U_{y^l, z^k} \end{vmatrix}, \quad \begin{vmatrix} U & U_w & U_y \\ U_x & U_{w,x} & 0 \\ U_z & 0 & U_{y,z} \end{vmatrix}, \tag{3.78}$$

$$i \neq l, \quad j \neq k, \quad i+j \geq 1, \quad i+k \geq 1, \quad j+l \geq 1, \quad k+l \geq 1.$$

However, all these differential polynomials also vanish under substitution of, e.g., $w + x$ for U, which represents an analytic function not belonging to S.

A finite generating set for $\mathscr{I}_{K\{U\}}(S)$ as a radical differential ideal is given by (3.77) and

$$\begin{vmatrix} U_w & U_{w,w} \\ U_{w,x} & U_{w,w,x} \end{vmatrix}, \quad \begin{vmatrix} U_x & U_{w,x} \\ U_{x,x} & U_{w,x,x} \end{vmatrix}, \quad \begin{vmatrix} U_y & U_{y,z} \\ U_{y,y} & U_{y,y,z} \end{vmatrix}, \quad \begin{vmatrix} U_z & U_{z,z} \\ U_{y,z} & U_{y,z,z} \end{vmatrix}, \quad \begin{vmatrix} U & U_w & U_y \\ U_x & U_{w,x} & 0 \\ U_z & 0 & U_{y,z} \end{vmatrix}.$$

In fact, Thomas decompositions of S are obtained in Examples 3.1.40, p. 143, and 3.1.41. They provide implicit descriptions of S in terms of differential equations and inequations. (The implicitization method that is applied in these examples will be discussed in Subsect. 3.3.2.) A Thomas decomposition of the differential system

defined by the above nine differential polynomials (using the degree-reverse lexicographical ranking satisfying $\partial_w > \partial_x > \partial_y > \partial_z$) can be computed which consists of
eight simple differential systems. One of them is trivially checked to be equivalent
to the generic simple system (3.11), p. 144. Using explicit solutions of the other
seven simple systems, which are easily determined as in Example 3.1.40, we check
that the closures of their solution sets are contained in the closure of the solution
set of the first simple system, i.e., in the closure of S. Hence, the solution set of the
above finite generating set equals the closure of S. We conclude that there exists
no differential polynomial in U which vanishes under substitution of every analytic
function in S and does not vanish under substitution of $w + x$.

Starting from the finite generating set for $\mathscr{I}_{K\{U\}}(S)$ as a radical differential ideal,
an alternative solution for the implicitization problem is to distinguish the cases
of vanishing or non-vanishing of the partial derivative $u_{w,x}$. By symmetry, we also
check the consequences of $u_{y,z} = 0$ and $u_{y,z} \neq 0$, respectively. We split the corresponding system of partial differential equations into four systems by adding the
equations or inequations with left hand sides $U_{w,x}$ or $U_{y,z}$.

The system which contains $U_{w,x} = 0$ and $U_{y,z} = 0$ admits analytic solutions of
the form $f_1(w) + f_2(x)$ which are not in S. In order to exclude these solutions, we
replace $U_{w,x} = 0$ with $U_w U_x = 0$, observing that we have

$$\mathrm{Sol}_\Omega(\{U_{w,x} = 0\}) = \mathrm{Sol}_\Omega(\{U_w U_x = 0\}) \uplus \mathrm{Sol}_\Omega(\{U_{w,x} = 0, U_w \neq 0, U_x \neq 0\}).$$

By symmetry, we replace $U_{y,z} = 0$ with $U_y U_z = 0$ in the same system. The equations
whose left hand sides are the generators of degree two and three are consequences
of $U_w U_x = 0$ and $U_y U_z = 0$ and can therefore be removed.

Let p denote the differential polynomial of degree three in the above generating
set of $\mathscr{I}_{K\{U\}}(S)$ as a radical differential ideal. If both $U_{w,x} \neq 0$ and $U_{y,z} = 0$ are
imposed, then the solution set is not changed when $p = 0$ is replaced with $U_y U_z = 0$.
Similarly, in the system containing both $U_{w,x} = 0$ and $U_{y,z} \neq 0$ we may replace $p = 0$
with $U_w U_x = 0$ without changing its solution set.

Finally, after adding $U_{w,x} \neq 0$ and $U_{y,z} \neq 0$ to the original system we may remove
the equations whose left hand sides are the generators of degree two. By symmetry, it
is enough to discuss the first two of these generators. Partial differentiation of $p = 0$
with respect to w or x yields the equation $p_w = 0$ or $p_x = 0$, respectively, if $U_{w,y} = 0$
and $U_{w,z} = 0$ or $U_{x,y} = 0$ and $U_{x,z} = 0$ are taken into account for simplification, where

$$p_w := \begin{vmatrix} U & U_{w,w} & U_y \\ U_x & U_{w,w,x} & 0 \\ U_z & 0 & U_{y,z} \end{vmatrix}, \qquad p_x := \begin{vmatrix} U & U_w & U_y \\ U_{x,x} & U_{w,x,x} & 0 \\ U_z & 0 & U_{y,z} \end{vmatrix}.$$

Now, $U_{w,x} \neq 0$ implies $U_w \neq 0$ and $U_x \neq 0$, so that the equations

$$U_x U_{y,z} \begin{vmatrix} U_w & U_{w,w} \\ U_{w,x} & U_{w,w,x} \end{vmatrix} = U_{w,x} p_w - U_{w,w,x} p,$$

$$U_w U_{y,z} \begin{vmatrix} U_x & U_{w,x} \\ U_{x,x} & U_{w,x,x} \end{vmatrix} = U_{w,x} p_x - U_{w,x,x} p$$

prove the redundancy of the equations with left hand sides of degree two.

We obtain a partition of S into four sets, which are the sets of analytic solutions of the systems of differential equations and inequations given by the combinations $((\Sigma_0), (\Sigma_1))$, $((\Sigma_0), (\Sigma_2))$, $((\Sigma_0), (\Sigma_3))$, and $((\Sigma_0), (\Sigma_4))$, where

$$U_{w,y} = U_{w,z} = U_{x,y} = U_{x,z} = 0, \tag{Σ_0}$$

$$U_w U_x = 0, \qquad U_y U_z = 0, \tag{Σ_1}$$

$$U_{y,z} \neq 0, \quad U_w U_x = 0, \quad \begin{vmatrix} U_y & U_{y,y} \\ U_{y,z} & U_{y,y,z} \end{vmatrix} = 0, \quad \begin{vmatrix} U_z & U_{y,z} \\ U_{z,z} & U_{y,z,z} \end{vmatrix} = 0, \tag{Σ_2}$$

$$U_{w,x} \neq 0, \quad U_y U_z = 0, \quad \begin{vmatrix} U_w & U_{w,w} \\ U_{w,x} & U_{w,w,x} \end{vmatrix} = 0, \quad \begin{vmatrix} U_x & U_{w,x} \\ U_{x,x} & U_{w,x,x} \end{vmatrix} = 0, \tag{Σ_3}$$

$$U_{w,x} \neq 0, \qquad U_{y,z} \neq 0, \qquad \begin{vmatrix} U & U_w & U_y \\ U_x & U_{w,x} & 0 \\ U_z & 0 & U_{y,z} \end{vmatrix} = 0. \tag{Σ_4}$$

The analytic solutions of the systems $((\Sigma_0), (\Sigma_1))$, $((\Sigma_0), (\Sigma_2))$, $((\Sigma_0), (\Sigma_3))$, and $((\Sigma_0), (\Sigma_4))$ are given, respectively, by the following four lines, where $f_1, \ldots, f_4$ denote arbitrary analytic functions of one argument:

$$\{ f_1(w) + f_3(y) \} \cup \{ f_1(w) + f_4(z) \} \cup \{ f_2(x) + f_3(y) \} \cup \{ f_2(x) + f_4(z) \},$$

$$\{ f_1(w) + f_3(y) f_4(z) \mid f_3' \neq 0 \neq f_4' \} \cup \{ f_2(x) + f_3(y) f_4(z) \mid f_3' \neq 0 \neq f_4' \},$$

$$\{ f_1(w) f_2(x) + f_3(y) \mid f_1' \neq 0 \neq f_2' \} \cup \{ f_1(w) f_2(x) + f_4(z) \mid f_1' \neq 0 \neq f_2' \},$$

$$\{ f_1(w) f_2(x) + f_3(y) f_4(z) \mid f_1' \neq 0, f_2' \neq 0, f_3' \neq 0, f_4' \neq 0 \}.$$

3.3.2 Elimination Methods for Multilinear Parametrizations

In this subsection, methods are presented which compute differential equations that are satisfied by all functions in a parametrized family $S(p; \alpha_1, \ldots, \alpha_k)$, where $p \in \mathcal{O}[F_1, \ldots, F_k] \subset K\{F_1, \ldots, F_k\}$ is a multilinear polynomial and $\alpha_1, \ldots, \alpha_k$ are analytic functions as described in Subsect. 3.3.1 (cf. also Def. 3.3.1).

Solutions to the implicitization problem which are constructed with the use of Thomas' algorithm are finite collections of simple differential systems of equations and inequations. In this case we obtain a partition of $S(p; \alpha_1, \ldots, \alpha_k)$ into the solution sets of the simple systems.

In many cases the elimination task which constructs a complete solution to the implicitization problem is computationally very difficult. Therefore, we discuss different strategies using techniques developed in previous sections. We proceed from the general case to more special cases, the type of implicitization method shifting from elimination procedures providing complete answers to algorithms returning more specific information. Another implicitization method, which is based on Kähler differentials, will be described in detail in Subsect. 3.3.3.

We start with some families of analytic functions admitting an implicit description in terms of differential equations and not requiring inequations. The next proposition extends the observation of d'Alembert mentioned in the introduction to this section. For variants dealing with different kinds of functions, we refer to, e.g., [Neu82], [GR89]. We stick to analytic functions.

Let $\Omega \subseteq \mathbb{C}^n$ be open and connected, $\mathcal{O}$ the ring of analytic functions on Ω and K its field of fractions.

Proposition 3.3.7. *Let* $n = 2$, $p := F_1 \cdot F_2$, *and* $\alpha_i(z_1,z_2) := z_i$, $i = 1$, 2. *For any analytic function* $u \colon \Omega \to \mathbb{C}$ *we have* $u \in S(p;\alpha_1,\alpha_2)$ *if and only if* u *satisfies the PDE*

$$\begin{vmatrix} U & U_{z_1} \\ U_{z_2} & U_{z_1,z_2} \end{vmatrix} = 0. \tag{3.79}$$

Proof. It is easy to check that every analytic function $u \in S(p;\alpha_1,\alpha_2)$ satisfies the PDE (3.79).

Conversely, let us assume that $u \colon \Omega \to \mathbb{C}$ is an analytic function which satisfies this PDE. If u is the zero function, then there is nothing to show. Otherwise,

$$\Omega_1 := u^{-1}(\mathbb{C} - \{z \in \mathbb{C} \mid \mathrm{Re}(z) \leq 0, \mathrm{Im}(z) = 0\}),$$

$$\Omega_2 := u^{-1}(\mathbb{C} - \{z \in \mathbb{C} \mid \mathrm{Re}(z) \geq 0, \mathrm{Im}(z) = 0\})$$

are open subsets of Ω, which are not both empty. Choosing appropriate branches of the logarithm on $u(\Omega_1)$ and $u(\Omega_2)$ (if not empty), respectively, we define

$$v_i := \log \circ u|_{\Omega_i}, \quad i = 1,2.$$

The assumption implies that v_i satisfies the linear partial differential equation

$$\frac{\partial^2 v}{\partial z_1 \partial z_2} = 0.$$

Applying the methods of Sect. 3.2, we conclude that there exist analytic functions $v_{i,1}$ and $v_{i,2}$ of one argument such that $v_i(z_1,z_2) = v_{i,1}(z_1) + v_{i,2}(z_2)$ on Ω_i. It follows

$$u(z_1,z_2) = \exp(v_{i,1}(z_1)) \cdot \exp(v_{i,2}(z_2)) \qquad \text{for all } (z_1,z_2) \in \Omega_i.$$

We define $f_{i,j} := \exp \circ v_{i,j}$ and $\Omega_{i,j} := \alpha_j(\Omega_i)$ for i, $j \in \{1,2\}$. If the intersection $\Omega_1 \cap \Omega_2$ is not empty, then, on each connected component,

$$\frac{f_{1,1}(z_1)}{f_{2,1}(z_1)} = \frac{f_{2,2}(z_2)}{f_{1,2}(z_2)}$$

is a non-zero constant, say λ. Hence, $f_{1,1}$ and $\lambda \cdot f_{2,1}$ coincide on $\Omega_{1,1} \cap \Omega_{2,1}$, as do $\lambda \cdot f_{1,2}$ and $f_{2,2}$ on $\Omega_{1,2} \cap \Omega_{2,2}$. If $\Omega_1 \cap \Omega_2$ is empty, we set $\lambda := 1$. Then

$$f_1(z_1) := \begin{cases} f_{1,1}(z_1), & z_1 \in \Omega_{1,1}, \\ \lambda \cdot f_{2,1}(z_1), & z_1 \in \Omega_{2,1}, \end{cases} \qquad f_2(z_2) := \begin{cases} \lambda \cdot f_{1,2}(z_2), & z_2 \in \Omega_{1,2}, \\ f_{2,2}(z_2), & z_2 \in \Omega_{2,2} \end{cases}$$

are analytic functions on $\Omega_{1,1} \cup \Omega_{2,1}$ and $\Omega_{1,2} \cup \Omega_{2,2}$, respectively, and we have $u(z_1,z_2) = f_1(z_1) \cdot f_2(z_2)$ for all $(z_1,z_2) \in \Omega_1 \cup \Omega_2 = \Omega - u^{-1}(\{0\})$. Since $\Omega_1 \cup \Omega_2$ is an open and dense subset of Ω, the claim follows. $\square$

More generally, we have the following characterization for certain families of products of analytic functions.

Theorem 3.3.8. *Let $m_1, \ldots, m_r \in \mathbb{N}$ satisfy $1 \le m_1 < m_2 < \ldots < m_r = n$, and let $p := F_1 \cdot \ldots \cdot F_r$, and $\alpha_i(z) := (z_{m_{i-1}+1}, \ldots, z_{m_i})$, $i = 1, \ldots, r$, where $m_0 := 0$. For any analytic function $u \colon \Omega \to \mathbb{C}$ we have $u \in S(p; \alpha_1, \ldots, \alpha_r)$, i.e., u is locally of the form*

$$u(z_1, \ldots, z_n) = f_1(z_1, \ldots, z_{m_1}) \cdot f_2(z_{m_1+1}, \ldots, z_{m_2}) \cdot \ldots \cdot f_r(z_{m_{r-1}+1}, \ldots, z_{m_r})$$

for analytic functions $f_1, \ldots, f_r$, if and only if u satisfies the following system of PDEs:

$$\begin{vmatrix} U & U_{z_a} \\ U_{z_b} & U_{z_a,z_b} \end{vmatrix} = 0, \qquad (a,b) \in \bigcup_{1 \le i < j \le r} \{m_{i-1}+1, \ldots, m_i\} \times \{m_{j-1}+1, \ldots, m_j\}.$$

Proof. The same method as used in the proof of Proposition 3.3.7 applies here. Note that the annihilator (in the sense of Sect. 3.2) of the $\mathbb{C}$-vector space of functions which are sums of analytic functions with arguments $z_1, \ldots, z_{m_1}$ and $z_{m_1+1}, \ldots, z_{m_2}$, etc., is generated by the differential operators $\partial_{z_a} \partial_{z_b}$, where (a,b) takes values in the above index set. $\square$

Example 3.3.9. Following the previous proposition, analytic functions of the form $f_1(z_1) f_2(z_2) f_3(z_3) f_4(z_4)$ are characterized by the system of nonlinear PDEs

$$\begin{vmatrix} U & U_{z_a} \\ U_{z_b} & U_{z_a,z_b} \end{vmatrix} = 0, \qquad \{a,b\} \subset \{1,2,3,4\}, \ a \ne b.$$

Analytic functions of the form $g_1(z_1,z_2) g_2(z_3,z_4)$ are the analytic solutions of

$$\begin{vmatrix} U & U_{z_1} \\ U_{z_3} & U_{z_1,z_3} \end{vmatrix} = \begin{vmatrix} U & U_{z_1} \\ U_{z_4} & U_{z_1,z_4} \end{vmatrix} = \begin{vmatrix} U & U_{z_2} \\ U_{z_3} & U_{z_2,z_3} \end{vmatrix} = \begin{vmatrix} U & U_{z_2} \\ U_{z_4} & U_{z_2,z_4} \end{vmatrix} = 0.$$

Omitting, say, the third condition, we enlarge the solution set to obtain precisely the analytic functions of the form $h_1(z_1,z_2) h_2(z_2,z_3) h_3(z_3,z_4)$.

The next algorithm effectively solves the implicitization problem for families of analytic functions that are locally a product of analytic functions with specified arguments. Since the families of analytic functions considered in this section are parametrized by multilinear polynomials, we restrict our attention to square-free monomials. Note that the logarithm is used only in formal computations here.

Algorithm 3.3.10 (Implicitization of products of analytic functions).

Input: A square-free monomial $M \neq 1$ in $\mathcal{O}[F_1, \ldots, F_k]$ and, for $i = 1, \ldots, k$, analytic functions $\alpha_i : \Omega \to \mathbb{C}^{\nu(i)}$ with Jacobian matrix of rank $\nu(i)$ at every point of Ω

Output: A finite system P of polynomial partial differential equations for an unknown function u of $z_1, \ldots, z_n$ such that $S(M; \alpha_1, \ldots, \alpha_k) = \mathrm{Sol}_\Omega(P)$

Algorithm:

1: compute a Janet basis J for the annihilating left ideal in $K\langle \partial_1, \ldots, \partial_n \rangle$ of the family of analytic functions of the form $f_{i_1}(\alpha_{i_1}(z)) + \ldots + f_{i_r}(\alpha_{i_r}(z))$, where $F_{i_1}, \ldots, F_{i_r}$ are the indeterminates dividing M

2: apply each operator in J to $\log(u(z))$ and let T be the set of these results after clearing denominators

3: **return** $\{t = 0 \mid t \in T\}$

Remark 3.3.11. The definition of the family of analytic functions dealt with in step 1 is to be understood as discussed in the section on linear differential elimination (cf. Subsect. 3.2.1, p. 161). Any method described in Sect. 3.2 which computes a Janet basis for the annihilating left ideal of this family may be applied in step 1. In the present case, the functions g_i are constant. Hence, Proposition 3.2.45 a), p. 185, implies that every monomial that appears with non-zero coefficient in any element of the Janet basis J is a differential operator of order at least one. Therefore, applying the operators in J to $\log(u(z))$, where u is a differential indeterminate, yields rational functions in the partial derivatives of u with coefficients in K whose denominators are powers of $u(z)$. Clearing denominators results in a system of polynomial PDEs for u. Note that, even though the zero function must be excluded as a candidate for the solutions of the intermediate system in step 2, it is a solution of the final PDE system because the numerators of the rational functions are polynomials in the partial derivatives of u without constant terms (by the chain rule of differentiation).

Example 3.3.12. Let Ω be a connected open subset of $\mathbb{C}^2$ with coordinates x, y. We would like to determine all analytic functions on Ω that can be represented locally

$$\text{both as} \quad f_1(x) \cdot f_2(y) \quad \text{and as} \quad f_3(x+y) \cdot f_4(x-y),$$

where f_1, f_2, f_3, f_4 are analytic functions of one variable (cf. also [RR92]).
Algorithm 3.3.10 translates the generators

$$\partial_x \partial_y, \qquad \partial_x^2 - \partial_y^2 = (\partial_x + \partial_y)(\partial_x - \partial_y) \tag{3.80}$$

of the annihilating left ideals of $K\langle\partial_x,\partial_y\rangle$ for the families of functions of the form $\widetilde{f}_1(x)+\widetilde{f}_2(y)$ and $\widetilde{f}_3(x+y)+\widetilde{f}_4(x-y)$, respectively, into the polynomial PDEs

$$u\,u_{x,y}-u_x\,u_y=0,\quad u\,(u_{x,x}-u_{y,y})-(u_x^2-u_y^2)=0.\tag{3.81}$$

In order to check that the set of analytic solutions of this PDE system coincides with the set of analytic functions which are representable in both ways specified above, we compute a Thomas decomposition for (3.81) and obtain

$$
\begin{array}{|c|}
\hline
u^2\,u_{y,y,y}-3\,u\,u_y\,u_{y,y}+2\,u_y^3=0\;\{\,*\,,\partial_y\} \\[4pt]
u\,(u_{x,x}-u_{y,y})-(u_x^2-u_y^2)=0\;\{\partial_x,\partial_y\} \\[4pt]
u\,u_{x,y}-u_x\,u_y=0\;\{\,*\,,\partial_y\} \\[4pt]
u\neq 0 \\
\hline
\end{array}
\qquad
\begin{array}{|c|}
\hline
u=0\;\{\partial_x,\partial_y\} \\
\hline
\end{array}
$$

The parametric jet variables defined by the first simple system are u, u_y, u_x, $u_{y,y}$. In fact, the analytic solutions of the first simple system are given by

$$\{\exp(a\,(x^2+y^2)+bx+dy+c)\mid a,b,c,d\in\mathbb{C}\}\tag{3.82}$$

(where the Taylor coefficients corresponding to the parametric jet variables are given by e^c, $d\,e^c$, $b\,e^c$, $(2a+d^2)\,e^c$). The $\mathbb{C}$-vector space of analytic functions which are representable both as $\widetilde{f}_1(x)+\widetilde{f}_2(y)$ and as $\widetilde{f}_3(x+y)+\widetilde{f}_4(x-y)$ can be determined as follows. Since

$$\partial_x^3=\partial_y\,\partial_x\,\partial_y+\partial_x\,(\partial_x^2-\partial_y^2),\qquad \partial_y^3=\partial_x\,\partial_x\,\partial_y-\partial_y\,(\partial_x^2-\partial_y^2)$$

are elements of the left ideal generated by (3.80), we may restrict our attention to polynomial functions of degree at most two. By applying the first annihilator in (3.80) to $\widetilde{f}_3(x+y)+\widetilde{f}_4(x-y)$, we conclude that the coefficients of $(x+y)^2$ and $(x-y)^2$ in $\widetilde{f}_3(x+y)$ and $\widetilde{f}_4(x-y)$, respectively, are equal. If this is taken into account, the representations of $\widetilde{f}_3(x+y)+\widetilde{f}_4(x-y)$ as $\widetilde{f}_1(x)+\widetilde{f}_2(y)$ are obtained by collecting terms in x and y, respectively, the constant terms of $\widetilde{f}_1(x)$ and $\widetilde{f}_2(y)$ being not uniquely determined by their sum. Hence, the above $\mathbb{C}$-vector space is given by

$$\widetilde{f}_1(x)=2\,\widetilde{a}x^2+(\widetilde{b}+\widetilde{d})x+\widetilde{g},\qquad \widetilde{f}_2(y)=2\,\widetilde{a}y^2+(\widetilde{b}-\widetilde{d})y+\widetilde{c}+\widetilde{e}-\widetilde{g},$$

$$\widetilde{f}_3(x+y)=\widetilde{a}(x+y)^2+\widetilde{b}(x+y)+\widetilde{c},\qquad \widetilde{f}_4(x-y)=\widetilde{a}(x-y)^2+\widetilde{d}(x-y)+\widetilde{e},$$

where $\widetilde{a}$, $\widetilde{b}$, $\widetilde{c}$, $\widetilde{d}$, $\widetilde{e}$, $\widetilde{g}$ are arbitrary constants. By composing the exponential function with $\widetilde{f}_1(x)+\widetilde{f}_2(y)$ or with $\widetilde{f}_3(x+y)+\widetilde{f}_4(x-y)$ in this parametrized form we obtain indeed the functions in (3.82) (with $a=2\,\widetilde{a}$, $b=\widetilde{b}+\widetilde{d}$, $d=\widetilde{b}-\widetilde{d}$, $c=\widetilde{c}+\widetilde{e}$).

Remark 3.3.13. The technique of Algorithm 3.3.10 can equally well be applied to families of analytic functions which are parametrized by a square-free monomial

M as above times arbitrary powers of a fixed non-zero analytic function $g(z)$. The annihilating left ideal computed in step 1 of Algorithm 3.3.10 is the intersection of the annihilators of the summands in $f_{i_1}(\alpha_{i_1}(z)) + \ldots + f_{i_r}(\alpha_{i_r}(z))$ (cf. Thm. 3.2.10, p. 164). The left ideal corresponding to the logarithm of $g(z)$ is the annihilator of a $\mathbb{C}$-vector space of dimension one (cf. also Rem. 3.2.50, p. 190). A generating set for this left ideal can be chosen as $\{\partial_1 \circ 1/\log(g(z)), \ldots, \partial_n \circ 1/\log(g(z))\}$.

Example 3.3.14. Let Ω be a simply connected subset of $\mathbb{C}^2 - \{0\}$ with coordinates x, y and K the field of meromorphic functions on Ω. We consider the family of analytic functions on Ω which are locally given by a power of y times an analytic function of x. The annihilating left ideal in $D = K\langle\partial_x, \partial_y\rangle$ of the family of analytic functions of the form $f(x)$ is generated by ∂_y. The annihilator of the $\mathbb{C}$-vector space generated by $\log(y)$ is the left ideal $_D\langle\partial_x, y\log(y)\partial_y - 1\rangle$. The intersection of these two left ideals is generated by $y\partial_y^2 + \partial_y$ and $\partial_x\partial_y$. After applying these operators to $\log(u(z))$ and clearing denominators, we obtain the left hand sides of the PDEs

$$y\, u\, u_{y,y} - y\, u_y^2 + u\, u_y = 0, \quad u\, u_{x,y} - u_x\, u_y = 0.$$

A Thomas decomposition of this PDE system is given by two simple differential systems, one containing the above PDEs and the inequation $u \neq 0$ and the other one consisting of the equation $u = 0$. It is a straightforward task to check that the above PDE system is an implicit description of the family of analytic functions under consideration.

We are going to present implicitization methods of different generality for families $S(p; \alpha_1, \ldots, \alpha_k)$ of analytic functions as defined above.

Remark 3.3.15. Following the construction in Remark 3.3.3, we define the differential polynomial ring $R = K\{F_1, \ldots, F_k, U\}$ in $k+1$ differential indeterminates. For $i = 1, \ldots, k$, let Z_i be the differential ideal of R which is generated by the linear differential polynomials in F_i which vanish for every analytic function of the form $f_i \circ \alpha_i$ (e.g., given by the special case $g_i(z) = 1$ of Lemma 3.2.6, p. 163). We define I to be the differential ideal of R which is generated by the polynomial $U - p(F_1, \ldots, F_k)$ and $Z_1, \ldots, Z_k$.

In order to compute the intersection of I with $K\{U\}$, we apply Thomas' algorithm to a finite generating set for the differential ideal I using a block ranking on R with blocks $\{F_1, \ldots, F_k\}$ and $\{U\}$. We refer to Subsect. 3.1.4 for more details about this elimination method.

In Example 3.1.40, p. 143, this method is applied to the family of analytic functions of the form $f_1(w) \cdot f_2(x) + f_3(y) \cdot f_4(z)$ (cf. also Ex. 3.3.6). Note that the intersections of the resulting simple systems with $K\{U\}$ have solution sets that are not disjoint, so that a postprocessing is necessary, which is, however, easily accomplished.

Note also that, for determining the vanishing ideal $\mathscr{I}_{K\{U\}}(S)$ of the given family $S := S(p; \alpha_1, \ldots, \alpha_k)$, it is sufficient to identify the generic simple system of a Thomas decomposition of the projection onto the component defining U (cf. Cor. 3.1.37, p. 141, and Cor. 2.2.66, p. 107).

Corollary 3.3.16. *Let L be a differential subfield of K. If $p \in L[F_1, \ldots, F_k]$ and the differential ideal Z_i of $K\{F_i\}$ is generated by a subset of $L\{F_i\}$ for all $i = 1, \ldots, k$, then the vanishing ideal $\mathscr{I}_{K\{U\}}(S)$ of $S = S(p; \alpha_1, \ldots, \alpha_k)$ is generated as a radical differential ideal by a finite subset of $L\{U\}$.*

Proof. We apply Thomas' algorithm as described in Remark 3.3.15, where the input consists of equations whose left hand sides are elements of $L\{F_1, \ldots, F_k, U\}$. If E is the differential ideal of $L\{U\}$ which is generated by the equations of the resulting generic simple system and q is the product of their initials and separants, then we have $E : q^\infty \subseteq L\{U\}$ and $E : q^\infty$ generates $\mathscr{I}_{K\{U\}}(S)$.   □

The following remark presents another variant of the method in Remark 3.3.15 which profits from the same idea as the technique developed in Subsect. 3.1.2, where a Veronese map is used to reduce the complexity of the elimination of variables (cf. also Rem. 3.1.17, p. 129).

Remark 3.3.17. Let $m_1, \ldots, m_r$ be some or all of the monomials in $\mathscr{O}[F_1, \ldots, F_k]$ which occur in the representation of p as sum of terms, and assume that after replacing $m_1, \ldots, m_r$ in p with new differential indeterminates $M_1, \ldots, M_r$, the indeterminates among $F_1, \ldots, F_k$ which occur in $m_1, \ldots, m_r$ do not appear in p anymore. If implicit descriptions of the families of analytic functions parametrized by the m_i are already known (e.g., provided by Thm. 3.3.8 or Alg. 3.3.10) as well as a generating set for the polynomial relations satisfied by the m_i (cf., e.g., Prop. 3.1.15, p. 129), we add the implicit descriptions, expressed in terms of the M_i, and the polynomial relations to the generators for I in the previous remark. Of course, the indeterminates M_i are adjoined to the block $\{F_1, \ldots, F_k\}$ of the block ranking. Moreover, those indeterminates F_i which are no longer present in the expression of $U - p$ in terms of the M_i can be ignored in the computation of the Thomas decomposition.

We refer to Example 3.1.41, p. 145, where we treat again the family of analytic functions of the form $f_1(w) \cdot f_2(x) + f_3(y) \cdot f_4(z)$ (cf. also Ex. 3.3.6) using this more refined implicitization method. In this case, all differential indeterminates F_i have been removed, and $F_1 \cdot F_2$ and $F_3 \cdot F_4$, which are represented by M_1 and M_2, respectively, are (differentially) algebraically independent.

Often applying Thomas' algorithm for the elimination is too time-consuming. In practice, partial information may be obtained if the algorithm is interrupted as soon as a certain number of simple differential systems have been constructed (cf. also Subsect. 2.2.6). Even if a Thomas decomposition has not been computed completely, partial results may already be of help to answer questions about the parametrized family, cf. Example 3.3.39 below.

As an alternative to the methods presented in the previous remarks, one may abandon differentiation in the elimination procedure and resort to purely algebraic techniques to obtain a partial result.

Remark 3.3.18. Forming all partial derivatives of $U = p(F_1, \ldots, F_k)$ up to some order ω, we consider the differences of the left and right hand sides as elements of the commutative polynomial algebra over K whose (algebraic) indeterminates are given by all jet variables in U and $F_1, \ldots, F_k$ that occur in these expressions. Let I be the ideal they generate. We can apply any method for elimination of variables (cf. Subsect. 3.1.1), in order to determine the intersection of I with the subalgebra generated by the relevant jet variables in U.

Since the vanishing ideal $\mathscr{I}_{K\{U\}}(S)$ is finitely generated as a radical differential ideal, we construct in this way a generating set for $\mathscr{I}_{K\{U\}}(S)$ if ω is large enough (cf. also Subsect. 3.3.3).

Among the techniques which produce such generators are Janet basis computations with respect to block orderings (cf. Subsect. 3.1.1), degree steering (cf. Rem. 3.1.9, p. 125), or resultants. If the parametrization p has a particular monomial structure, the elimination procedure of Subsect. 3.1.2 using a Veronese map can be applied.

In Example 3.1.19, p. 130, and Example 3.1.20, the latter method is used to compute implicit equations for the families of analytic functions of the form

$$f_1(x) \cdot f_2(y) + f_3(y) \cdot f_4(z)$$

and

$$f_1(v) \cdot f_2(w) + f_3(w) \cdot f_4(x) + f_5(x) \cdot f_6(y) + f_7(y) \cdot f_8(z),$$

respectively, by restricting to differentiation order $\omega = 3$.

The following method lends itself when elements of the vanishing ideal of specific degree and differentiation order and constant coefficients are to be determined.

Remark 3.3.19. Note that $u \in S := S(p; \alpha_1, \ldots, \alpha_k)$ implies that we have

$$c \cdot u \in S \qquad \text{for all } c \in \mathbb{C}.$$

Let us assume that all components of each α_i are homogeneous polynomials in z (not necessarily of the same degree), that $0 \in \Omega$, and that

$$X(\Omega) := \{ c \in \mathbb{C} \mid c \cdot z \in \Omega \text{ for all } z \in \Omega \}$$

is an infinite set. Then $u \in S$ also implies

$$u(c \cdot -) \in S \qquad \text{for all } c \in X(\Omega),$$

where the analytic function $u(c \cdot -)$ on Ω is defined by

$$u(c \cdot -)(z_1, \ldots, z_n) := u(c \cdot z_1, \ldots, c \cdot z_n), \qquad (z_1, \ldots, z_n) \in \Omega.$$

By Lemmas A.3.14, p. 253, and A.3.20 the vanishing ideal $\mathscr{I}_{K\{U\}}(S)$ of S in $K\{U\}$ has a generating set consisting of differential polynomials that are homogeneous with respect to the standard grading of $K\{U\}$ and isobaric around 0.

If all components of each α_i are homogeneous polynomials in z of degree one, then the annihilating operators of $f_i \circ \alpha_i$ given by Lemma 3.2.6, p. 163, may be chosen to have constant coefficients (i.e., $b_j \in \mathbb{C}^{n \times 1}$ for all $j = 1, \ldots, n - \nu(i)$).

Hence, the differential ideal Z_i referred to in Remark 3.3.15 is generated by a subset of $\mathbb{C}\{F_i\}$, $i = 1, \ldots, k$. Moreover, if $p \in \mathbb{C}[F_1, \ldots, F_k]$, then, by Corollary 3.3.16, $\mathscr{I}_{K\{U\}}(S)$ is generated as a radical differential ideal by a finite subset of $\mathbb{C}\{U\}$.

Although the homogeneous component of degree d of $\mathbb{C}\{U\}$ has infinite dimension as a $\mathbb{C}$-vector space, the subspace $\mathbb{C}\{U\}_{d,\omega}$ of differential polynomials which are isobaric of weight ω is finite dimensional. In case $\mathbb{C}\{U\}$ has only one derivation (i.e., $n = 1$) this dimension equals the number of ways of writing ω as a sum of exactly d non-negative integers up to permutation of the summands. In general, $\mathbb{C}\{U\}_{1,\omega}$ is a $\mathbb{C}$-vector space of dimension $o(\omega) := \binom{\omega+n-1}{n-1}$, and we have

$$\dim_{\mathbb{C}} \mathbb{C}\{U\}_{d,\omega} = \sum_{i=0}^{\omega} \prod \binom{n_i + o(i) - 1}{o(i) - 1},$$

where the sum is taken over all $n_0, \ldots, n_\omega \in \mathbb{Z}_{\geq 0}$ such that $n_0 + n_1 + \ldots + n_\omega = d$ and $n_0 \cdot 0 + n_1 \cdot 1 + \ldots + n_\omega \cdot \omega = \omega$.

Therefore, in principle, for each given pair $(d, \omega) \in (\mathbb{Z}_{\geq 0})^2$, the $\mathbb{C}$-vector space $\mathscr{I}_{K\{U\}}(S) \cap \mathbb{C}\{U\}_{d,\omega}$ can be determined by substituting $p(f_1(\alpha_1(z)), \ldots, f_k(\alpha_k(z)))$ for U into a linear combination of the monomial vector space basis of $\mathbb{C}\{U\}_{d,\omega}$ with unknown coefficients, equating the result with zero, and solving the system of linear equations in the unknowns.

Example 3.3.20. Let x, y be coordinates of $\mathbb{C}^2$. Applying the method discussed in the previous remark to the implicitization problem of the family S of analytic functions on $\mathbb{C}^2$ of the form

$$f_1(x) \cdot f_2(y) + f_3(x+y),$$

it is a straightforward task to determine the $\mathbb{C}$-vector spaces

$$V_{d,\omega} := \mathscr{I}_{K\{U\}}(S) \cap \mathbb{C}\{U\}_{d,\omega}$$

for $d \in \{1, \ldots, 4\}$ and $\omega \in \{1, \ldots, 12\}$, say. The dimensions are given by the following matrix:

$$\begin{pmatrix} 0 & 0 & 0 & 0 & 0 & 0 & 0 & 0 & 0 & 0 & 0 & 0 \\ 0 & 0 & 0 & 0 & 0 & 1 & 2 & 5 & 8 & 14 & 20 & 30 \\ 0 & 0 & 0 & 0 & 0 & 1 & 4 & 12 & 29 & 58 & 107 & 183 \\ 0 & 0 & 0 & 0 & 0 & 1 & 4 & 15 & 41 & 101 & 214 & 425 \end{pmatrix}.$$

Of course, we obtain a much clearer picture by considering the factor space $W_{d,\omega}$ of $V_{d,\omega}$ modulo the subspace generated by the partial derivatives of appropriate order of polynomial multiples of degree d of elements of all $V_{d',\omega'}$ for $d' \leq d$, $\omega' \leq \omega$, $(d', \omega') \neq (d, \omega)$. The $\mathbb{C}$-dimensions of the factor spaces $W_{d,\omega}$ are then given by:

$$\begin{pmatrix} 0 & 0 & 0 & 0 & 0 & 0 & 0 & 0 & 0 & 0 & 0 & 0 \\ 0 & 0 & 0 & 0 & 0 & 1 & 0 & 2 & 0 & 3 & 0 & 4 \\ 0 & 0 & 0 & 0 & 0 & 0 & 0 & 1 & 0 & 1 & 1 \\ 0 & 0 & 0 & 0 & 0 & 0 & 0 & 0 & 0 & 0 & 0 \end{pmatrix}.$$

For instance, the differential polynomial

$$(U_{x,x,y} - U_{x,y,y})(U_{x,x,x} - U_{y,y,y}) + (U_{x,y,y,y} - U_{x,x,x,y})(U_{x,x} - U_{y,y}) +$$
$$(U_{x,x,x,y,y} - U_{x,x,y,y,y})(U_x - U_y)$$

represents a generator for $W_{2,6}$, while a representative for a generator of $W_{3,9}$ may be chosen as the Wronskian determinant

$$\begin{vmatrix} U_x - U_y & U_{x,x} - U_{y,y} & U_{x,x,y} - U_{x,y,y} \\ U_{x,x} - U_{x,y} & U_{x,x,x} - U_{x,y,y} & U_{x,x,x,y} - U_{x,x,y,y} \\ U_{x,x,x} - U_{x,x,y} & U_{x,x,x,x} - U_{x,x,y,y} & U_{x,x,x,x,y} - U_{x,x,x,y,y} \end{vmatrix}$$

(cf. also Ex. 3.3.39, p. 219, for another interpretation of the latter polynomial).

The next subsection introduces another implicitization method, which uses Kähler differentials.

3.3.3 Implicitization using Kähler Differentials

The elimination method using linear algebra discussed in Subsect. 3.2.2 will now be generalized to the multilinear case. In particular, we define and apply an analog of the derivative matrix. The main algebraic tool is the module of Kähler differentials.

Let F be a field, A a commutative F-algebra and M an A-module. We denote by $\mathrm{der}_F(A,M)$ the set of (F-linear) derivations from A to M, i.e., F-linear maps $D \colon A \to M$ satisfying the Leibniz rule

$$D(a_1 a_2) = a_1 D(a_2) + a_2 D(a_1)$$

for all $a_1, a_2 \in A$. Recall that the functor $M \mapsto \mathrm{der}_F(A,M)$ from the category of A-modules to itself is representable, i.e., there exists an A-module $\Omega_F(A)$ such that for every A-module M we have

$$\mathrm{der}_F(A,M) \cong \mathrm{hom}_A(\Omega_F(A),M).$$

More precisely, we define the *module of Kähler differentials of A over F*, denoted by $\Omega_F(A)$, as the quotient of the free A-module generated by the symbols da, $a \in A$, modulo the relations which express that $d \colon a \mapsto da$ is an F-linear derivation. Then $\Omega_F(A)$ has the universal property that for any A-module M, every F-linear derivation $D \colon A \to M$ factors as $D = \varphi \circ d$ for a unique homomorphism $\varphi \colon \Omega_F(A) \to M$ of A-modules, and $\Omega_F(A)$ and $d \colon A \to \Omega_F(A)$ are uniquely determined up to unique isomorphism (cf., e.g., [Mat86]).

The following proposition is of crucial importance for the elimination method discussed in this subsection.

Proposition 3.3.21 (cf., e.g., [Coh91], Thm. 5.4.6, or [Eis95], Thm. 16.14). *Let F be a field of characteristic zero, E an extension field of F, and $x_1, \ldots, x_m$ elements of the field E. Then $x_1, \ldots, x_m$ are algebraically independent over F if and only if $dx_1, \ldots, dx_m$ are linearly independent over E.*

We apply Kähler differentials in our context as explained next.

Remark 3.3.22. Let K be the differential field of meromorphic functions on the open and connected subset Ω of $\mathbb{C}^n$. We consider the family $S := S(p; \alpha_1, \ldots, \alpha_k)$ of analytic functions parametrized by $p \in \mathscr{O}[F_1, \ldots, F_k]$. By Proposition 3.3.4, the differential ideal $\mathscr{I}_{K\{U\}}(S)$ of all differential polynomials in $K\{U\}$ which vanish under substitution of $p(f_1(\alpha_1(z)), \ldots, f_k(\alpha_k(z)))$ for U is prime. It is the kernel of the homomorphism

$$\Phi \colon K\{U\} \longrightarrow K\{F_1, \ldots, F_k\}/Z \colon U \longmapsto p + Z$$

of differential algebras over K, where the prime differential ideal Z is generated by the annihilators of $f_1(\alpha_1(z)), \ldots, f_k(\alpha_k(z))$, cf. Remark 3.3.3.

In order to detect polynomial relations (with coefficients in K) that hold for any function of the family and certain of its partial derivatives, we check K-linear dependence of the Kähler differentials of the images under Φ of jet variables in U representing these partial derivatives.

Let

$$F = K, \quad E = \mathrm{Quot}(K\{F_1, \ldots, F_k\}/Z),$$

i.e., E is the field of fractions of the integral domain $K\{F_1, \ldots, F_k\}/Z$. By Proposition 3.3.21, elements $x_1, \ldots, x_m$ of E are algebraically independent over K if and only if $dx_1, \ldots, dx_m$ are linearly independent over E, where d denotes the universal derivation[8] $E \to \Omega_K(E)$. We apply this to elements of $\mathrm{Quot}(\mathrm{im}(\Phi))$, which is an intermediate differential field of the extension E/K. Enumerating by $x_1, \ldots, x_m$ the elements of E represented by $\Phi(U_I)$ for jets U_I up to differentiation order ω and choosing ω large enough, a finite generating set for $\mathscr{I}_{K\{U\}}(S) = \ker(\Phi)$ as radical differential ideal (cf. Prop. 3.3.4 and the preceding paragraph) is obtained using algebraic elimination methods (cf. Sect. 3.1). (It is not known a priori how large ω has to be chosen.)

In order to define an analog of the derivative matrix used in the linear case, we assume that orderly rankings $>$ have been chosen on $K\{U\}$ and on $K\{F_1, \ldots, F_k\}$. For $\omega \in \mathbb{Z}_{\geq 0}$ we set

$$\zeta(\omega) := \binom{n + \omega}{\omega}, \quad \sigma(\omega) := \sum_{i=1}^{k} \binom{\nu(i) + \omega}{\omega}$$

and define $I_1, \ldots, I_m \in (\mathbb{Z}_{\geq 0})^n$, where $m := \zeta(\omega)$, such that

---

[8] The module of Kähler differentials $\Omega_K(E)$ has a structure of differential module over E defined by $\partial_i(de) = d(\partial_i e)$ for all $e \in E$ and $i = 1, \ldots, n$ (cf. [Joh69b]). We will not, however, make use of this differential module structure.

$$U_{I_m} > \ldots > U_{I_2} > U_{I_1} = U$$

are the $\zeta(\omega)$ lowest jet variables of $K\{U\}$ with respect to the chosen ranking.

Recall that the differential ideal Z of $K\{F_1,\ldots,F_k\}$ is generated by first order linear differential polynomials in F_i, which annihilate $f_i(\alpha_i(z))$. Since the Jacobian matrix of α_i is assumed to have rank $\nu(i)$ at every point of Ω, there exist subsets $\Delta^{(i)}$ of $\{\partial_1,\ldots,\partial_n\}$ of cardinality $\nu(i)$, $i = 1, \ldots, k$, such that

$$P := \{\, \theta_i F_i + Z \mid \theta_i \in \mathrm{Mon}(\Delta^{(i)}), i = 1,\ldots,k \,\}$$

is a maximal subset of $K\{F_1,\ldots,F_k\}/Z$ which is algebraically independent over K. Let J_i be a Janet basis for the annihilator of $f_i(\alpha_i(z))$ in the skew polynomial ring $K\langle \partial_1,\ldots,\partial_n \rangle$ with respect to the term ordering determined by the chosen ranking on $K\{F_1,\ldots,F_k\}$ restricted to linear differential polynomials in F_i. Clearly, the operators in J_i have degree one. The set $\Delta^{(i)}$ is uniquely determined as the complement in $\{\partial_1,\ldots,\partial_n\}$ of the derivations occurring in $\mathrm{lm}(J_i)$ (i.e., as the set of derivations which are not leading monomials of operators in J_i; cf. also Def. 2.1.24, p. 22).

The fact that P is a maximal subset of E which is algebraically independent over K implies that $\Omega_K(E)$ is an E-vector space with basis $d(P)$.

Sorting the elements of $\Delta^{(i)}$ with respect to the ranking of the jet variables in F_i, we use $(\overline{F_i})_L$ as a name for $\theta_i F_i + Z$, where $\theta_i \in \mathrm{Mon}(\Delta^{(i)})$ has exponent tuple $L \in (\mathbb{Z}_{\geq 0})^{\nu(i)}$.

Now define $l_j \in \{1,\ldots,k\}$ and $L_j \in (\mathbb{Z}_{\geq 0})^{\nu(l_j)}$, $j = 1, \ldots, \sigma(\omega)$, such that

$$(\overline{F_{l_j}})_{L_j}, \quad j = 1,\ldots,\sigma(\omega),$$

is the sequence of pairwise distinct residue classes in P with least possible jet variables in $F_1, \ldots, F_k$ as representatives, increasingly sorted with respect to the chosen ranking. We will use $\overline{F_i}$ as a synonym for $(\overline{F_i})_{(0,\ldots,0)}$, $i = 1, \ldots, k$.

Definition 3.3.23. Let $p \in \mathcal{O}[F_1,\ldots,F_k]$, let $\alpha_i \colon \Omega \to \mathbb{C}^{\nu(i)}$ be analytic with Jacobian matrix of rank $\nu(i)$ at every point of Ω, $i = 1, \ldots, k$, and let $\omega \in \mathbb{Z}_{\geq 0}$. Using the homomorphism Φ and the universal derivation $d \colon E \to \Omega_K(E)$ defined in Remark 3.3.22, the *derivative matrix* of order ω

$$\Delta_\omega(p,\alpha) \in E^{\zeta(\omega) \times \sigma(\omega)}$$

is defined to be the $\zeta(\omega) \times \sigma(\omega)$ matrix whose entry at position (i,j) is the coefficient of $d((\overline{F_{l_j}})_{L_j})$ in $d\Phi(U_{I_i})$.

Remark 3.3.24. The derivative matrix $\Delta_\omega(p,\alpha)$ depends, of course, on the chosen rankings, which is not indicated in the notation. Since p is a polynomial of zero order jets in $K\{F_1,\ldots,F_k\}$, $d\Phi(U_{I_i})$ is a linear combination of $d((\overline{F_{l_j}})_{L_j})$, $j = 1,\ldots,\sigma(\omega)$, with coefficients in E, for every $i = 1, \ldots, \zeta(\omega)$; i.e., $d\Phi(U_{I_i})$ can be represented as a row in the derivative matrix. The coefficients are uniquely determined because $d(P)$ is a basis for the E-vector space $\Omega_K(E)$.

Remark 3.3.25. Definition 3.3.23 extends the notion of derivative matrix defined in Subsect. 3.2.2 in the following sense. The families of analytic functions discussed in Sect. 3.2 are parametrized by linear polynomials $p \in \mathscr{O}[F_1, \ldots, F_k]$. Note that Z is generated by linear differential polynomials in any case. Identifying $d((\overline{F_{l_j}})_{L_j})$ with the corresponding jet variable $(F_{l_j})_{L_j}$ in the context of Definition 3.2.12, p. 166, yields the derivative matrix in the sense of Subsect. 3.2.2 as an instance of $\Delta_\omega(p, \alpha)$ with entries in K (due to linearity).

Remark 3.3.26. The assumption that the rankings on $K\{U\}$ and on $K\{F_1, \ldots, F_k\}$ are orderly implies that the derivative matrix has a lower block triangular structure similar to the linear case (cf. Rem. 3.2.16, p. 167).

The following two examples illustrate the notion of derivative matrix.

Example 3.3.27. Let $\mathscr{O}$ be the ring of analytic functions on $\mathbb{C}^2$ with coordinates x, y, and K the field of fractions of $\mathscr{O}$. Let the family $S(F_1 F_2; \alpha_1, \alpha_2)$ of analytic functions on $\mathbb{C}^2$ be defined by $\alpha_1(x, y) := x$, $\alpha_2(x, y) := y$, i.e., we consider analytic functions of the form

$$u(x, y) = f_1(x) \cdot f_2(y).$$

The annihilators of $f_1 \circ \alpha_1$ and $f_2 \circ \alpha_2$ in $K\langle \partial_x, \partial_y \rangle$ are generated by ∂_y and ∂_x, respectively, and the differential ideal Z of $K\{F_1, F_2\}$ is generated by $\partial_y F_1$ and $\partial_x F_2$. The submatrix of $\Delta_2(p, \alpha)$ whose rows express $d\Phi(U), d\Phi(U_x), d\Phi(U_y), d\Phi(U_{x,y})$, respectively, as linear combination of $d\overline{F_1}, d\overline{F_2}, d(\overline{F_1}'), d(\overline{F_2}')$ with coefficients in E, where $\overline{F_i}' := (\overline{F_i})_{(1)}$, $i = 1, 2$, is given by

$$\begin{pmatrix} \overline{F_2} & \overline{F_1} & 0 & 0 \\ 0 & \overline{F_1}' & \overline{F_2} & 0 \\ \overline{F_2}' & 0 & 0 & \overline{F_1} \\ 0 & 0 & \overline{F_2}' & \overline{F_1}' \end{pmatrix}.$$

The linear relations which are satisfied by the rows of this matrix are generated by the coefficient vector $(\overline{F_1}'\,\overline{F_2}', -\overline{F_1}\,\overline{F_2}', -\overline{F_1}'\,\overline{F_2}, \overline{F_1}\,\overline{F_2})$, which we recognize as

$$(\Phi(U_{x,y}), -\Phi(U_y), -\Phi(U_x), \Phi(U)).$$

(A different generator for the linear relations would have required a multiplication by a suitable element of E to arrive at the same conclusion.) Hence, we have

$$0 = \Phi(U_{x,y})\,d\Phi(U) - \Phi(U_y)\,d\Phi(U_x) - \Phi(U_x)\,d\Phi(U_y) + \Phi(U)\,d\Phi(U_{x,y})$$

$$= d\Phi(U\,U_{x,y} - U_x\,U_y).$$

In fact,

$$U\,U_{x,y} - U_x\,U_y \in \ker(\Phi)$$

is the Wronskian determinant given in Proposition 3.3.7.

Example 3.3.28. Let x, y, z be coordinates of $\mathbb{C}^3$, $\mathcal{O}$ the ring of analytic functions on $\mathbb{C}^3$ and K its field of fractions. We consider the family of analytic functions on $\mathbb{C}^3$ which are locally of the form

$$u(x,y,z) = f_1(x) \cdot f_2(y,x^2 - z) + f_3(y - z^2) \cdot f_4(x - y^2, z),$$

i.e., we choose the polynomial $p := F_1 F_2 + F_3 F_4 \in \mathcal{O}[F_1,\ldots,F_4]$ and the functions

$$\alpha_1(x,y,z) := x, \qquad \alpha_2(x,y,z) := (y,x^2 - z),$$
$$\alpha_3(x,y,z) := y - z^2, \qquad \alpha_4(x,y,z) := (x - y^2, z).$$

The annihilators in $K\langle \partial_x, \partial_y, \partial_z \rangle$ of $f_1 \circ \alpha_1, \ldots, f_4 \circ \alpha_4$ are generated by

$$\{ \partial_y, \partial_z \}, \quad \{ 2x\partial_z + \partial_x \}, \quad \{ 2z\partial_y + \partial_z, \partial_x \}, \quad \{ 2y\partial_x + \partial_y \},$$

respectively, and the differential ideal Z of $K\{F_1,\ldots,F_4\}$ is generated by

$$\{ \partial_y F_1, \partial_z F_1, (2x\partial_z + \partial_x) F_2, (2z\partial_y + \partial_z) F_3, \partial_x F_3, (2y\partial_x + \partial_y) F_4 \}.$$

Choosing $\omega = 1$, the rows of $\Delta_1(p,\alpha)$ express

$$d\Phi(U), \quad d\Phi(U_x), \quad d\Phi(U_y), \quad d\Phi(U_z),$$

respectively, as linear combination of

$$d\overline{F_1}, \quad d\overline{F_2}, \quad d\overline{F_3}, \quad d\overline{F_4},$$
$$d(\overline{F_1}'), \quad d((\overline{F_2})_{(1,0)}), \quad d((\overline{F_2})_{(0,1)}), \quad d(\overline{F_3}'), \quad d((\overline{F_4})_{(1,0)}), \quad d((\overline{F_4})_{(0,1)})$$

with coefficients in the field E, where $\overline{F_i}' := (\overline{F_i})_{(1)}$, $i \in \{1,3\}$. More precisely, $\Delta_1(p,\alpha)$ equals

$$\begin{pmatrix}
\overline{F_2} & \overline{F_1} & \overline{F_4} & \overline{F_3} & 0 & 0 & 0 & 0 & 0 & 0 \\
2x(\overline{F_2})_{(0,1)} & \overline{F_1}' & (\overline{F_4})_{(1,0)} & 0 & \overline{F_2} & 0 & 2x\overline{F_1} & 0 & \overline{F_3} & 0 \\
(\overline{F_2})_{(1,0)} & 0 & -2y(\overline{F_4})_{(1,0)} & \overline{F_3}' & 0 & \overline{F_1} & 0 & \overline{F_4} & -2y\overline{F_3} & 0 \\
-(\overline{F_2})_{(0,1)} & 0 & (\overline{F_4})_{(0,1)} & -2z\overline{F_3}' & 0 & 0 & -\overline{F_1} & -2z\overline{F_4} & 0 & \overline{F_3}
\end{pmatrix}.$$

We summarize the discussion following Proposition 3.3.21.

Proposition 3.3.29. Let $S := S(p; \alpha_1,\ldots,\alpha_k)$ be a parametrized family of analytic functions as above. Every non-zero polynomial relation $r \in \mathscr{I}_{K\{U\}}(S)$ gives rise to a non-zero linear relation with coefficients in E satisfied by the rows of $\Delta_\omega(p,\alpha)$, where ω is an upper bound for the differentiation order of jet variables occurring in r. Conversely, every such linear dependency gives rise to a non-zero polynomial partial differential equation satisfied by every function in S.

Corollary 3.3.30. *If* $v(i) < n$ *for every* $i = 1, \ldots, k$, *then there exists a non-zero polynomial partial differential equation which is satisfied by every function in* $S(p; \alpha_1, \ldots, \alpha_k)$.

Proof. The shape of the derivative matrix $\Delta_\omega(p, \alpha)$ is $\zeta(\omega) \times \sigma(\omega)$, where

$$\zeta(\omega) = \binom{n + \omega}{\omega}, \qquad \sigma(\omega) = \sum_{i=1}^{k} \binom{v(i) + \omega}{\omega}.$$

Hence, $\zeta(\omega)$ is a polynomial in ω of degree n, and $\sigma(\omega)$ is a polynomial in ω of degree $\max\{v(1), \ldots, v(k)\}$. Under the assumption of the corollary, the number of rows of $\Delta_\omega(p, \alpha)$ exceeds the number of columns if ω is large enough. Then the rows of $\Delta_\omega(p, \alpha)$ are linearly dependent over E, which proves the claim.   □

Example 3.3.31. Let Ω be an open and connected subset of $\mathbb{C}^2$ with coordinates x, y and consider the family S of analytic functions on Ω which are locally of the form

$$f_1(x) \cdot f_2(y) + f_3(x + y) \cdot f_4(x - y).$$

Then the ranks of the derivative matrices up to order 6 are given by:

ω		0	1	2	3	4	5	6
$\zeta(\omega) = \binom{\omega+2}{\omega}$		1	3	6	10	15	21	28
$\sigma(\omega) = 4\binom{\omega+1}{\omega}$		4	8	12	16	20	24	28
$\mathrm{rank}(\Delta_\omega(p, \alpha))$		1	3	6	10	15	21	26

Hence, the minimal ω such that $\Delta_\omega(p, \alpha)$ does not have full row rank is 6. Since the derivative matrix $\Delta_6(p, \alpha)$ is of shape 28×28 and has rank 26, there exist two independent generators of $\mathscr{I}_{K\{U\}}(S)$ which are polynomials involving jet variables in U up to order 6. Since the arguments of $f_1, \ldots, f_4$ are homogeneous polynomials in z of degree one, these two generators may be chosen as homogeneous differential polynomials with constant coefficients which are isobaric around 0, if $0 \in \Omega$ and $X(\Omega)$ is infinite (cf. Lemma A.3.20). Applying the elimination method of Remark 3.3.19, it is a straightforward task to check that

$$\dim_{\mathbb{C}}(\mathscr{I}_{K\{U\}}(S) \cap \mathbb{C}\{U\}_{d,\omega}) = 0$$

for all $d \leq 20$ and $\omega \leq 20$, say. We expect these generators to consist of too many terms to be useful here.

Using Kähler differentials, the following algorithm computes a finite generating set for the vanishing ideal $\mathscr{I}_{K\{U\}}(S)$ of a parametrized family $S = S(p; \alpha_1, \ldots, \alpha_k)$ of analytic functions, where $p \in \mathscr{O}[F_1, \ldots, F_k]$, if primality of radical differential ideals defined by simple differential systems over $K\{U\}$ can be decided and if the Hilbert series counting the parametric jet variables defined by the generic simple system of a Thomas decomposition of $\mathscr{I}_{K\{U\}}(S)$ is known (cf. Def. 2.2.80, p. 113). In what follows, we refer to the latter as the Hilbert series of $\mathscr{I}_{K\{U\}}(S)$.

Algorithm 3.3.32 (Implicitization using Kähler differentials).

Input: A polynomial $p \in \mathcal{O}[F_1, \ldots, F_k]$, analytic functions $\alpha_i \colon \Omega \to \mathbb{C}^{v(i)}$ with Jacobian matrix of rank $v(i)$ at every point of Ω, $i = 1, \ldots, k$, and $H_\Delta \in \mathbb{Z}[[\lambda]]$

Output: A simple differential system T over $K\{U\}$ such that

$$\langle T^= \rangle : q^\infty = \mathscr{I}_{K\{U\}}(S) = \ker(\Phi), \tag{3.83}$$

where q is the product of the initials and separants of all elements of $T^=$ (provided that H_Δ equals the Hilbert series of $\mathscr{I}_{K\{U\}}(S)$)

Algorithm:

1: $\omega \leftarrow 0;\ H_{\overline{T}} \leftarrow 0$

2: **repeat**

3:     **if** $\operatorname{rank}(\Delta_\omega(p, \alpha)) < \zeta(\omega)$ **then**   // $\Delta_\omega(p, \alpha)$ *not of full row rank*

4:          $G \leftarrow$ a generating set for $\ker(\Phi) \cap K[U_I \mid |I| \leq \omega]$

5:          // *i.e., generating set for the polynomial relations satisfied by the partial derivatives up to order ω of analytic functions $p(f_1 \circ \alpha_1, \ldots, f_k \circ \alpha_k)$*

6:          choose a simple system T of a Thomas decomposition of the differential system defined by G satisfying $\langle T^= \rangle : q^\infty \subseteq \ker(\Phi)$, where q is the product of the initials and separants of all elements of $T^=$

7:          // *the inclusion is decided by checking whether substitution of the expression $p(f_1 \circ \alpha_1, \ldots, f_k \circ \alpha_k)$ for U in $T^=$ yields $\{0\}$ (where f_i are symbols representing analytic functions such that $f_i \circ \alpha_i$ is defined)*

8:          $H_{\overline{T}} \leftarrow$ Hilbert series (in λ) counting the parametric jet variables of T

9:     **end if**

10:     $\omega \leftarrow \omega + 1$

11: **until** $\langle T^= \rangle : q^\infty$ is prime **and** $H_{\overline{T}} = H_\Delta$

12: **return** T

Proof. First we show that a simple system T as in step 6 exists. Clearly, the radical differential ideal of $K\{U\}$ which is generated by G in step 4 is a subset of $\ker(\Phi)$. Let $S_1, \ldots, S_r$ be a Thomas decomposition of the differential system defined by G. Using the notation of Proposition 2.2.72, we have

$$\ker(\Phi) \supseteq \sqrt{\langle G \rangle} = (E^{(1)} : (q^{(1)})^\infty) \cap \ldots \cap (E^{(r)} : (q^{(r)})^\infty).$$

We consider the minimal representations as intersections of prime differential ideals of the radical differential ideals $E^{(i)} : (q^{(i)})^\infty$. Among these prime ideals there exists one which is contained in $\ker(\Phi)$, because $\ker(\Phi)$ is prime by Proposition 3.3.4. If $j \in \{1, \ldots, r\}$ satisfies that such a prime ideal occurs in the minimal representation of $E^{(j)} : (q^{(j)})^\infty$, then we may choose T to be S_j.

If $\sqrt{\langle G \rangle} = \ker(\Phi)$ holds, then T is the generic simple system of the Thomas decomposition $S_1, \ldots, S_r$ (cf. Cor. 2.2.66, p. 107). Under the assumption that H_Δ equals the Hilbert series of $\mathcal{I}_{K\{U\}}(S)$ the condition in step 11 is then satisfied.

Let us consider the case $\langle T^= \rangle : q^\infty \subsetneq \ker(\Phi)$, where T and q are defined in step 6. Moreover, let us assume that $\langle T^= \rangle : q^\infty$ is a prime differential ideal. Then we have $\mathrm{ld}(\langle T^= \rangle : q^\infty) \subsetneq \mathrm{ld}(\ker(\Phi))$ by Proposition 2.2.69, p. 107. We conclude that $H_{\overline{T}}$ is greater than H_Δ when the sequences of their coefficients are compared lexicographically.

Since $\mathcal{I}_{K\{U\}}(S)$ is finitely generated as a radical differential ideal (cf. Prop. 3.3.4) and polynomial relations satisfied by all functions in the family $S(p; \alpha_1, \ldots, \alpha_k)$ and their derivatives up to order ω are detected as linear dependencies of the rows of $\Delta_\omega(p, \alpha)$ by Proposition 3.3.29, the equality $\sqrt{\langle G \rangle} = \ker(\Phi)$ is satisfied after finitely many steps. $\square$

Remarks 3.3.33. We comment on a few details concerning Algorithm 3.3.32.

a) Clearly, the generating set G can be constructed in an accumulative fashion. The corresponding update in step 4 may be realized by using any elimination method which computes the relevant intersection (e.g., the methods discussed in Sect. 3.1) and may take the previous generating set G into account.
b) A more efficient variant is to perform the computation of the Thomas decomposition in step 6 in such a way that only the candidates for the generic simple system are returned, i.e., the computation of simple systems with inappropriate Hilbert series is abandoned as soon as possible (cf. Prop. 2.2.82, p. 114).
c) The Hilbert series $H_{\overline{T}}(\lambda)$ counting the parametric jet variables of the simple differential system T depends, in general, on the ranking which is chosen for the computation of the Thomas decomposition. We prefer a ranking which respects the differentiation order. Then a comparison of the Taylor coefficient of order ω of $H_{\overline{T}}(\lambda) \cdot \frac{1}{1-\lambda}$ with the rank of the derivative matrix $\Delta_\omega(p, \alpha)$ indicates whether enough generating equations have been found up to differential order ω.
d) The primality of the radical differential ideal $\langle T^= \rangle : q^\infty$ needs to be checked only because the equality test using the Hilbert series (i.e., Proposition 2.2.69) requires the included differential ideal to be prime. By this means, we exclude the possibility that $\langle T^= \rangle : q^\infty$ is the intersection of $\ker(\Phi)$ with a radical ideal which is not contained in $\langle T^= \rangle : q^\infty$, but such that $\mathrm{ld}(\langle T^= \rangle : q^\infty) = \mathrm{ld}(\ker(\Phi))$. In that case we would have $\sqrt{\langle G \rangle} \subsetneq \ker(\Phi)$.

The following remark provides a partial description of the kernel of the linear map induced by $\Delta_\omega(p, \alpha)$ on column vectors. Knowledge of this kernel can be used to estimate the Hilbert series of $\mathcal{I}_{K\{U\}}(S)$.

Remark 3.3.34. Assume that the monomial $F_{i_1} \cdot \ldots \cdot F_{i_r}$ is present in the multilinear polynomial $p \in \mathcal{O}[F_1, \ldots, F_k]$. For every $a \in \{1, \ldots, r\}$ we define

$$J(i_a) := \{ j \in \{1, \ldots, \sigma(\omega)\} \mid l_j = i_a \}.$$

Then for each pair $(a, b) \in \mathbb{N}^2$ such that $1 \le a < b \le r$, the coefficients of

$$\sum_{j\in J(i_a)} (\overline{F_{i_a}})_{L_j}\, d(\overline{F_{i_a}})_{L_j} - \sum_{j\in J(i_b)} (\overline{F_{i_b}})_{L_j}\, d(\overline{F_{i_b}})_{L_j}$$

define a non-zero column vector $v_{a,b}$ in the kernel of the linear map induced by the derivative matrix $\Delta_\omega(p,\alpha)$. In fact, for every $I\in(\mathbb{Z}_{\geq 0})^n$, the differential monomials which occur in $\partial^I(F_{i_1}\cdot\ldots\cdot F_{i_r})$ are square-free products involving exactly one jet variable in F_{i_a} for each $a\in\{1,\ldots,r\}$. Hence, we have

$$\left(\sum_{\theta\in \mathrm{Mon}(\{\partial_1,\ldots,\partial_n\})} \theta F_{i_a}\,\frac{\partial}{\partial\,\theta F_{i_a}}\right)\partial^I(F_{i_1}\cdot\ldots\cdot F_{i_r}) = \partial^I(F_{i_1}\cdot\ldots\cdot F_{i_r})$$

for all $I\in(\mathbb{Z}_{\geq 0})^n$ and $a\in\{1,\ldots,r\}$. Therefore, the claim follows by taking residue classes modulo the differential ideal Z on both sides and noticing that the coefficient of $d(\overline{F_{i_a}})_{L_j}$ in $d\Phi(\partial^I(F_{i_1}\cdot\ldots\cdot F_{i_r}))$ is the residue class of $\frac{\partial}{\partial\,\theta_i F_i}\partial^I(F_{i_1}\cdot\ldots\cdot F_{i_r})$, if we have $(\overline{F_{i_a}})_{L_j}=\theta_i F_i + Z$ (cf. Def. 3.3.23).

If $1\leq a<b<c\leq r$ then $v_{a,b}-v_{a,c}+v_{b,c}=0$. It is easily checked that the vector space which is generated by the vectors obtained in this way is of dimension $r-1$.

Example 3.3.35. Let Ω be an open and connected subset of $\mathbb{C}^4$ with coordinates w, x, y, z and consider the family S of analytic functions on Ω of the form

$$u(w,x,y,z) = f_1(w,x)\cdot f_2(y) + f_3(x,y)\cdot f_4(z).$$

We summarize here information about the corresponding derivative matrices up to order 9.

ω	0	1	2	3	4	5	6	7	8	9
$\zeta(\omega) = \binom{\omega+4}{\omega}$	1	5	15	35	70	126	210	330	495	715
$\sigma(\omega) = 2\binom{\omega+2}{\omega}+2\binom{\omega+1}{\omega}$	4	10	18	28	40	54	70	88	108	130
$\mathrm{rank}(\Delta_\omega(p,\alpha))$	1	5	14	24	36	50	66	84	104	126

By performing Algorithm 3.3.32 for $\omega=0,\ldots,3$, we obtain the following partial differential equations, whose left hand sides are elements of $\mathscr{I}_{K\{U\}}(S)$ and which, in fact, generate $\mathscr{I}_{K\{U\}}(S)$ as a radical differential ideal:

$$U_{w,z}=0,$$

$$\begin{vmatrix} U_z & U_{y,z}\\ U_{z,z} & U_{y,z,z}\end{vmatrix} = \begin{vmatrix} U_z & U_{x,z}\\ U_{z,z} & U_{x,z,z}\end{vmatrix} = \begin{vmatrix} U_w & U_{w,x}\\ U_{w,y} & U_{w,x,y}\end{vmatrix} = \begin{vmatrix} U_w & U_{w,w}\\ U_{w,y} & U_{w,w,y}\end{vmatrix}=0,$$

$$\begin{vmatrix} U & U_w & U_z\\ U_y & U_{w,y} & U_{y,z}\\ U_{y,y} & U_{w,y,y} & U_{y,y,z}\end{vmatrix}=0,$$

$$\begin{vmatrix} U_w & U_x\\ U_{w,y} & U_{x,y}\end{vmatrix}\cdot\begin{vmatrix} U_w & U_z\\ U_{w,y} & U_{y,z}\end{vmatrix} + \begin{vmatrix} U & U_y\\ U_w & U_{w,y}\end{vmatrix}\cdot\begin{vmatrix} U_w & U_{x,z}\\ U_{w,y} & U_{x,y,z}\end{vmatrix}=0.$$

Let T be the generic simple system of a Thomas decomposition of the differential system given by the above equations. Then the Hilbert series counting the parametric jet variables of T is

$$H_{\overline{T}}(\lambda) = 1 + 4\lambda + 9\lambda^2 + \lambda^3 \left(\frac{8}{1-\lambda} + \frac{2}{(1-\lambda)^2} \right).$$

The Taylor coefficient c_ω of

$$\sum_{i \geq 0} c_i \lambda^i := H_{\overline{T}}(\lambda) \cdot \frac{1}{1-\lambda}$$

counts the parametric jet variables of T up to order ω. The list of the first ten coefficients coincides with the corresponding ranks of the derivative matrices:

ω	0	1	2	3	4	5	6	7	8	9
c_ω	1	5	14	24	36	50	66	84	104	126

Knowledge of the column rank of $\Delta_\omega(p, \alpha)$ for all ω allows an asymptotic comparison. In fact, for $\omega \geq 2$, Remark 3.3.34 provides two linearly independent vectors of the kernel of $\Delta_\omega(p, \alpha)$, namely those determined by

$$\sum (\overline{F_a})_{(i_1, i_2)} d(\overline{F_a})_{(i_1, i_2)} - \sum (\overline{F_b})_{(j)} d(\overline{F_b})_{(j)}, \qquad (a, b) \in \{(1,2), (3,4)\},$$

where the sums are taken over all $(i_1, i_2) \in (\mathbb{Z}_{\geq 0})^2$, $i_1 + i_2 \leq \omega$, and over $0 \leq j \leq \omega$, respectively, and two more independent vectors of the kernel arise due to the fact that not enough derivatives of u are taken into account (otherwise requiring a derivative matrix with infinitely many rows).

3.3.4 Constructing a Representation

Similarly to Subsect. 3.2.3, which deals with the case of linear parametrizations, we comment here on the problem of finding parameters $f_1, \ldots, f_k$ which realize a given analytic function u as an element of $S(p; \alpha_1, \ldots, \alpha_k)$.

Remark 3.3.36. In order to determine suitable analytic functions $f_1, \ldots, f_k$ such that $p(f_1 \circ \alpha_1, \ldots, f_k \circ \alpha_k)$ equals a given analytic function u, we try to solve the system of polynomial equations given by $U = p(F_1, \ldots, F_k)$ and its partial derivatives for each of the F_i. Due to the ambiguity of such a representation, the extent to which this is possible varies with the given parametrization (cf. also the case of linear parametrizations). At best, each F_i can be expressed as a rational function in the jet variables in U up to some order. Clearly, this task is an elimination problem. The methods of Subsect. 3.3.2 can be modified so as to search for expressions as above. Moreover, a particular structure of the monomials $m_1, \ldots, m_r$ which occur in p may allow splitting the problem in two parts: first expressing $m_1, \ldots, m_r$ as rational functions in the jet variables in U, and subsequently determining the parameters $f_1, \ldots, f_k$ from their products $m_1, \ldots, m_r$.

We are going to treat an earlier example along the lines of Remark 3.3.18, exemplifying an efficient procedure to search for rational functions which express (monomials in) the functional parameters.

Example 3.3.37. Let x, y, z be coordinates of $\mathbb{C}^3$ and consider the family of analytic functions of the form (cf. also Ex. 3.1.19, p. 130)

$$u(x,y,z) = f_1(x) \cdot f_2(y) + f_3(y) \cdot f_4(z). \tag{3.84}$$

Recall that $UU_{x,y} - U_x U_y = 0$ and $UU_{y,z} - U_y U_z = 0$ are implicit descriptions of the families of analytic functions defined by the summands on the right hand side of (3.84) (cf. Prop. 3.3.7). We form all partial derivatives of first order of these two equations. Let $P, Q_1, \ldots, Q_4, R_1, \ldots, R_4$ be new indeterminates which represent U, the left hand sides of the two equations above and their partial derivatives of first order, respectively. Adjoining the equation $U_{x,z} = 0$ and its partial derivatives of first order, which are relations satisfied by all analytic functions u of the form (3.84), yields the following algebraic system:

$$
\begin{cases}
\quad\quad\quad\quad\quad\quad\quad U - P = 0, \\
\quad\quad\quad UU_{x,y} - U_x U_y - Q_1 = 0, \\
\quad\quad\quad UU_{x,x,y} - U_{x,x} U_y - Q_2 = 0, \\
\quad\quad\quad UU_{x,y,y} - U_x U_{y,y} - Q_3 = 0, \\
U_z U_{x,y} + UU_{x,y,z} - U_{x,z} U_y - U_x U_{y,z} - Q_4 = 0, \\
\quad\quad\quad UU_{y,z} - U_y U_z - R_1 = 0, \\
U_x U_{y,z} + UU_{x,y,z} - U_{x,y} U_z - U_y U_{x,z} - R_2 = 0, \\
\quad\quad\quad UU_{y,y,z} - U_{y,y} U_z - R_3 = 0, \\
\quad\quad\quad UU_{y,z,z} - U_y U_{z,z} - R_4 = 0,
\end{cases}
\quad
\begin{aligned}
U_{x,z} &= 0, \\
U_{x,x,z} &= 0, \\
U_{x,y,z} &= 0, \\
U_{x,z,z} &= 0.
\end{aligned}
\tag{3.85}
$$

We compute a Janet basis for the above system with respect to a term ordering which ranks the jet variables in U higher than $P, Q_1, \ldots, Q_4, R_1, \ldots, R_4$. In principle, a block ordering can be used to find polynomials involving the new indeterminates only. In the present case, the use of a degree-reverse lexicographical ordering is sufficient to produce the consequence

$$U_z Q_1 - U_x R_1 + R_2 P = 0,$$

which may be solved for P. Using the equations in (3.85) as rewriting rules, we obtain

$$U = \frac{U_x \begin{vmatrix} U & U_z \\ U_y & U_{y,z} \end{vmatrix}}{\begin{vmatrix} U_x & U_z \\ U_{x,y} & U_{y,z} \end{vmatrix}} + \frac{U_z \begin{vmatrix} U & U_x \\ U_y & U_{x,y} \end{vmatrix}}{\begin{vmatrix} U_z & U_x \\ U_{y,z} & U_{x,y} \end{vmatrix}}, \tag{3.86}$$

where the summands on the right hand side express $f_1 \cdot f_2$ and $f_3 \cdot f_4$, respectively. By the well-known property of the Wronskian determinant, the vanishing of the denominator characterizes linear dependence of u_x and u_z as functions of y, which is equivalent to the non-uniqueness of a representation of u as $f_1 \cdot f_2 + f_3 \cdot f_4$, if it exists. For instance, the function $u(x,y,z) = (x+1)y + yz = xy + y(z+1)$ has two different representations, and $u_x = y$ and $u_z = y$ are linearly dependent. In that case, applying formula (3.86) to a suitable perturbation of u and canceling the effect of the perturbation in $f_1 \cdot f_2$ and $f_3 \cdot f_4$ yields a representation of u. Defining $\widetilde{u} = u + z$ in the above example, we obtain

$$
\widetilde{u} = \frac{\widetilde{u}_x \begin{vmatrix} \widetilde{u} & \widetilde{u}_z \\ \widetilde{u}_y & \widetilde{u}_{y,z} \end{vmatrix}}{\begin{vmatrix} \widetilde{u}_x & \widetilde{u}_z \\ \widetilde{u}_{x,y} & \widetilde{u}_{y,z} \end{vmatrix}} + \frac{\widetilde{u}_z \begin{vmatrix} \widetilde{u} & \widetilde{u}_x \\ \widetilde{u}_y & \widetilde{u}_{x,y} \end{vmatrix}}{\begin{vmatrix} \widetilde{u}_z & \widetilde{u}_x \\ \widetilde{u}_{y,z} & \widetilde{u}_{x,y} \end{vmatrix}} = (x+1)y + (y+1)z,
$$

and thus $u(x,y,z) = \widetilde{u}(x,y,z) - z = (x+1)y + yz$.

Note that it is necessary to introduce new indeterminates P, Q_i, R_j as names for certain relations in the above procedure, because multiplying (3.86) by the denominator and collecting all terms on one side results in the zero polynomial.

For the sake of completeness, we give a partition of the given family into five sets, which is obtained in a similar way as the one in Example 3.3.6, p. 196. The partition is formed by the sets of analytic solutions of the systems of differential equations and inequations given by the combinations $((\Sigma_0), (\Sigma_1))$, $((\Sigma_0), (\Sigma_2))$, $((\Sigma_0), (\Sigma_3))$, $((\Sigma_0), (\Sigma_4))$, and $((\Sigma_0), (\Sigma_5))$, where

$$U_{x,z} = 0, \tag{Σ_0}$$

$$U_x U_z = 0, \qquad U_{x,y} = U_{y,z} = 0, \tag{Σ_1}$$

$$U_{y,z} \neq 0, \quad U_{x,y} = 0, \quad \begin{vmatrix} U_y & U_{y,y} \\ U_{y,z} & U_{y,y,z} \end{vmatrix} = 0, \quad \begin{vmatrix} U_z & U_{y,z} \\ U_{z,z} & U_{y,z,z} \end{vmatrix} = 0, \tag{Σ_2}$$

$$U_{x,y} \neq 0, \quad U_{y,z} = 0, \quad \begin{vmatrix} U_x & U_{x,x} \\ U_{x,y} & U_{x,x,y} \end{vmatrix} = 0, \quad \begin{vmatrix} U_y & U_{x,y} \\ U_{y,y} & U_{x,y,y} \end{vmatrix} = 0, \tag{Σ_3}$$

$$U_x \neq 0, \quad U_z \neq 0, \quad \begin{vmatrix} U & U_x \\ U_y & U_{x,y} \end{vmatrix} = 0, \quad \begin{vmatrix} U & U_y \\ U_z & U_{y,z} \end{vmatrix} = 0, \tag{Σ_4}$$

$$
\left.
\begin{aligned}
U_{x,y} \neq 0, \quad U_{y,z} \neq 0, \quad & \begin{vmatrix} U_x & U_{x,x} \\ U_{x,y} & U_{x,x,y} \end{vmatrix} = 0, \quad \begin{vmatrix} U_z & U_{y,z} \\ U_{z,z} & U_{y,z,z} \end{vmatrix} = 0, \\[2mm]
& \begin{vmatrix} U & U_y & U_{y,y} \\ U_x & U_{x,y} & U_{x,y,y} \\ U_z & U_{y,z} & U_{y,y,z} \end{vmatrix} = 0, \quad \begin{vmatrix} U_x & U_{x,y} \\ U_z & U_{y,z} \end{vmatrix} \neq 0.
\end{aligned}
\right\} \tag{Σ_5}
$$

The analytic solutions of $((\Sigma_0), (\Sigma_1))$, $((\Sigma_0), (\Sigma_2))$, $((\Sigma_0), (\Sigma_3))$, $((\Sigma_0), (\Sigma_4))$, and $((\Sigma_0), (\Sigma_5))$ are given, respectively, by the following five lines, where $f_1, \ldots, f_4$ denote arbitrary analytic functions of one argument:

$$\{f_1(x)+f_3(y)\} \cup \{f_2(y)+f_4(z)\},$$

$$\{f_1(x)+f_3(y)f_4(z) \mid f_3' \neq 0 \neq f_4'\},$$

$$\{f_1(x)f_2(y)+f_4(z) \mid f_1' \neq 0 \neq f_2'\},$$

$$\{f_2(y)(f_1(x)+f_4(z)) \mid f_1' \neq 0 \neq f_4'\},$$

$$\{f_1(x)f_2(y)+f_3(y)f_4(z) \mid f_1' \neq 0, f_2' \neq 0, f_3' \neq 0, f_4' \neq 0, f_2/f_3 \neq \text{const.}\}.$$

Example 3.3.38. In Example 3.1.20, p. 132, implicit equations for the family of analytic functions of the form

$$u(v,w,x,y,z) = f_1(v) \cdot f_2(w) + f_3(w) \cdot f_4(x) + f_5(x) \cdot f_6(y) + f_7(y) \cdot f_8(z)$$

are computed. We equate the generator (3.6), p. 133, with zero and solve for U. It turns out that the four summands which define U as a rational function in the other jet variables in U correspond to the four summands of the parametrization. Given an analytic function u, we obtain products $f_1 \cdot f_2, \ldots, f_7 \cdot f_8$ realizing u in the above form (if possible) as follows:

$$f_1 \cdot f_2 = \frac{u_v \cdot \begin{vmatrix} u_w & u_{w,x} \\ u_{w,w} & u_{w,w,x} \end{vmatrix}}{\begin{vmatrix} u_{v,w} & u_{w,x} \\ u_{v,w,w} & u_{w,w,x} \end{vmatrix}}, \qquad f_3 \cdot f_4 = \frac{u_{w,x} \cdot \begin{vmatrix} u_{v,w} & u_w \\ u_{v,w,w} & u_{w,w} \end{vmatrix} \cdot \begin{vmatrix} u_x & u_{x,y} \\ u_{x,x} & u_{x,x,y} \end{vmatrix}}{\begin{vmatrix} u_{v,w} & u_{w,x} \\ u_{v,w,w} & u_{w,w,x} \end{vmatrix} \cdot \begin{vmatrix} u_{w,x} & u_{x,y} \\ u_{w,x,x} & u_{x,x,y} \end{vmatrix}},$$

$$f_5 \cdot f_6 = \frac{u_{x,y} \cdot \begin{vmatrix} u_{w,x} & u_x \\ u_{w,x,x} & u_{x,x} \end{vmatrix} \cdot \begin{vmatrix} u_y & u_{y,z} \\ u_{y,y} & u_{y,y,z} \end{vmatrix}}{\begin{vmatrix} u_{w,x} & u_{x,y} \\ u_{w,x,x} & u_{x,x,y} \end{vmatrix} \cdot \begin{vmatrix} u_{x,y} & u_{y,z} \\ u_{x,y,y} & u_{y,y,z} \end{vmatrix}}, \qquad f_7 \cdot f_8 = \frac{u_z \cdot \begin{vmatrix} u_{x,y} & u_y \\ u_{x,y,y} & u_{y,y} \end{vmatrix}}{\begin{vmatrix} u_{x,y} & u_{y,z} \\ u_{x,y,y} & u_{y,y,z} \end{vmatrix}}.$$

Note that, similarly to the previous example, the vanishing of a denominator in the above expressions is equivalent to the linear dependence of u_v and u_x as functions of w or of u_w and u_y as functions of x or of u_x and u_z as functions of y.

Even if the implicitization problem is too hard to solve in acceptable time using the elimination methods discussed in the previous subsections, intermediate results of the elimination process may still be valuable, as the following example shows.

Example 3.3.39. Let x, y be coordinates of $\mathbb{C}^2$ and consider the family of analytic functions on $\mathbb{C}^2$ of the form

$$f_1(x) \cdot f_2(y) + f_3(x+y). \tag{3.87}$$

The left hand sides of the following equations occur as intermediate results of a computation which eliminates the jets of F_1, F_2, F_3 representing derivatives of f_1, f_2, f_3, respectively (e.g., when applying the method discussed in Remark 3.3.18):

$$\begin{cases} F_1''(x) \cdot (U_x - U_y) - F_1'(x) \cdot (U_{x,x} - U_{y,y}) + F_1(x) \cdot (U_{x,x,y} - U_{x,y,y}) = 0, \\[2mm] F_2''(y) \cdot (U_x - U_y) - F_2'(y) \cdot (U_{x,x} - U_{y,y}) + F_2(y) \cdot (U_{x,x,y} - U_{x,y,y}) = 0, \\[2mm] F_3''(x+y) \cdot (U_x - U_y) - F_3'(x+y) \cdot (U_{x,x} - U_{y,y}) + \\[1mm] F_3(x+y) \cdot (U_{x,x,y} - U_{x,y,y}) + U_{x,x} U_y - U_x U_{y,y} - U U_{x,x,y} + U U_{x,y,y} = 0. \end{cases} \qquad (3.88)$$

Note that each equation involves derivatives of only one F_i. Assuming that u is a given analytic function, solving the above ordinary differential equations may produce explicit representations of u in the form $f_1(x) \cdot f_2(y) + f_3(x+y)$ (if possible).

For instance, given $u(x,y) = x^2 + y^2$, the equations (3.88) yield the necessary conditions

$$F_1''(x) = 0, \quad F_2''(y) = 0, \quad F_3''(x+y) = 2. \qquad (3.89)$$

Substituting a parametrization of the solutions of (3.89) into (3.87) and comparing with $x^2 + y^2$ shows that all representations of $x^2 + y^2$ as $f_1(x) \cdot f_2(y) + f_3(x+y)$ are given by

$$f_1(x) = ax + b,$$
$$f_2(y) = -\tfrac{2}{a}\left(y + \tfrac{b}{a}\right),$$
$$f_3(x+y) = (x+y)^2 + 2\tfrac{b}{a}(x+y) + 2\tfrac{b^2}{a^2},$$

where a and b are constants and $a \neq 0$. For

$$u(x,y) = \sin(x)\cos(y) + \cos(x+y) = \sin(x)\cos(y) + \cos(x)\cos(y) - \sin(x)\sin(y)$$

we obtain $F_i'' = -F_i$ for $i = 1, 2, 3$, from which all analytic functions f_1, f_2, f_3 representing u can be determined as $\mathbb{C}$-linear combinations of sin and cos.

Note also that the Wronskian determinants (cf. also Ex. 3.3.20, p. 206)

$$\begin{vmatrix} U_x - U_y & U_{x,x} - U_{y,y} & U_{x,x,y} - U_{x,y,y} \\ U_{x,x} - U_{x,y} & U_{x,x,x} - U_{x,y,y} & U_{x,x,x,y} - U_{x,x,y,y} \\ U_{x,x,x} - U_{x,x,y} & U_{x,x,x,x} - U_{x,x,y,y} & U_{x,x,x,x,y} - U_{x,x,x,y,y} \end{vmatrix},$$

$$\begin{vmatrix} U_x - U_y & U_{x,x} - U_{y,y} & U_{x,x,y} - U_{x,y,y} \\ U_{x,y} - U_{y,y} & U_{x,x,y} - U_{y,y,y} & U_{x,x,y,y} - U_{x,y,y,y} \\ U_{x,y,y} - U_{y,y,y} & U_{x,x,y,y} - U_{y,y,y,y} & U_{x,x,y,y,y} - U_{x,y,y,y,y} \end{vmatrix},$$

together, characterize linear dependence of the analytic functions which are defined by the coefficients of F_i'', F_i', F_i in each equation of (3.88) after substitution of an analytic function u for U. The observation that these differential polynomials in U vanish for all analytic functions of the form (3.87) furnishes another method to compute elements of the vanishing ideal of the family parametrized by (3.87).

Additional pieces of information about a representation of a given function as element of the family under consideration may be used for constructing a complete representation. Note that the differential operator $\partial_x - \partial_y$ annihilates $f_3(x+y)$. Consequently, we obtain

$$u_x - u_y = f_1'(x) \cdot f_2(y) - f_1(x) \cdot f_2'(y),$$

if u is of the form (3.87). Let us assume that f_2 is known. Then we interpret the above equation as an inhomogeneous linear ordinary differential equation for f_1 and use the variation of constants formula. Knowledge of both f_1 and f_2, of course, yields f_3 by subtraction.

For instance, if

$$u(x,y) = \sin(x)\cos(y) + \cos(x+y)$$

and if f_2 is known, we obtain

$$\varphi_1(x,y) = e^{\frac{f_2'(y)}{f_2(y)} \cdot x} \left(c(y) + \int_0^x e^{-\frac{f_2'(y)}{f_2(y)} \cdot t} \frac{1}{f_2(y)} \left(\sin(t)\sin(y) + \cos(t)\cos(y) \right) dt \right),$$

where $c(y)$ has to be chosen appropriately such that $\varphi_1(x,y)$ does not depend on y and $u(x,y) - \varphi_1(x,y) f_2(y)$ is of the form $f_3(x+y)$, i.e., satisfies the differential equation $v_x - v_y = 0$. In fact, if $f_2(y) = \cos(y)$, then the above integral equals

$$\exp\left(\frac{\sin(y)}{\cos(y)} x \right) \cdot \sin(x),$$

and choosing the zero function for $c(y)$, we obtain, as expected, $\varphi_1(x,y) = \sin(x)$.

3.3.5 Applications

In this subsection we outline some applications of the results of the previous subsections to symbolic solving of PDE systems. In particular, an implicit description of a family of analytic functions of a special form in terms of differential equations may allow to construct explicit solutions of a given PDE system, which are of this form (cf. also [OR86], where this approach was examined from the symmetry point of view, and [HE71] for an approach using exterior differential forms). This technique can be understood as a generalization of the method of separation of variables, cf., e.g., [Mil77], [JRR95], [GR89, Sect. 6].

At the end of this subsection we investigate the Korteweg-de Vries equation and the Navier-Stokes equations from this point of view (cf. Ex. 3.3.49 and Ex. 3.3.50, respectively).

The following example demonstrates how primary decomposition of ideals in polynomial rings of differential operators with constant coefficients can be used to solve PDE systems whose solution sets are certain families of analytic functions as discussed in this section. (We refer to [GTZ88], [EHV92], [SY96], [DGP99] for algorithmic approaches to primary decomposition and comparisons.)

Example 3.3.40. Let us consider the following system of partial differential equations for one unknown function u of w, x, y, z (in jet notation):

$$
\left\{
\begin{aligned}
u_{w,y} - u_{y,z} &= 0, \\
u_{x,y} - u_{y,y} &= 0, \\
u_{w,x} - u_{x,x} - 2u_{w,y} + u_{y,y} + u_{x,z} &= 0, \\
u_{w,w} - 2u_{w,x} + u_{x,x} + 2u_{w,y} - u_{y,y} - u_{z,z} &= 0, \\
(u_{w,y} + u_{y,z}) \cdot (u_{w,x} + u_{x,x} - u_{x,y} - u_{x,z}) \cdot u & \\
-(u_{w,y} + u_{y,z}) \cdot (u_w + u_x - u_y - u_z) \cdot (u_x - u_y) & \\
-(u_{w,x} + u_{x,x} - u_{x,y} - u_{x,z}) \cdot u_y \cdot (u_w - u_x + u_y + u_z) &= 0.
\end{aligned}
\right. \tag{3.90}
$$

First of all, we examine the linear equations of this system. The differential operators occurring on the left hand sides of the first four equations generate an ideal I of the commutative polynomial algebra $\mathbb{Q}[\partial_w, \partial_x, \partial_y, \partial_z]$, which has the following primary decomposition:

$$
I = \langle \partial_w - \partial_x + \partial_z, \partial_y \rangle \cap \langle \partial_w - \partial_z, \partial_x - \partial_y \rangle.
$$

By Proposition 3.2.42 (viz. the implication b) $\Rightarrow$ a)), we conclude that the set of analytic solutions of (3.90) is a subset of

$$
\{ g_1(w - z, x + z) + g_2(w + z, x + y) \mid g_1, g_2 \text{ analytic functions} \}
$$

(with appropriate domains of definition). We expect that the coordinate change

$$
W = w - z, \quad X = x + z, \quad Y = w + z, \quad Z = x + y
$$

results in a simplification of the above system. In fact, we obtain[9]

$$
\left\{
\begin{aligned}
2U_{W,Z} - U_{X,Z} &= 0, \\
U_{X,Z} &= 0, \\
2U_{X,Y} - 2U_{W,Z} - U_{X,Z} &= 0, \\
4U_{W,Y} - 4U_{X,Y} + 2U_{X,Z} &= 0, \\
-4(U_{W,Z}U_Y U_Z + U_W U_X U_{Y,Z} + U_{W,X} U_Y U_Z - U U_{W,X} U_{Y,Z} & \\
-U U_{W,Z} U_{Y,Z}) - 2(U_W U_X U_{X,Z} - U U_{W,X} U_{X,Z} - U U_{W,Z} U_{X,Z}) &= 0.
\end{aligned}
\right. \tag{3.91}
$$

---

[9] The transformation of jet expressions under this coordinate change can be performed, e.g., by applying ichjet($[W = w - z, X = x + z, Y = w + z, Z = x + y, U = u], \ldots, [w, x, y, z], [u]$) using the Maple package jets [Bar01].

A Janet basis for the four linear PDEs in (3.91) is given by

$$U_{W,Y} = 0, \qquad U_{X,Y} = 0, \qquad U_{W,Z} = 0, \qquad U_{X,Z} = 0. \tag{3.92}$$

Accordingly, the fifth equation in (3.91) simplifies to

$$\begin{vmatrix} U & U_W & U_Y \\ U_X & U_{W,X} & 0 \\ U_Z & 0 & U_{Y,Z} \end{vmatrix} = 0. \tag{3.93}$$

We assume that the solutions of the combined system ((3.92), (3.93)) are known due to an investigation of such systems in another context, cf. the combined system $((\Sigma_0), (\Sigma_4))$ in Example 3.3.6, p. 196 (without the inequations). By examining the consequences of ((3.92), (3.93)) together with $U_{W,X} = 0$ and $U_{W,X} \neq 0$, respectively, we conclude that, in the original coordinates, the set of analytic solutions of (3.90) is given in parametrized form as

$$\{ f_1(w-z) f_2(x+z) + f_3(w+z) f_4(x+y) \}$$

$$\cup \{ f_1(w-z) + g_2(w+z, x+y) \}$$

$$\cup \{ f_2(x+z) + g_2(w+z, x+y) \}$$

$$\cup \{ g_1(w-z, x+z) + f_3(w+z) \}$$

$$\cup \{ g_1(w-z, x+z) + f_4(x+y) \}$$

$$\cup \{ f_1(w-z) + f_2(x+z) + f_3(w+z) + f_4(x+y) \},$$

where $f_1, \ldots, f_4$ and g_1, g_2 are arbitrary analytic functions (with appropriate domains of definition) of one argument and two arguments, respectively.

A generalization of the method exemplified above to PDE systems whose linear equations have (non-constant) polynomial or rational function coefficients, requires an analog of primary decomposition for left ideals of the skew polynomial ring $D = K\langle \partial_1, \ldots, \partial_n \rangle$ of partial differential operators. In particular, effective techniques for factoring elements of D are necessary. In trying to construct the left ideals $\mathscr{I}(\alpha_i, g_i)$ in Theorem 3.2.10, p. 164, we restrict ourselves to left ideals of D which are generated by linear differential operators of order one (cf. Lemma 3.2.6). For the purpose of this subsection, we only outline here a naive method to find generators for left ideals of this kind, and we refer to, e.g., [GS04], [Gri09], [LST03], [vdPS03, Chap. 4] for more details on factoring differential operators.

Remark 3.3.41. Let $D = K\langle \partial_1, \ldots, \partial_n \rangle$ and $P \in D$ of differential order $d > 0$. The differential operator P has a right factor $Q \in D$ of differential order d', i.e., we have $P = \widetilde{Q} \cdot Q$ for some $\widetilde{Q} \in D$, if and only if the annihilator of the residue class $Q + _D\langle P \rangle$ in $D/_D\langle P \rangle$ is generated by an element of D of differential order $d - d'$.

Algorithm 3.3.42 (Finding right factors of order one).

Input: An element P of $D = K\langle \partial_1, \ldots, \partial_n \rangle$ of differential order d

Output: A set of elements of D of differential order one, each of which represents
an element in $D/_D\langle P \rangle$ whose annihilator is generated by an element of D of
differential order $d - 1$

Algorithm:

1: let J be a Janet basis for the left ideal of D generated by P with respect to a term
   ordering which is compatible with the differentiation order

2: let $\partial^{l_1}, \ldots, \partial^{l_r}$ be the parametric derivatives defined by J up to order $d - 1$ and
   $\partial^{l_1}, \ldots, \partial^{l_r}, \ldots, \partial^{l_s}$ those up to order d (cf. Rem. 2.1.67, p. 51).

3: define $Q := a_0 + a_1 \partial_1 + \ldots + a_n \partial_n$ with symbols $a_0, a_1, \ldots, a_n$ representing
   elements of K to be determined

4: for $i = 1, \ldots, r$ compute the Janet normal form $\sum_{j=1}^{s} M_{i,j} \partial^{l_j}$ of $\partial^{l_i} Q$ modulo J,
   where each $M_{i,j}$ is a differential expression in $a_0, \ldots, a_n$

5: solve the system of partial differential equations for $a_0, \ldots, a_n$ whose left hand
   sides are given by the $r \times r$ minors of the $r \times s$ matrix M

6: **return** $\{ a_0 + a_1 \partial_1 + \ldots + a_n \partial_n \mid (a_0, \ldots, a_n)$ is a solution obtained in step 6$\}$

We comment on a few aspects of the previous algorithm.

Remarks 3.3.43. a) The Janet basis J defined in step 1 consists of the generator P
only, but the leading monomial of P depends on the choice of the term ordering
in general.

b) The vanishing of all $r \times r$ minors of the matrix M is equivalent to the condition
that M (with suitably specialized $a_0, \ldots, a_n$) has rank less than r. Since M represents the left action of $\partial^{l_1}, \ldots, \partial^{l_r}$ by left multiplication on the cyclic submodule
of $D/_D\langle P \rangle$ generated by Q, the previous condition characterizes the existence
of a left K-linear combination of $\partial^{l_1}, \ldots, \partial^{l_r}$ of differential order $d - 1$ which
annihilates the residue class $Q +_D \langle P \rangle$.

c) Success of Algorithm 3.3.42 obviously depends on solving a system of nonlinear
PDEs. Applying Thomas' algorithm to this system results in a finite collection of
simple differential systems which partition the solution set. If not all annihilators
need to be determined, it may be sufficient to consider only one simple system.
The PDE system to be solved quickly gets large as the number n, the order of
the operators, and the number of parametric derivatives grow. Still, examining
some particular $r \times r$ minors of M and solving the arising algebraic systems can
already construct some right factors (cf. the next example). (The method used in
step 5 to solve differential equations should, however, not use Algorithm 3.3.42
again in a way that leads to an infinite recursion.)

d) The product rule of differentiation implies that the term of differentiation order
zero in $\partial^{l_i} Q$ is $\partial^{l_i} a_0$. If P does not involve any term of differential order zero,
then one can set $a_0 = 0$ beforehand.

Example 3.3.44. We would like to find right factors of order one of

$$P := 2y^2 \partial_x^2 \partial_y + y \partial_x \partial_y^2 - \partial_x \partial_y \in D := K\langle \partial_x, \partial_y \rangle$$

using Algorithm 3.3.42. We choose the degree-reverse lexicographical ordering on $\mathrm{Mon}(D)$ satisfying $\partial_x > \partial_y$. Of course, the Janet basis J for the left ideal of D which is generated by P consists of P only. The generalized Hilbert series enumerating the parametric derivatives of the corresponding PDE system is

$$\frac{1}{1 - \partial_y} + \frac{\partial_x}{1 - \partial_y} + \frac{\partial_x^2}{1 - \partial_x}.$$

The parametric derivatives up to order three are

$$1, \quad \partial_y, \quad \partial_x, \quad \partial_y^2, \quad \partial_x \partial_y, \quad \partial_x^2, \quad \partial_y^3, \quad \partial_x \partial_y^2, \quad \partial_x^3. \tag{3.94}$$

In order to determine all operators of the form

$$Q = a_1(x,y)\partial_x + a_2(x,y)\partial_y,$$

which represent a residue class in $D/{}_D\langle P \rangle$ whose annihilator is generated by an element of D of differential order 2, we compute the Janet normal forms of Q, $\partial_y Q$, $\partial_x Q$, $\partial_y^2 Q$, $\partial_x \partial_y Q$, $\partial_x^2 Q$ modulo J and collect in each case the coefficients $M_{i,j}$ of the operators in (3.94) as described in step 4 of Algorithm 3.3.42. The submatrix of M which is formed by its columns indexed by ∂_x, $\partial_x \partial_y$, ∂_x^2, ∂_y^3, $\partial_x \partial_y^2$, ∂_x^3 is

$$\begin{pmatrix} a_1 & 0 & 0 & 0 & 0 & 0 \\ (a_1)_y & a_1 & 0 & 0 & 0 & 0 \\ (a_1)_x & a_2 & a_1 & 0 & 0 & 0 \\ (a_1)_{y,y} & 2(a_1)_y & 0 & a_2 & a_1 & 0 \\ (a_1)_{x,y} & \frac{1}{2y^2}a_1 + (a_1)_x + (a_2)_y & (a_1)_y & 0 & a_2 - \frac{1}{2y}a_1 & 0 \\ (a_1)_{x,x} & \frac{1}{2y^2}a_2 + 2(a_2)_x & 2(a_1)_x & 0 & -\frac{1}{2y}a_2 & a_1 \end{pmatrix}.$$

The determinant of this matrix is $a_1^4 a_2 (\frac{1}{2y}a_1 - a_2)$. For $a_1(x,y) = 0$, $a_2(x,y) = 1$, and $a_1(x,y) = 1$, $a_2(x,y) = 0$, and $a_1(x,y) = 2y$, $a_2(x,y) = 1$ we obtain respectively

$$\partial_y + {}_D\langle P \rangle \text{ with annihilator } {}_D\langle 2y^2 \partial_x^2 + y \partial_x \partial_y - \partial_x \rangle,$$

$$\partial_x + {}_D\langle P \rangle \text{ with annihilator } {}_D\langle 2y^2 \partial_x \partial_y + y \partial_y^2 - \partial_y \rangle,$$

$$2y \partial_x + \partial_y + {}_D\langle P \rangle \text{ with annihilator } {}_D\langle y \partial_x \partial_y - \partial_x \rangle.$$

These pairs of residue classes and annihilators give rise to the factorizations

$$P = (2y^2 \partial_x^2 + y \partial_x \partial_y - \partial_x)\partial_y = (2y^2 \partial_x \partial_y + y \partial_y^2 - \partial_y)\partial_x = (y \partial_x \partial_y - \partial_x)(2y \partial_x + \partial_y).$$

We outline an algorithm which solves PDE systems whose solution sets are certain families of products of analytic functions.

Algorithm 3.3.45 (Recognizing products of analytic functions).

Input: A finite subset L of $K\{u\}$

Output: Analytic functions $\alpha_i \colon \Omega \to \mathbb{C}^{v(i)}$ with Jacobian matrix of rank $v(i)$ at every point of Ω, where $i = 1, \ldots, k$, such that

$$\mathrm{Sol}_\Omega(\{p = 0 \mid p \in L\}) = S(F_1 \cdot \ldots \cdot F_k; \alpha_1, \ldots, \alpha_k),$$

or "FAIL"

Algorithm:

1:    substitute $\partial^I(\exp(v))$ for $\partial^I u$ in each element of L, where v is a differential indeterminate, $I \in (\mathbb{Z}_{\geq 0})^n$, and let E be the set of the resulting polynomials, in which the terms are collected that involve the same power of $\exp(v)$

2:    **if** $E = \{\exp(v)^{d_i} \cdot p_i \mid i = 1, \ldots, r\}$ for some homogeneous differential polynomials $p_1, \ldots, p_r \in K\{v\}$ of degree one and $d_1, \ldots, d_r \in \mathbb{Z}_{\geq 0}$ **then**

3:      let $P_1, \ldots, P_r \in K\langle \partial_1, \ldots, \partial_n \rangle$ be such that $p_i = P_i u$ for $i = 1, \ldots, r$

4:      **if** $P_1, \ldots, P_r \in \mathbb{C}[\partial_1, \ldots, \partial_n]$ **then**

5:        compute an irredundant primary decomposition of $\langle P_1, \ldots, P_r \rangle$, i.e.,

$$\langle P_1, \ldots, P_r \rangle = Q_1 \cap \ldots \cap Q_k, \quad Q_i = \langle q_1^{(i)}, \ldots, q_{m_i}^{(i)} \rangle, \quad i = 1, \ldots, k$$

6:        **if** $\deg(q_j^{(i)}) = 1$ for all $j = 1, \ldots, m_i, i = 1, \ldots, k$ **then**

7:          for $i = 1, \ldots, k$, compute a basis $v_1^{(i)}, \ldots, v_{t_i}^{(i)}$ for the kernel of the linear map of column vectors induced by $\left(\partial q_a^{(i)} / \partial \partial_b \right)_{1 \leq a \leq m_i, 1 \leq b \leq n}$

8:          **return** $(\alpha_1, \ldots, \alpha_k)$, where $\alpha_i(z) := \left(\sum_{j=1}^n (v_l^{(i)})_j z_j \right)_{1 \leq l \leq t_i}$

9:        **else**

10:          **return** "FAIL"

11:        **end if**

12:      **else**

13:        try iteratively to find right factors of each P_i (e.g., using Alg. 3.3.42)

14:        **if** in this way the left ideal $\langle P_1, \ldots, P_r \rangle$ of $K\langle \partial_1, \ldots, \partial_n \rangle$ is identified as an irredundant intersection of left ideals as in Lemma 3.2.6, p. 163, **then**

15:          **return** $(\alpha_1, \ldots, \alpha_k)$ such that the kernels of the Jacobian matrices of α_i give rise to the above left ideals as in Lemma 3.2.6

16:        **else**

17:          **return** "FAIL"

18:        **end if**

19:      **end if**

20: **else**

21:    **return** "FAIL"

22: **end if**

Remarks 3.3.46. a) We assume that input to Algorithm 3.3.45 is given in such a way that its steps can be performed effectively, e.g., the computation of a primary decomposition. The input L should preferably be a subset of a simple system in a Thomas decomposition.

b) Usually, the subset Ω of $\mathbb{C}^n$ and the field K of meromorphic functions are determined a posteriori (cf. also the comments before Def. 2.2.53, p. 98).

c) Note that $P_1, \ldots, P_r$ are well-defined in step 3 because $p_1, \ldots, p_r$ are homogeneous of degree one. An element of the input L yields as a result of step 1 an expression of the form $\exp(v)^{d_i} \cdot p_i$ for some differential polynomial p_i in v and $d_i \in \mathbb{Z}_{\geq 0}$ if and only if it is a homogeneous differential polynomial in u. Since a family of analytic functions $S = S(F_1 \cdot \ldots \cdot F_k; \alpha_1, \ldots, \alpha_k)$ as aimed for in Algorithm 3.3.45 is closed under multiplication by arbitrary constants, Lemma A.3.14, p. 253, shows that the vanishing ideal $\mathscr{I}_R(S)$ of S in $R = K\{u\}$ is indeed a homogeneous differential ideal. Clearly, not every generating set of a differential ideal of R with solution set S consists of homogeneous differential polynomials. Therefore, a preprocessing of the input (e.g., the computation of a Thomas decomposition) is required in general.

d) For an explanation of steps 5 to 8, let $D = K\langle \partial_1, \ldots, \partial_n \rangle$ and let $\langle P_1, \ldots, P_r \rangle$ be the ideal of $\mathbb{C}[\partial_1, \ldots, \partial_n]$ which is generated by $P_1, \ldots, P_r$. According to Proposition 3.2.42, p. 184, the existence of analytic functions $g_i \colon \Omega \to \mathbb{C}, \, g_i \neq 0$, and $\alpha_i \colon \Omega \to \mathbb{C}^{v(i)}$ with Jacobian matrix of rank $v(i)$ at every point of Ω, $i = 1, \ldots, k$, such that $D \cdot \langle P_1, \ldots, P_r \rangle$ equals the annihilating left ideal of $S(\alpha_1, g_1; \ldots; \alpha_k, g_k)$, is equivalent to the condition that $\langle P_1, \ldots, P_r \rangle$ is a radical ideal and that its prime components are generated by polynomials of degree one.

Example 3.3.47. Let us consider the following partial differential equation for one unknown function u of x, y:

$$(u_{x,x,y} - u_{x,y,y}) u^2 - (u_{x,x} u_y + 2 u_x u_{x,y} - 2 u_{x,y} u_y - u_x u_{y,y}) u + 2 u_x^2 u_y - 2 u_x u_y^2 = 0. \tag{3.95}$$

The result of substituting $\exp(v)$ for u (and correspondingly for the derivatives) is

$$\exp(v)^3 (v_{x,x,y} - v_{x,y,y}) = 0.$$

Hence, we obtain $P_1 = \partial_x^2 \partial_y - \partial_x \partial_y^2$ in step 3 of Algorithm 3.3.45 (where we write ∂_x and ∂_y instead of ∂_1 and ∂_2, respectively). An irredundant primary decomposition of the ideal of $\mathbb{C}[\partial_x, \partial_y]$ which is generated by P_1 is given by

$$\langle P_1 \rangle = \langle \partial_y \rangle \cap \langle \partial_x \rangle \cap \langle \partial_x - \partial_y \rangle.$$

In step 7 the kernels of the linear maps which are induced by the matrices

$$\begin{pmatrix} 0 & 1 \end{pmatrix}, \quad \begin{pmatrix} 1 & 0 \end{pmatrix}, \quad \begin{pmatrix} 1 & -1 \end{pmatrix},$$

respectively, are computed, and we define

$$\alpha_1(x,y) := x, \qquad \alpha_2(x,y) := y, \qquad \alpha_3(x,y) := x+y.$$

Therefore, the set of analytic solutions of (3.95) is

$$\{f(x)\,g(y)\,h(x+y) \mid f,g,h \text{ analytic}\}$$

(with appropriate domains of definition).

Example 3.3.48. Let us consider the following system of partial differential equations for one unknown function u of x, y, z:

$$\begin{cases} u_{x,y,z}\,u^2 - (u_{x,y}\,u_z + u_{x,z}\,u_y + u_{y,z}\,u_x)\,u + 2\,u_x\,u_y\,u_z = 0, \\ (2y^2\,u_{x,x,y} + y\,u_{x,y,y} - u_{x,y})\,u^2 - (2y^2\,u_{x,x}\,u_y + 2y\,u_{x,y}\,u_y + 4y^2\,u_x\,u_{x,y} + \qquad (3.96) \\ \quad y\,u_x\,u_{y,y} - u_x\,u_y)\,u + 2y\,u_x\,u_y^2 + 4y^2\,u_x^2\,u_y = 0. \end{cases}$$

Substituting $\exp(v)$ for u (and correspondingly for the derivatives) and collecting terms involving the same power of $\exp(v)$, we obtain

$$\exp(v)^3\,v_{x,y,z} = 0, \qquad \exp(v)^3\,(2y^2\,v_{x,x,y} + y\,v_{x,y,y} - v_{x,y}) = 0,$$

hence,

$$p_1 = v_{x,y,z}, \qquad p_2 = 2y^2\,v_{x,x,y} + y\,v_{x,y,y} - v_{x,y}$$

and

$$P_1 = \partial_x\,\partial_y\,\partial_z, \qquad P_2 = 2y^2\,\partial_x^2\,\partial_y + y\,\partial_x\,\partial_y^2 - \partial_x\,\partial_y.$$

Applying Algorithm 3.3.42, p. 224, to P_2, we obtain the right factors ∂_x, ∂_y, and $2y\,\partial_x + \partial_y$ (cf. also Ex. 3.3.44) and deduce that the left ideal of $K\langle \partial_x, \partial_y, \partial_z \rangle$ which is generated by P_1 and P_2 has the following representation as intersection:

$$\langle P_1, P_2 \rangle = \langle \partial_y \rangle \cap \langle \partial_x \rangle \cap \langle \partial_z, 2y\,\partial_x + \partial_y \rangle.$$

Hence, the set of analytic solutions of (3.96) is

$$\{f(x,z)\,g(y,z)\,h(x-y^2) \mid f,g,h \text{ analytic}\}$$

(with appropriate domains of definition).

The following examples apply implicit descriptions of certain families of analytic functions to symbolic solving of PDEs. Having solutions of a specific form in mind (which may often be motivated by the geometry of the problem), we add the implicit equations for functions of this form to the original system. Applying Thomas' algorithm to the combined system determines the algebraic consequences of this combination and, in particular, allows to decide whether there exist solutions of the prescribed type. This approach can produce differential equations that are easier to solve than the original equations due to possible algebraic simplifications.

Example 3.3.49. Let us consider the Korteweg-de Vries (KdV) equation (cf., e.g., [BC80]), which is used to describe shallow water waves.

$$\frac{\partial u}{\partial t} - 6u\frac{\partial u}{\partial x} + \frac{\partial^3 u}{\partial x^3} = 0. \tag{3.97}$$

We also write the left hand side of (3.97) as

$$p := u_t - 6uu_x + u_{x,x,x}.$$

The command `pdsolve` in Maple 13 returns solutions

$$u(t,x) = 2c_3^2\tanh^2(c_3 x + c_2 t + c_1) + \frac{c_2 - 8c_3^3}{6c_3}, \tag{3.98}$$

where c_1, c_2, c_3 are arbitrary constants. We are going to compute exact solutions to (3.97) of the form

$$u(t,x) = f(t)g(x), \tag{3.99}$$

which clearly amounts to a separation of variables. Substitution of (3.99) into (3.97) gives no hint how to obtain more information about f or g. We note that analytic functions of the form (3.99) are characterized as solutions of (cf. Prop. 3.3.7)

$$\begin{vmatrix} u & u_x \\ u_t & u_{t,x} \end{vmatrix} = 0. \tag{3.100}$$

A Thomas decomposition of the differential system

$$\{p = 0,\ uu_{t,x} - u_t u_x = 0\}$$

is given by (cf. Ex. 2.2.61, p. 103):

	$\underline{u_t - 6uu_x} = 0\ \{\partial_t,\partial_x\}$	$u_t = 0\ \{\partial_t,\partial_x\}$
		$\underline{u_{x,x,x}} - 6uu_x = 0\ \{*,\partial_x\}$
	$u_{x,x} = 0\ \{*,\partial_x\}$	$u_{x,x} \neq 0$
$u = 0\ \{\partial_t,\partial_x\}$	$u \neq 0$	$u \neq 0$

The analytic solutions of the second simple system can be determined as follows. By assumption we have $u(t,x) = f(t)g(x)$. Now $u_{x,x} = 0$ yields

$$g(x) = C_1 x + C_2, \quad \text{where } C_1, C_2 \in \mathbb{R}.$$

At least one of C_1 and C_2 is non-zero because $u \neq 0$. Then $u_t - 6uu_x = 0$ implies that f is a non-zero constant in case $C_1 = 0$ and otherwise

$$f(t) = 1/(-6C_1 t + C_0), \quad \text{where } C_0 \in \mathbb{R}.$$

Hence, the solutions are

$$u(t,x) = c, \qquad u(t,x) = \frac{x+c_1}{-6t+c_2}, \tag{3.101}$$

where $c \in \mathbb{R} - \{0\}$, $c_1, c_2 \in \mathbb{R}$ (determined by $c_1 = C_2/C_1$, $c_2 = C_0/C_1$).

The general solution $u(x)$ of the third system is given implicitly by

$$x = \pm \int_{u(0)}^{u(x)} \frac{dz}{\sqrt{2z^3 + az + b}} = \pm \int_0^{u(x)} \frac{dz}{\sqrt{2z^3 + az + b}} + c, \qquad a, b, c \text{ constant.}$$

This can be seen as follows. Considering $u_{x,x,x} - 6uu_x = 0$, we have that $u_{x,x} - 3u^2$ is constant, say equal to $\tilde{a}$. Multiplying by $u_x \neq 0$ we obtain $u_x u_{x,x} - 3u^2 u_x = \tilde{a} u_x$. Integration with respect to x yields $\frac{u_x^2}{2} = u^3 + \tilde{a}u + \tilde{b}$ with a constant $\tilde{b}$. Hence, we have $u_x = \pm\sqrt{2u^3 + au + b}$ for some constants a and b, and therefore

$$x = \int_0^x dy = \int_0^x \frac{\pm u'(y)\,dy}{\sqrt{2u(y)^3 + au(y) + b}} = \pm \int_{u(0)}^{u(x)} \frac{dz}{\sqrt{2z^3 + az + b}}.$$

Some solutions of the third system are

$$u(t,x) = 2c_1^2 \tanh^2(c_1 x + c_2) - \frac{4c_1^2}{3}, \qquad c_1 \neq 0.$$

We have

$$u_{x,x} = 4c_1^4 (\tanh(c_1 x + c_2) - 1)(\tanh(c_1 x + c_2) + 1)(3\tanh^2(c_1 x + c_2) - 1) \neq 0$$

because $c_1 \neq 0$.

In any case, substituting (3.98) into (3.100) yields a rational function in cosh which vanishes identically only if $c_2 = 0$ or $c_3 = 0$ (using that cosh is transcendental). Hence, the set of functions of the form (3.98), proposed by `pdsolve`, does not include any function of the form $f(t)g(x)$; in particular, the solutions (3.101) of (3.97) are not found by `pdsolve`.

Example 3.3.50. We consider the Navier-Stokes equations, i.e., the equations of motion of a viscous fluid (cf., e.g., [LL66, p. 54], [Bem06]), in the case of an incompressible flow:

$$\begin{cases} \dfrac{\partial v}{\partial t} + (v \cdot \nabla)v = -\dfrac{1}{\rho} \operatorname{grad}(p) + \dfrac{\eta}{\rho} \Delta v + f, \\[2mm] \nabla \cdot v = 0. \end{cases} \tag{3.102}$$

Incompressibility means that the fluid density ρ is assumed to be constant in time and space. The constant η/ρ is the kinematic viscosity and is abbreviated by the Greek letter v. We choose the force f to be zero.

In order to find some exact solutions of a certain form of the Navier-Stokes equations, say in cylindrical coordinates r, φ, z, we add an implicit description of the family of analytic functions of this form in terms of polynomial partial differential equations to (3.102) and compute a Thomas decomposition of the combined system. Here we choose the assumption that each component of the velocity field is a product of a function of r, a function of φ, and a function of z, without any assumption on the dependence of t. An implicit description of functions of this type is given by

$$
\begin{vmatrix} u & u_z \\ u_r & u_{r,z} \end{vmatrix} = 0, \quad
\begin{vmatrix} u & u_\varphi \\ u_r & u_{r,\varphi} \end{vmatrix} = 0, \quad
\begin{vmatrix} u & u_z \\ u_\varphi & u_{\varphi,z} \end{vmatrix} = 0, \quad u \in \{v_1, v_2, v_3\}.
$$

A Thomas decomposition for the combined system consists of too many systems to be computable completely in reasonable time. Therefore, we make use of the feature of the Maple package `DifferentialThomas` (cf. Subsect. 2.2.6), which allows to stop the computation of a Thomas decomposition as soon as one simple system is produced and to continue the computation later. Collecting simple differential systems in this way and applying the Maple command `pdsolve` to them, we get, for instance, the following family of exact solutions of (3.102):

$$
\begin{cases}
v_1(t,r,\varphi,z) = -\dfrac{(t+c_1)\,F_1(t)}{r} - \dfrac{r}{2\,(t+c_1)}, \\[2ex]
v_2(t,r,\varphi,z) = \dfrac{(\varphi+c_2)\,r}{t+c_1}, \\[2ex]
v_3(t,r,\varphi,z) = 0, \\[2ex]
p(t,r,\varphi,z) = (t+c_1)\ln(r)\dfrac{dF_1(t)}{dt} - \dfrac{(t+c_1)^2\,F_1(t)^2}{2\,r^2} \\[2ex]
\qquad\qquad + (\ln(r) + (\varphi+c_2)^2)\,F_1(t) + F_2(t) \\[2ex]
\qquad\qquad - \dfrac{2\,v\ln(r)}{t+c_1} + \dfrac{((\varphi+c_2)^2 - \frac{3}{4})\,r^2}{2\,(t+c_1)^2},
\end{cases}
$$

where c_1, c_2 are arbitrary constants and F_1, F_2 are arbitrary analytic functions of t (and where no assumption is made about the constant v).

Appendix A
Basic Principles and Supplementary Material

Abstract This appendix reviews elementary concepts of module theory, homological algebra, and differential algebra and presents additional basic material that is relevant for the previous chapters. Fundamental notions of homological algebra, which are used for the study of systems of linear functional equations, are to be found in the first section. In the second section the chain rule for higher derivatives is deduced, which is employed in the earlier section on linear differential elimination. The description of Thomas' algorithm and the development of nonlinear differential elimination rely on the principles of differential algebra that are recalled in the third section. Two notions of homogeneity for differential polynomials are discussed, and the analogs of Hilbert's Basis Theorem and Hilbert's Nullstellensatz in differential algebra are outlined.

A.1 Module Theory and Homological Algebra

In this section some basic notions of module theory and homological algebra are collected. General references for this material are, e.g., [Eis95], [Lam99], [Rot09].

Let R be a (not necessarily commutative) ring with multiplicative identity element 1. In this section all modules are understood to be left R-modules, and 1 is assumed to act as identity. For every statement about left R-modules in this section an analogous statement about right R-modules holds, which will not be mentioned in what follows.

A.1.1 Free Modules

Definition A.1.1. An R-module M is said to be *free*, if it is a direct sum of copies of R. If $M = \bigoplus_{i \in I} R b_i$ and $R b_i \cong R$ for all $i \in I$, where I is an index set, then $(b_i)_{i \in I}$ is a *basis* for M. If n denotes the cardinality of I, then M is said to be *free of rank n*.

© Springer International Publishing Switzerland 2014   233
D. Robertz, *Formal Algorithmic Elimination for PDEs*,
Lecture Notes in Mathematics 2121, DOI 10.1007/978-3-319-11445-3

Example A.1.2. Let R be a field (or a skew-field). Then every R-module is free because, as a vector space, it has a basis (by the axiom of choice).

Remark A.1.3 (Universal property of a free module). A homomorphism from a free R-module M to another R-module N is uniquely determined by specifying the images of the elements of a basis $B = (b_i)_{i \in I}$ of M. In other words, every map $B \to N$ defines a unique homomorphism $M \to N$ by R-linear extension.

Remark A.1.4. Every R-module M is a factor module of a free module. Indeed, let $G \subseteq M$ be any generating set for M. We define the free (left) R-module

$$F = \bigoplus_{g \in G} R\hat{g},$$

where the $\hat{g}$ for $g \in G$ are pairwise different new symbols, and we define the homomorphism $\varphi \colon F \to M$ by $\varphi(\hat{g}) = g$. By the universal property of a free module, φ exists and is uniquely determined. Since G is a generating set for M, the homomorphism φ is surjective. Defining $N := \ker(\varphi)$, the homomorphism theorem for R-modules states that

$$M = \mathrm{im}(\varphi) \cong F/N. \tag{A.1}$$

Whereas in an arbitrary representation of M it may be difficult to address elements $m \in M$ or to tell such elements apart, the isomorphism in (A.1) is advantageous in the sense that the free module F allows a very explicit representation of its elements (e.g., coefficient vectors with respect to a chosen basis) and the ambiguity of addressing an element in M in that way is captured by N.

Definition A.1.5. By definition, an equivalent formulation for the statement in (A.1) is:

$$0 \longleftarrow M \overset{\varphi}{\longleftarrow} F \longleftarrow N \longleftarrow 0 \quad \text{is a short exact sequence}, \tag{A.2}$$

where the homomorphism $N \to F$ is the inclusion map.

More generally, a *chain complex* of R-modules is defined to be a family $(M_i)_{i \in \mathbb{Z}}$ of R-modules with homomorphisms $d_i \colon M_i \to M_{i-1}$, $i \in \mathbb{Z}$, denoted by

$$M_\bullet : \qquad \cdots \overset{d_0}{\longleftarrow} M_0 \overset{d_1}{\longleftarrow} M_1 \overset{d_2}{\longleftarrow} \cdots \overset{d_{n-1}}{\longleftarrow} M_{n-1} \overset{d_n}{\longleftarrow} M_n \overset{d_{n+1}}{\longleftarrow} \cdots$$

such that the composition of each two consecutive homomorphisms is the zero map, i.e.,

$$\mathrm{im}(d_{i+1}) \subseteq \ker(d_i) \qquad \text{for all } i \in \mathbb{Z}.$$

If $\mathrm{im}(d_{i+1}) = \ker(d_i)$ holds, then $M_\bullet$ is said to be *exact at M_i*. The factor module

$$H_i(M_\bullet) := \ker(d_i)/\mathrm{im}(d_{i+1}), \qquad i \in \mathbb{Z},$$

is the *homology at M_i* or the *defect of exactness at M_i*.

A *cochain complex* of R-modules is defined to be a family $(N^i)_{i \in \mathbb{Z}}$ of R-modules with homomorphisms $d^i \colon N^i \to N^{i+1}$, $i \in \mathbb{Z}$, denoted by

$$N^\bullet: \qquad \cdots \longrightarrow N^0 \xrightarrow{d^0} N^1 \xrightarrow{d^1} \cdots \xrightarrow{d^{n-2}} N^{n-1} \xrightarrow{d^{n-1}} N^n \xrightarrow{d^n} \cdots$$

such that

$$\mathrm{im}(d^{i-1}) \subseteq \ker(d^i) \qquad \text{for all } i \in \mathbb{Z}.$$

If $\mathrm{im}(d^{i-1}) = \ker(d^i)$ holds, then $N^\bullet$ is said to be *exact at* N^i. The factor module

$$H^i(N^\bullet) := \ker(d^i)/\mathrm{im}(d^{i-1}), \qquad i \in \mathbb{Z},$$

is the *cohomology at* N^i or the *defect of exactness at* N^i.

The term *exact sequence* is a synonym for a (co)chain complex with trivial (co)homology groups.

A short exact sequence as in (A.2), where F is a free module, is also called a *presentation* of M. If F and N are finitely generated, then (A.1) is called a *finite presentation* of M. If there exists a finite presentation of an R-module M, then M is said to be *finitely presented*.

A.1.2 Projective Modules and Injective Modules

Definition A.1.6. An R-module M is said to be *projective*, if for every epimorphism $\beta \colon B \to C$ and every homomorphism $\alpha \colon M \to C$ there exists a homomorphism $\gamma \colon M \to B$ satisfying $\alpha = \beta \circ \gamma$, i.e., such that the following diagram is commutative:

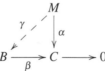

Remark A.1.7. Let M be a free R-module. Then M is projective. Indeed, knowledge of the images $\alpha(m_i)$ in C of elements in a basis $(m_i)_{i \in I}$ of M allows to choose preimages $b_i \in B$ of the $\alpha(m_i)$ under β, and by the universal property of free modules, there exists a unique homomorphism $\gamma \colon M \to B$ such that $\gamma(m_i) = b_i$ for all $i \in I$.

Remark A.1.8. The condition in Definition A.1.6 is equivalent to the right exactness of the (covariant) functor $\hom_R(M, -)$, i.e., the condition that for every exact sequence of R-modules

$$A \longrightarrow B \longrightarrow C \longrightarrow 0 \qquad\qquad (A.3)$$

the complex of abelian groups

$$\hom_R(M,A) \longrightarrow \hom_R(M,B) \longrightarrow \hom_R(M,C) \longrightarrow 0$$

(with homomorphisms which compose with the corresponding homomorphisms in (A.3)) is exact. (The functor $\hom_R(M, -)$ is left exact for every R-module M, i.e., for every exact sequence of R-modules

$$0 \longrightarrow A \longrightarrow B \longrightarrow C \tag{A.4}$$

the complex of abelian groups

$$0 \longrightarrow \hom_R(M,A) \longrightarrow \hom_R(M,B) \longrightarrow \hom_R(M,C)$$

(with homomorphisms which compose with the corresponding homomorphisms in (A.4)) is exact.)

Remark A.1.9. Let M be a projective R-module and let

$$0 \longleftarrow M \overset{\pi}{\longleftarrow} F \overset{\varepsilon}{\longleftarrow} N \longleftarrow 0$$

be a presentation of M with a free R-module F. By choosing the modules $B = F$, $C = M$ and the homomorphisms $\alpha = \mathrm{id}_M$, $\beta = \pi$, we conclude that there exists a homomorphism $\sigma\colon M \to F$ (viz. $\sigma = \gamma$) satisfying $\pi \circ \sigma = \mathrm{id}_M$. The short exact sequence is said to be *split* in this case (and the following discussion is essentially the Splitting Lemma in homological algebra).

Let $\varphi\colon F \to F$ be the homomorphism defined by

$$\varphi = \mathrm{id}_F - \sigma \circ \pi.$$

For $f \in F$ we have

$$\varphi(f) = f \quad \Longleftrightarrow \quad f \in \ker(\pi) = \mathrm{im}(\varepsilon)$$

because σ is a monomorphism. Since the homomorphism ε is injective, we get a homomorphism $\rho\colon F \to N$ such that $\rho \circ \varepsilon = \mathrm{id}_N$, and we have (cf., e.g., [BK00, Thm. 2.4.5])

$$\mathrm{id}_F = \varepsilon \circ \rho + \sigma \circ \pi.$$

This shows that

$$F = \varepsilon(N) \oplus \sigma(M)$$

and ρ, π are the projections $F \to N$ and $F \to M$ onto the respective summands (up to isomorphism) of this direct sum.

Conversely, let M be a direct summand of a free R-module F, i.e., there exists an R-module N such that $M \oplus N = F$. Then every element of F has a unique representation as sum of an element of M and an element of the complement N. In particular, for the elements of a basis $(f_i)_{i \in I}$ of F we have $f_i = m_i + n_i$, where $m_i \in M$ and $n_i \in N$ are uniquely determined by f_i. In the situation of Definition A.1.6 we may choose preimages $b_i \in B$ of the $\alpha(m_i)$ under β and use the universal property of free modules to define a homomorphism $F \to B$ by $f_i \mapsto b_i$, $i \in I$, whose restriction to M is a homomorphism γ as in Definition A.1.6.

Therefore, an R-module is projective if and only if it is (isomorphic to) a direct summand of a free R-module.

Examples A.1.10. a) Let R be a (commutative) principal ideal domain. Then every projective R-module is free, because every submodule of a free R-module is free.

b) (Quillen-Suslin Theorem, the resolution of Serre's Problem)

Every finitely generated projective module over $K[x_1,\ldots,x_n]$, where K is a field or a (commutative) principal ideal domain, is free. (We also refer to [Lam06] for details on Serre's Problem and generalizations, and to [FQ07] for constructive aspects and applications.)

c) Let R be a commutative local ring (with 1). Then, by using Nakayama's Lemma, every finitely generated projective R-module is free.

We refer to Example A.1.17 below for an example of a projective module which is not free.

Definition A.1.11. An R-module M is said to be *injective*, if for every monomorphism $\beta\colon A \to B$ and every homomorphism $\alpha\colon A \to M$ there exists a homomorphism $\gamma\colon B \to M$ satisfying $\alpha = \gamma \circ \beta$, i.e., such that the following diagram is commutative:

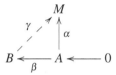

The notion of an injective module is clearly dual to the notion of projective module in the sense that all arrows are reversed.

Remark A.1.12. The condition in Definition A.1.11 is equivalent to the right exactness of the contravariant functor $\hom_R(-,M)$, i.e., the condition that for every exact sequence of R-modules

$$C \longleftarrow B \longleftarrow A \longleftarrow 0 \tag{A.5}$$

the complex of abelian groups

$$\hom_R(C,M) \longrightarrow \hom_R(B,M) \longrightarrow \hom_R(A,M) \longrightarrow 0$$

(with homomorphisms which pre-compose with the corresponding homomorphisms in (A.5)) is exact. (The contravariant functor $\hom_R(-,M)$ is left exact for every R-module M, i.e., for every exact sequence of R-modules

$$0 \longleftarrow C \longleftarrow B \longleftarrow A \tag{A.6}$$

the complex of abelian groups

$$0 \longrightarrow \hom_R(C,M) \longrightarrow \hom_R(B,M) \longrightarrow \hom_R(A,M)$$

(with homomorphisms which pre-compose with the corresponding homomorphisms in (A.6)) is exact.)

If the monomorphism $\beta : A \to B$ in Definition A.1.11 is actually an inclusion of modules, then injectivity can be understood as the possibility to extend every homomorphism $A \to M$ to a homomorphism $B \to M$. Moreover, choosing $A = M$ and $\alpha = \mathrm{id}_M$, we conclude that an injective R-module is a direct summand of every R-module which contains it. Another particular case of the above condition is that $A = I$ is a (left) ideal of $B = R$. It turns out that the restriction of the condition to arbitrary (left) ideals of R is sufficient for injectivity.

Theorem A.1.13 (Baer's criterion). *A (left) R-module M is injective if and only if for every (left) ideal I of R every homomorphism $I \to M$ can be extended to a homomorphism $R \to M$.*

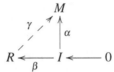

Example A.1.14. Let A be an abelian group (which is the same as a $\mathbb{Z}$-module). Then A is injective if and only if it is *divisible*, i.e., for every $a \in A$ and every integer $n \in \mathbb{Z} - \{0\}$ there exists $b \in A$ such that $a = nb$.

Definition A.1.15. An R-module M is said to be *stably free of rank $n \in \mathbb{Z}_{\geq 0}$*, if there exist $m \in \mathbb{Z}_{\geq 0}$ and a free R-module F of rank m such that $M \oplus F$ is isomorphic to a free R-module of rank $m + n$.

Remark A.1.16. Every free R-module of rank $n \in \mathbb{Z}_{\geq 0}$ is stably free of the same rank. Every stably free R-module is projective because it is a direct summand of a free module (cf. Rem. A.1.9). Hence, the following chain of implications holds for finitely generated R-modules M:

$$M \text{ free} \Rightarrow M \text{ stably free} \Rightarrow M \text{ projective}.$$

Example A.1.17. Let $R = \mathbb{R}[x_1, \ldots, x_n]/\langle 1 - x_1^2 - \ldots - x_n^2 \rangle$ and define the left R-module $M = R^{1 \times n}/R(x_1, \ldots, x_n)$. Then M is stably free of rank $n - 1$, but not free if $n \notin \{1, 2, 4, 8\}$ (because the tangent bundle to the $(n-1)$-sphere is trivial only if $n \in \{1, 2, 4, 8\}$, cf. [Eis95, Ex. 19.17] and the references therein).

For an example of a projective, but not stably free module, cf. Example A.1.32.

A.1.3 Syzygies and Resolutions

As mentioned earlier, a presentation of a module M as $M \cong F/N$ with a free module F furnishes a very concrete description of M. We may treat N in the same way as M, in order to get an ever clearer picture of M.

For more details about the following material, we refer to, e.g., [Eis95], [Rot09]. A proof of the next proposition can be found, e.g., in [Lam99, Chap. 2, § 5A].

Proposition A.1.18 (Schanuel's Lemma). *Let P and Q be projective R-modules and*

$$0 \longleftarrow M \longleftarrow P \longleftarrow K \longleftarrow 0,$$

$$0 \longleftarrow M \longleftarrow Q \longleftarrow L \longleftarrow 0$$

short exact sequences of R-modules. Then we have $K \oplus Q \cong L \oplus P$ as R-modules.

The conclusion of Schanuel's Lemma suggests to define the following equivalence relation on the class of all R-modules.

Definition A.1.19. Two R-modules K and L are said to be *projectively equivalent* if there exist projective R-modules P and Q such that $K \oplus Q \cong L \oplus P$.

Remark A.1.20. Let M be an R-module with generating sets G_1 and G_2. Let

$$F_i := \bigoplus_{g \in G_i} R\hat{g} \quad \text{and} \quad \varphi_i \colon F_i \longrightarrow M \colon \hat{g} \longmapsto g, \qquad i = 1, 2,$$

be the homomorphisms of R-modules discussed in Remark A.1.4, where all $\hat{g}$ for $g \in G_1 \cup G_2$ are pairwise different new symbols. The *syzygy module* of G_i is defined to be the kernel of φ_i, $i = 1, 2$. Every element of $\ker(\varphi_i)$ is also called a *syzygy* of G_i. By Schanuel's Lemma, $\ker(\varphi_1)$ and $\ker(\varphi_2)$ are projectively equivalent. Therefore, we may associate with M the equivalence class of $\ker(\varphi_1)$ (or of $\ker(\varphi_2)$, whose equivalence class is the same) under projective equivalence.

By iterating the syzygy module construction we get a *free resolution*

$$F_0 \longleftarrow F_1 \longleftarrow \dots \longleftarrow F_{n-1} \longleftarrow F_n \longleftarrow \dots \tag{A.7}$$

of M, i.e., a chain complex of free R-modules that is exact at each F_i except at F_0, where the homology (i.e., the cokernel of the leftmost map) is isomorphic to M.

If the modules F_i are assumed to be projective rather than free, then (A.7) is called a *projective resolution* of M.

Similarly, a cochain complex of injective modules

$$I^0 \longrightarrow I^1 \longrightarrow \dots \longrightarrow I^{n-1} \longrightarrow I^n \longrightarrow \dots$$

which is exact at each I^i except at I^0, where the cohomology (i.e., the kernel of the leftmost map) is isomorphic to M, is called an *injective resolution* of M.

If a resolution is *finite*, i.e., if there exists a non-negative integer n such that the i-th module is zero for all $i \geq n$, then the number of non-zero homomorphisms in the resolution is called the *length* of the resolution.

Theorem A.1.21. *Every R-module has a free resolution and an injective resolution.*

For examples of free resolutions, cf. Examples A.1.31 and A.1.32 below, or Subsect. 3.1.5, e.g., p. 152.

Example A.1.22. Let $R = \mathbb{Z}$ and $M = \mathbb{Z}$. Then

$$0 \longrightarrow \mathbb{Z} \longrightarrow \mathbb{Q} \longrightarrow \mathbb{Q}/\mathbb{Z} \longrightarrow 0$$

is a short exact sequence of $\mathbb{Z}$-modules, and

$$\mathbb{Q} \longrightarrow \mathbb{Q}/\mathbb{Z} \longrightarrow 0$$

is an injective resolution of $\mathbb{Z}$. Note that $\mathbb{Q}$ is not finitely generated as $\mathbb{Z}$-module.

For particular kinds of rings it is known that finite free resolutions exist for every finitely generated module. For instance, every finitely generated module over a commutative polynomial algebra in n variables over a field has a free resolution of length at most n with finitely generated free modules, as ensured by Hilbert's Syzygy Theorem, cf. Corollary 3.1.47, p. 151, or, e.g., [Eis95, Cor. 19.7].

Definition A.1.23. The *left projective dimension* of a left R-module M is defined to be the smallest length of a projective resolution of M if such a finite resolution exists and ∞ otherwise. The *left global dimension* of a ring R is defined to be the supremum of the left projective dimensions of its left modules. In an analogous way the notions of right projective dimension and right global dimension are defined.

Remark A.1.24. The left projective dimension of a left R-module M is zero if and only if M is projective.

Proposition A.1.25 (cf. [MR01], Subsect. 7.1.11). *If R is left and right Noetherian, then its left and right global dimensions are equal.*

Since we only deal with Noetherian rings, we denote the left and right global dimension of R by $\mathrm{gld}(R)$. For notation concerning skew polynomial rings $R[\partial; \sigma, \delta]$, cf. Subsect. 2.1.2.

Theorem A.1.26 ([MR01], Thm. 7.5.3). *If $\mathrm{gld}(R) < \infty$, σ an automorphism of R and δ a σ-derivation of R, then we have $\mathrm{gld}(R) \leq \mathrm{gld}(R[\partial; \sigma, \delta]) \leq \mathrm{gld}(R) + 1$.*

Remark A.1.27 (cf. [QR07], Cor. 21). Let D be a ring and

$$D^{1 \times r_0} \xleftarrow{\ d_1\ } D^{1 \times r_1} \xleftarrow{\ d_2\ } \ldots \xleftarrow{d_{m-1}} D^{1 \times r_{m-1}} \xleftarrow{\ d_m\ } D^{1 \times r_m} \longleftarrow 0 \qquad \text{(A.8)}$$

be a finite free resolution of a left D-module M. If $m \geq 3$ and there exists a homomorphism $s \colon D^{1 \times r_{m-1}} \to D^{1 \times r_m}$ such that $s \circ d_m$ is the identity on $D^{1 \times r_m}$, then a free resolution of M of length $m - 1$ is given by

$$D^{1 \times r_0} \xleftarrow{\ d_1\ } \ldots \xleftarrow{d_{m-3}} D^{1 \times r_{m-3}} \xleftarrow{\widetilde{d}_{m-2}} D^{1 \times r_{m-2}} \oplus D^{1 \times r_m} \xleftarrow{\widetilde{d}_{m-1}} D^{1 \times r_{m-1}} \longleftarrow 0,$$

where $\widetilde{d}_{m-1}$ combines the values of d_{m-1} and s to a pair and $\widetilde{d}_{m-2}$ applies d_{m-2} to the first component of such a pair. In case $m = 2$ the same reduction is possible, where now the canonical projection onto M is to be modified instead of d_{m-2}.

Theorem A.1.28. *Let* $D = R[\partial_1; \sigma_1, \delta_1][\partial_2; \sigma_2, \delta_2] \dots [\partial_l; \sigma_l, \delta_l]$ *be an Ore algebra (cf. Def. 2.1.14, p. 17), where R is either a field, or $\mathbb{Z}$, or a commutative polynomial algebra over a field or $\mathbb{Z}$ with finitely many indeterminates, and where every σ_i is an automorphism. Moreover, let M be a finitely generated (left or right) D-module. Then there exists a finite free resolution (A.8) of M, where either $m = 1$ and there exists a homomorphism $s\colon D^{1\times r_{m-1}} \to D^{1\times r_m}$ such that $s \circ d_m$ is the identity on $D^{1\times r_m}$, or where $m \geq 1$ and there exists no such s. In the first case, M is stably free; in the second case, M is not projective.*

An iteration of Theorem A.1.26 shows that a finite free resolution of M exists. A free resolution of M as claimed in Theorem A.1.28 is obtained by a repeated use of Remark A.1.27. The assertion that M is stably free in the first case is immediate, whereas an application of Schanuel's Lemma (Prop. A.1.18) proves that M is not projective in the second case. We refer to the proof of [Rob14, Thm. 3.23] for more details. The relevant techniques can also be found in [Lam99, Chap. 2, § 5A].

Remark A.1.27 allows to compute the projective dimension of a finitely generated D-module.

Theorem A.1.28 yields the following corollary (cf. also [MR01, Cor. 12.3.3]).

Corollary A.1.29. *Let D be an Ore algebra as in Theorem A.1.28. Then every finitely generated projective D-module is stably free.*

By applying Corollary A.1.29 and induction on the projective dimension one can prove the following corollary (cf. also [Rot09, Lem. 8.42], [CQR05, Prop. 8], [Rob14, Cor. 3.25]).

Corollary A.1.30. *Let D be an Ore algebra as in Theorem A.1.28. Then every finitely generated D-module has a free resolution with finitely generated free modules of length at most* $\mathrm{gld}(D) + 1$.

Example A.1.31 ([QR07], Ex. 50). Let K be a field of characteristic zero and either $D = A_1(K) = K[z][\partial; \sigma, \delta]$ be the Weyl algebra or $D = B_1(K) = K(z)[\partial; \sigma, \delta]$ be the algebra of differential operators with rational function coefficients (cf. Ex. 2.1.18 b), p. 19). Moreover, let $R = (\partial \quad -z^k) \in D^{1\times 2}$ for some $k \in \mathbb{Z}_{\geq 0}$ and $M = D^{1\times 2}/DR$. Then the left D-module M is stably free of rank 1 because the short exact sequence

$$0 \longleftarrow M \xleftarrow{\ \pi\ } D^{1\times 2} \xleftarrow{\ \varepsilon\ } D \longleftarrow 0,$$

where $\pi\colon D^{1\times 2} \to M$ is the canonical projection and $\varepsilon\colon D \to D^{1\times 2}$ is induced by R, is split (cf. Rem. A.1.9), the homomorphism $\rho\colon D^{1\times 2} \to D$ represented with respect to the standard bases by the matrix

$$\begin{pmatrix} \displaystyle\sum_{j=1}^{k} (-1)^{j-1} \frac{z^j}{j!} \partial^{j-1} \\[2ex] (-1)^{k+1} \frac{1}{k!} \partial^k \end{pmatrix} \in D^{2\times 1}$$

satisfying $\rho \circ \varepsilon = \mathrm{id}_D$. In order to show that the left D-module M is not free, we consider the exact sequence of left D-modules

$$D \xleftarrow{\ \psi\ } D^{1\times 2} \xleftarrow{\ \varepsilon\ } D \xleftarrow{\qquad} 0,$$

where $\psi \colon D^{1\times 2} \to D$ is induced by

$$Q = \begin{pmatrix} z^{k+1} \\ z\partial + k + 1 \end{pmatrix} \in D^{2\times 1}.$$

The homomorphism of left D-modules

$$\phi \colon M \longrightarrow D \colon r + DR \longmapsto rQ, \qquad r \in D^{1\times 2},$$

is well-defined due to $RQ = 0$ and injective because of $\ker(\psi) = \mathrm{im}(\varepsilon)$. Hence, we have $M \cong \mathrm{im}(\phi) = D^{1\times 2}Q$, which is a left ideal of D. A Janet basis or Gröbner basis computation proves that this left ideal is not principal. Therefore, M is not free.

(The matrix Q is referred to as a *minimal parametrization* of M in [CQR05, Thm. 8], cf. also [PQ99].)

In contrast to the previous example we note that non-principal ideals of Dedekind domains are projective modules which are not stably free (cf. also [MR01, Ex. 11.1.4 (i)] for the general statement or [Rot09, Ex. 4.92 (iii)] for a different example).

Example A.1.32. Let $D = \mathbb{Z}[\sqrt{-5}]$. The left D-modules $M_i = D^{1\times 2}R_i$, $i = 1, 2$, where

$$R_1 = \begin{pmatrix} -2 & 1 - \sqrt{-5} \\ -1 - \sqrt{-5} & 3 \end{pmatrix}, \qquad R_2 = \begin{pmatrix} 3 & -1 + \sqrt{-5} \\ 1 + \sqrt{-5} & -2 \end{pmatrix},$$

are projective because the D-module homomorphisms $\pi_i \colon D^{1\times 2} \to D^{1\times 2}$ induced by R_i, $i = 1, 2$, satisfy

$$\pi_1 \circ \pi_1 = \pi_1, \quad \pi_2 \circ \pi_2 = \pi_2, \quad \pi_1 \circ \pi_2 = \pi_2 \circ \pi_1 = 0, \quad \pi_1 + \pi_2 = \mathrm{id}_{D^{1\times 2}},$$

which implies $M_1 \oplus M_2 = D^{1\times 2}$. However, M_1 (and M_2) is not free, because the ideal of D which is generated by 2 and $1 + \sqrt{-5}$ (by 3 and $1 + \sqrt{-5}$, respectively) is isomorphic to M_1 (to M_2, respectively) and is not principal. The left D-modules M_1 and M_2 are not stably free either, because an isomorphism $M_i \oplus D^{1\times m} \cong D^{1\times n}$ of finitely generated D-modules for some m, $n \in \mathbb{N}$ would allow cancelation of free D-modules, i.e., would imply that M_i is free (as a consequence of Steinitz' Theorem, cf., e.g., [BK00, Cor. 6.1.8]). A free resolution of M_1 is given by the periodic resolution

$$0 \longleftarrow M_1 \xleftarrow{\ \pi_1\ } D^{1\times 2} \xleftarrow{\ \pi_2\ } D^{1\times 2} \xleftarrow{\ \pi_1\ } D^{1\times 2} \xleftarrow{\ \pi_2\ } \cdots,$$

and by exchanging the roles played by π_1 and π_2, we obtain one for M_2.

A.2 The Chain Rule of Differentiation

In this section we deduce the chain rule for higher derivatives, which is applied in Subsect. 3.2.2. For more details on the basic definitions, we refer to, e.g., [Die69], and for an equivariant viewpoint on the tensors discussed below, cf. [KMS93].

In what follows let E, E_1, ..., E_n, and F be Banach spaces over either $\mathbb{R}$ or $\mathbb{C}$, and let A, A_1, ..., A_n be non-empty open subsets of E, E_1, ..., E_n, respectively.

If $f\colon A \to F$ is a continuously differentiable map, then the *derivative Df* of f associates to each $x \in A$ a certain continuous linear map $E \to F$, the linearization of f at x. We also write Df for the continuous map

$$A \times E \longrightarrow F\colon (x,v) \longmapsto (Df)(x)(v),$$

which is linear in its second argument.

Let $f\colon A_1 \times \ldots \times A_n \to F$ be a map, $j \in \{1,\ldots,n\}$, and define

$$A_j^{(x_1,\ldots,x_{j-1},x_{j+1},\ldots,x_n)} := \{x_1\} \times \ldots \times \{x_{j-1}\} \times A_j \times \{x_{j+1}\} \times \ldots \times \{x_n\}$$

for $x_i \in A_i$, $1 \leq i \leq n$, $i \neq j$. If the restriction of f to $A_j^{(x_1,\ldots,x_{j-1},x_{j+1},\ldots,x_n)}$ is continuously differentiable, then the derivative of this restriction associates to each $(x_1,\ldots,x_n) \in A_j^{(x_1,\ldots,x_{j-1},x_{j+1},\ldots,x_n)}$ a certain continuous linear map $E_j \to F$ as above. If all restrictions of f to $A_j^{(x_1,\ldots,x_{j-1},x_{j+1},\ldots,x_n)}$, $x_i \in A_i$, $1 \leq i \leq n$, $i \neq j$, are continuously differentiable, then the *j-th partial derivative D_jf* of f associates to each $(x_1,\ldots,x_n) \in A_1 \times \ldots \times A_n$ the continuous linear map $E_j \to F$ given by the derivative of the corresponding restriction of f. We also write D_jf for the continuous map

$$A_1 \times \ldots \times A_n \times E_j \longrightarrow F\colon (x_1,\ldots,x_n,v) \longmapsto (D_jf)(x_1,\ldots,x_n)(v),$$

which is linear in its last argument.

Let $f\colon A \to F$ be a continuously differentiable map and assume that the first partial derivative of $Df\colon A \times E \to F$ exists and is continuous. Then

$$D^2f\colon A \times E^2 \longrightarrow F\colon (x,v_1,v_2) \longmapsto D_1(Df)(x,v_1,v_2) = D_1(Df)(x,v_1)(v_2)$$

is a continuous map, which is bilinear and symmetric in its last two arguments. It is called the *second derivative* of f. More generally, if the $(k-1)$-st derivative of $f\colon A \to F$ is defined and if its first partial derivative exists and is continuous, then

$$D^kf\colon A \times E^k \longrightarrow F\colon (x,v_1,\ldots,v_k) \longmapsto D_1(D^{k-1}f)(x,v_1,\ldots,v_k)$$

is a continuous map, which is multilinear and symmetric in its last k arguments. It is called the *k-th derivative* of f.

Remark A.2.1. If all partial derivatives $D_j f$ of $f: A_1 \times \ldots \times A_n \to F$ exist and if $\alpha_i: A \to A_i$ is a continuously differentiable map, $i = 1, \ldots, n$, then

$$D(f \circ (\alpha_1, \ldots, \alpha_n)): A \times E \longrightarrow F$$

is given by

$$(D(f \circ \alpha))(x, v) = \sum_{i=1}^{n} (D_i f)(\alpha(x), (D\alpha_i)(x, v)), \qquad x \in A, \quad v \in E,$$

where $\alpha := (\alpha_1, \ldots, \alpha_n)$.

Remark A.2.2. Let $j \in \{1, \ldots, n\}$. If

$$g: A_1 \times \ldots \times A_{j-1} \times E_j \times A_{j+1} \times \ldots \times A_n \longrightarrow F$$

is a map such that

$$E_j \longrightarrow F: x_j \longmapsto g(a_1, \ldots, a_{j-1}, x_j, a_{j+1}, \ldots, a_n)$$

is a continuous linear map for all $a_i \in A_i$, $1 \le i \le n$, $i \ne j$, then the j-th partial derivative of g exists and we have

$$(D_j g)(x_1, \ldots, x_n, v) = g(x_1, \ldots, x_{j-1}, v, x_{j+1}, \ldots, x_n),$$

$$(x_1, \ldots, x_n) \in A_1 \times \ldots \times A_{j-1} \times E_j \times A_{j+1} \times \ldots \times A_n, \qquad v \in E_j$$

(i.e., the j-th partial derivative of g associates to each

$$(x_1, \ldots, x_n) \in A_1 \times \ldots \times A_{j-1} \times E_j \times A_{j+1} \times \ldots \times A_n$$

the continuous linear map

$$E_j \longrightarrow F: v \longmapsto g(x_1, \ldots, x_{j-1}, v, x_{j+1}, \ldots, x_n),$$

which does not depend on x_j).

Remark A.2.3. For every $(i_1, \ldots, i_j) \in \{1, \ldots, k\}^j$ we define

$$\varepsilon^k_{i_1, \ldots, i_j}: E^k \longrightarrow E^j: (v_1, \ldots, v_k) \longmapsto (v_{i_1}, \ldots, v_{i_j}).$$

In what follows, $(i_1, \ldots, i_j)$ will be chosen to have pairwise distinct entries. We have $\varepsilon^k_{1, \ldots, k} = \mathrm{id}_{E^k}$.

Let us assume that the k-th derivatives of the maps $f: A_1 \times \ldots \times A_n \to F$ and $\alpha_i: A \to A_i$, $i = 1, \ldots, n$, exist, and write $\alpha := (\alpha_1, \ldots, \alpha_n)$. Applying Remark A.2.1 repeatedly, we obtain the *chain rule* for higher derivatives

$$D^k(f \circ \alpha) = \sum_{i=1}^{k} \sum_{(I_1,\dots,I_i)} D^i f \left(D^{|I_1|} \alpha \circ \varepsilon_{I_1}, \dots, D^{|I_i|} \alpha \circ \varepsilon_{I_i} \right), \qquad (A.9)$$

where the dependencies of the derivatives $D^{|I_j|}\alpha$ on $x \in A$ and of the derivatives $D^i f$ on $\alpha(x) = (\alpha_1(x),\dots,\alpha_n(x))$ are suppressed, and where the inner sum is taken over all partitions

$$I_1 \uplus \dots \uplus I_i = \{1,\dots,k\}$$

with $\min I_l < \min I_m$ for all $l < m$, and

$$\varepsilon_{I_j} := \varepsilon_{i_1,\dots,i_r}^k, \qquad I_j = \{i_1,\dots,i_r\}, \qquad i_1 < \dots < i_r$$

(where r depends on j). Formula (A.9) is a generalization of Faà di Bruno's formula, where $n = 1$ and $A, A_1 \subseteq \mathbb{R}$ (cf., e.g., [KP02, Sect. 1.3]).

Example A.2.4. For $k = 2$, the chain rule (A.9) specializes to

$$D^2(f \circ \alpha) = D^2 f(D\alpha \circ \varepsilon_1^2, D\alpha \circ \varepsilon_2^2) + Df(D^2\alpha \circ \varepsilon_{1,2}^2).$$

Remark A.2.5. Note that each (inner) summand on the right hand side of (A.9) is given by composing the j-th argument of $D^i f$ with a certain derivative of α, evaluated at the appropriate arguments of $D^k(f \circ \alpha)$. Moreover, the summand for $i = k$ involves the highest derivative of f which occurs in (A.9), and the arguments of this derivative are composed with the first derivative of α.

For use in Subsect. 3.2.2, we investigate more closely the summand for $i = k$ in the chain rule (A.9) in case $D\alpha$ is represented by a Jacobian matrix of square shape. First a general remark is made about the determinant of the k-th symmetric tensor power of a square matrix.

Remark A.2.6. Let V and W be vector spaces over $\mathbb{R}$ or $\mathbb{C}$ as above. Every linear map $\varphi \colon V \to W$ induces a linear map

$$\bigotimes^k \varphi \colon \bigotimes^k V \longrightarrow \bigotimes^k W \colon v_1 \otimes \dots \otimes v_k \longmapsto \varphi(v_1) \otimes \dots \otimes \varphi(v_k)$$

between the k-th tensor powers of V and W and via symmetrization a linear map

$$S^k \varphi \colon S^k V \longrightarrow S^k W$$

between the k-th symmetric tensor powers of V and W.

Let V and W be of finite dimension with bases $b_1, \dots, b_s$ and $b_1', \dots, b_t'$, respectively. We recall that if the $s \times t$ matrix M represents φ with respect to these bases, then the Kronecker product $M \otimes M$ represents $\varphi \otimes \varphi$ with respect to the bases $b_1 \otimes b_1, b_1 \otimes b_2, \dots, b_s \otimes b_{s-1}, b_s \times b_s$ and $b_1' \otimes b_1', b_1' \otimes b_2', \dots, b_t' \otimes b_{t-1}', b_t' \otimes b_t'$ of $V \otimes V$ and $W \otimes W$.

Moreover, let us assume that $s = t$. Then $S^k \varphi$ is of shape $\binom{s+k-1}{k} \times \binom{s+k-1}{k}$ and we have

$$\det(S^k \varphi) = \det(\varphi)^N, \qquad N := \binom{s+k-1}{s}. \tag{A.10}$$

On the Zariski dense subset of diagonalizable matrices with pairwise distinct eigenvalues this may be checked by writing the determinant as product of the eigenvalues. Symmetry implies that the degree of $\det(S^k \varphi)$ is the same in each of the s eigenvalues, namely

$$k \binom{s+k-1}{k} \bigg/ s = \frac{k}{s} \frac{(s+k-1)!}{k!\,(s-1)!} = \frac{(s+k-1)!}{(k-1)!\,s!} = \binom{s+k-1}{s}.$$

By continuity, formula (A.10) for the determinant of a k-th symmetric tensor power holds for all matrices.

Remark A.2.7. Let us assume that the k-th derivatives of $f : A_1 \times \ldots \times A_n \to F$ and $\alpha_i : A \to A_i$, $i = 1, \ldots, n$, exist and are continuous. We consider the map $D^k f$, which is k-linear and symmetric on $(E_1 \oplus \ldots \oplus E_n)^k$, as a map which is linear on the k-th symmetric tensor power $S^k(E_1 \oplus \ldots \oplus E_n)$.

If the vector spaces $E, E_1, \ldots, E_n$ are finite dimensional and if we have

$$s := \dim E_1 + \ldots + \dim E_n = n \cdot \dim E,$$

then the Jacobian matrix of α, which represents $D\alpha$, is of shape $s \times s$. If F is of finite dimension r, then the summand for $i = k$ in (A.9) may be represented as the product of two matrices of shape $r \times \binom{s+k-1}{k}$ and $\binom{s+k-1}{k} \times \binom{s+k-1}{k}$, respectively, the latter matrix representing the symmetrization of the k-th tensor power of the Jacobian matrix of α. By Remark A.2.6, the determinant of this symmetrization is the $\binom{s+k-1}{s}$-th power of the functional determinant of α.

Example A.2.8. Let $K \in \{\mathbb{R}, \mathbb{C}\}$ and let A be an open subset of $E := K^2$. Moreover, let $\alpha_i : A \to A_i$, $i = 1, 2$, and $f : A_1 \times A_2 \to F$ be maps, where A_i is an open subset of $E_i := K$, $i = 1, 2$, and $F := K$. Assume that the k-th derivatives of f and α_1, α_2 exist and are continuous. Then the summand for $i = k$ in the chain rule (A.9) may be represented as the product of two matrices of shape $1 \times (k+1)$ and $(k+1) \times (k+1)$, respectively. The determinant of the latter matrix equals the $\binom{k+1}{2}$-th power of the functional determinant of $\alpha = (\alpha_1, \alpha_2)$. Denoting the partial derivative $D_i D_j f$ by $D_{i,j} f$, the summand in question for $k = 2$ may be written as

$$\begin{pmatrix} D_{1,1}f & D_{1,2}f & D_{2,2}f \end{pmatrix} \begin{pmatrix} (D_1\alpha_1)^2 & (D_1\alpha_1)(D_2\alpha_1) & (D_2\alpha_1)^2 \\ 2(D_1\alpha_1)(D_1\alpha_2) & (D_1\alpha_1)(D_2\alpha_2)+(D_2\alpha_1)(D_1\alpha_2) & 2(D_2\alpha_1)(D_2\alpha_2) \\ (D_1\alpha_2)^2 & (D_1\alpha_2)(D_2\alpha_2) & (D_2\alpha_2)^2 \end{pmatrix}.$$

A.3 Differential Algebra

A.3.1 Basic Notions of Differential Algebra

In this subsection we recall a few basic definitions from differential algebra. Standard references are [Rit34], [Rit50], [Kol73], [Kol99], [Kap76].

A *differential ring* R is a ring together with a certain number of *derivations* acting on R, i.e., maps $\partial : R \to R$ satisfying

$$\partial(r_1 + r_2) = \partial(r_1) + \partial(r_2), \quad \partial(r_1 \cdot r_2) = r_1 \cdot \partial(r_2) + \partial(r_1) \cdot r_2, \qquad r_1, r_2 \in R.$$

Let $\partial_1, \ldots, \partial_n$ be the derivations of the differential ring R. We usually assume that each two of $\partial_1, \ldots, \partial_n$ commute, i.e., $\partial_i \circ \partial_j = \partial_j \circ \partial_i$ for all $i, j = 1, \ldots, n$. For $r \in R$ and $i \in \{1, \ldots, n\}$, the element $\partial_i(r)$ of R is called a *derivative* of r. An element of R all of whose derivatives are zero is called a *constant*.

An ideal of R that is closed under the action of $\partial_1, \ldots, \partial_n$ is called a *differential ideal*. The homomorphism theorem for rings holds in an analogous way for differential rings. A differential ring which is a field is called a *differential field*.

Examples of differential fields are fields of rational functions or fields of (formal or convergent) Laurent series in n variables, or fields of meromorphic functions on connected open subsets of $\mathbb{C}^n$, where in each case the n commuting derivations are given by partial differentiation with respect to the n variables. (Choosing all derivations to be zero yields a trivial differential structure for any field.)

Let R be a differential ring with derivations $\partial_1, \ldots, \partial_n$ and S a differential ring with derivations $\delta_1, \ldots, \delta_n$. Then S is called a *differential algebra over R* if S is an algebra over R and

$$\delta_i(r \cdot s) = \partial_i(r) \cdot s + r \cdot \delta_i(s), \qquad r \in R, \quad s \in S, \quad i = 1, \ldots, n.$$

Let S_1 and S_2 be differential algebras over R with derivations $\partial_1, \ldots, \partial_n$ on S_1 and $\delta_1, \ldots, \delta_n$ on S_2. A *homomorphism $\varphi : S_1 \to S_2$ of differential algebras over R* is a homomorphism of algebras over R which satisfies

$$\varphi \circ \partial_i = \delta_i \circ \varphi, \qquad i = 1, \ldots, n.$$

All algebras are assumed to be associative and unital; algebra homomorphisms map the multiplicative identity element to the multiplicative identity element.

We assume that K is a differential field of characteristic zero with commuting derivations $\partial_1, \ldots, \partial_n$. The *differential polynomial ring* $K\{u_1, \ldots, u_m\}$ over K in the *differential indeterminates* $u_1, \ldots, u_m$ is by definition the commutative polynomial algebra $K[(u_k)_J \mid 1 \leq k \leq m, J \in (\mathbb{Z}_{\geq 0})^n]$ with infinitely many, algebraically independent indeterminates $(u_k)_J$, also called *jet variables*, which represent the partial derivatives

$$\frac{\partial^{J_1+\ldots+J_n} U_k}{\partial z_1^{J_1} \ldots \partial z_n^{J_n}}, \qquad k = 1, \ldots, m, \quad J \in (\mathbb{Z}_{\geq 0})^n,$$

of smooth functions $U_1, \ldots, U_m$ of $z_1, \ldots, z_n$. We use u_k as a synonym for $(u_k)_{(0,\ldots,0)}$, $k = 1, \ldots, m$. The ring $K\{u_1, \ldots, u_m\}$ is considered as differential ring with commuting derivations $\delta_1, \ldots, \delta_n$ defined by extending

$$\delta_i u_k := (u_k)_{1_i}, \qquad i = 1, \ldots, n, \quad k = 1, \ldots, m,$$

additively, respecting the product rule of differentiation, and restricting to the derivation ∂_i on K. (Here 1_i denotes the multi-index $(0, \ldots, 0, 1, 0, \ldots, 0)$ of length n with 1 at position i.) More generally, the differential polynomial ring may be constructed with coefficients in a differential ring rather than in a differential field in the same way.

A jet variable $(u_k)_J$ has *(differential) order*

$$\mathrm{ord}((u_k)_J) := |J| := J_1 + \ldots + J_n, \qquad k = 1, \ldots, m, \quad J \in (\mathbb{Z}_{\geq 0})^n.$$

The *(differential) order* of a non-constant differential polynomial p is defined to be the maximum of the orders of jet variables occurring in p.

We also use an alternative notation for jet variables. Denoting, for simplicity, by x, y, z the first three coordinates and by u one differential indeterminate,

$$\underbrace{u_{x,\ldots,x}}_{i},\underbrace{{}_{,y,\ldots,y}}_{j},\underbrace{{}_{,z,\ldots,z}}_{k} \quad \text{and} \quad u_{x^i,y^j,z^k}$$

are used as synonyms for the jet variable $u_{(i,j,k)}$.

Let F and E be differential fields such that E is a differential algebra over F and $F \subseteq E$. Then $F \subseteq E$ is called a *differential field extension*. A family of elements e_i, $i \in I$, of E, where I is an index set, is said to be *differentially algebraically independent over F* if the family of all derivatives of e_i, $i \in I$, is algebraically independent over F (which is a condition that does not involve the differential structure of F).

The differential polynomial ring $K\{u_1, \ldots, u_m\}$ is the free differential algebra over K generated by $u_1, \ldots, u_m$, in the sense that $u_1, \ldots, u_m$ are differentially algebraically independent over K, which gives rise to the following universal property of $K\{u_1, \ldots, u_m\}$. Let A be any differential algebra over K. For any $a_1, \ldots, a_m \in A$ there exists a unique homomorphism $\varphi \colon K\{u_1, \ldots, u_m\} \to A$ of differential algebras over K satisfying $\varphi(u_k) = a_k$, $k = 1, \ldots, m$.

Let R be a differential ring of characteristic zero and I a differential ideal of R. Then the radical of I, i.e.,

$$\sqrt{I} := \{ r \in R \mid r^e \in I \text{ for some } e \in \mathbb{N} \},$$

is a differential ideal of R. A differential ideal which equals its radical is said to be *radical* (also called a *perfect differential ideal*). For any subset G of R, the *radical differential ideal generated by* G is defined to be the smallest radical differential ideal of R which contains G. For the importance of radical differential ideals, cf. Subsect. A.3.4.

The differential polynomial ring $K\{u_1,\ldots,u_m\}$ contains the polynomial algebra $K[u_1,\ldots,u_m]$. Kolchin's Irreducibility Theorem (cf. [Kol73, Sect. IV.17, Prop. 10]) states that the radical differential ideal of $K\{u_1,\ldots,u_m\}$ which is generated by a prime ideal of $K[u_1,\ldots,u_m]$ is a prime differential ideal. However, the radical differential ideal generated by an (algebraically) irreducible differential polynomial (involving proper derivatives) may not be prime, as the example of Clairaut's equation (cf. also [Inc56, pp. 39-40])

$$u - x u_x - f(u_x) = 0, \quad \text{e.g., with} \quad f(u_x) = -\frac{1}{4}u_x^2,$$

shows (cf. [Kol99, p. 575]).

A.3.2 Characteristic Sets

The basic notions of the theory of characteristic sets (cf., e.g., [Rit50], [Wu00]) are recalled in this section; cf. also [ALMM99], [Hub03a, Hub03b], [Wan01]. Differential algebra in general and the methods introduced by J. M. Thomas (cf. Sect. 2.2) in particular, make extensive use of the concepts which are discussed below.

Let K be a differential field of characteristic zero with n commuting derivations.

Definition A.3.1. Let $R := K\{u_1,\ldots,u_m\}$ be the differential polynomial ring with commuting derivations $\partial_1, \ldots, \partial_n$, let $\Delta := \{\partial_1,\ldots,\partial_n\}$, and denote by $\mathrm{Mon}(\Delta)$ the (commutative) monoid of monomials in $\partial_1, \ldots, \partial_n$. A *ranking* $>$ on R is a total ordering on

$$\mathrm{Mon}(\Delta)u := \{\, (u_k)_J \mid 1 \le k \le m, J \in (\mathbb{Z}_{\geq 0})^n \,\} = \{\, \partial^J u_k \mid 1 \le k \le m, J \in (\mathbb{Z}_{\geq 0})^n \,\}$$

which satisfies the following two conditions.

a) For all $1 \le k \le m$ and all $1 \le j \le n$ we have $\partial_j u_k > u_k$.
b) For all $1 \le k_1, k_2 \le m$ and all $J_1, J_2 \in (\mathbb{Z}_{\geq 0})^n$ we have

$$(u_{k_1})_{J_1} > (u_{k_2})_{J_2} \quad \Longrightarrow \quad \partial_j (u_{k_1})_{J_1} > \partial_j (u_{k_2})_{J_2} \qquad \text{for all } j = 1,\ldots,n.$$

A ranking $>$ on $K\{u_1,\ldots,u_m\}$ is said to be *orderly* if

$$|J_1| > |J_2| \quad \Longrightarrow \quad (u_{k_1})_{J_1} > (u_{k_2})_{J_2} \qquad \text{for all } 1 \le k_1, k_2 \le m, J_1, J_2 \in (\mathbb{Z}_{\geq 0})^n.$$

Remark A.3.2. Every ranking is a well-ordering, i.e., every non-empty subset of $\mathrm{Mon}(\Delta)u$ has a least element. Equivalently, every descending sequence of elements of $\mathrm{Mon}(\Delta)u$ terminates.

Example A.3.3. Let $R := K\{u\}$ be the differential polynomial ring with one differential indeterminate u and commuting derivations $\partial_1, \ldots, \partial_n$. A ranking $>$ on R which is analogous to the degree-reverse lexicographical ordering of monomials (cf. Ex. 2.1.27, p. 23) is defined for jet variables u_J, $u_{J'}$, $J = (j_1, \ldots, j_n)$, $J' = (j'_1, \ldots, j'_n) \in (\mathbb{Z}_{\geq 0})^n$, by

$$u_J > u_{J'} \quad :\Longleftrightarrow \quad \begin{cases} j_1 + \ldots + j_n > j'_1 + \ldots + j'_n \quad \text{or} \\[2mm] (\ j_1 + \ldots + j_n = j'_1 + \ldots + j'_n \quad \text{and} \quad J \neq J' \quad \text{and} \\[2mm] j_i < j'_i \quad \text{for} \quad i = \max\{1 \leq k \leq n \mid j_k \neq j'_k\}\ ). \end{cases}$$

In this case the ranking of the first order jet variables is given by

$$\partial_1 u > \partial_2 u > \ldots > \partial_n u.$$

The ranking can be extended to more than one differential indeterminate in ways analogous to the term-over-position or position-over-term orderings (cf. Ex. 2.1.28, p. 23). In the former case the ranking is orderly, in the latter case it is not.

In what follows, we fix a ranking $>$ on $R = K\{u_1, \ldots, u_m\}$.

Remark A.3.4. With respect to the chosen ranking, for any non-constant differential polynomial $p \in R - K$ the *leader* of p is defined to be the greatest jet variable with respect to $>$ which occurs in p. It is denoted by $\mathrm{ld}(p)$. Any such differential polynomial p may be viewed as a univariate polynomial in $\mathrm{ld}(p)$ and its coefficients may be considered recursively in the same way (if not constant).

We recall the pseudo-reduction process of differential polynomials.

Definition A.3.5. Let $p \in R$ and $q \in R - K$. The differential polynomial p is said to be *partially reduced with respect to* q if no proper derivative of $\mathrm{ld}(q)$ occurs in p. It is said to be *reduced with respect to* q if it is constant or if it is partially reduced with respect to q and $\deg_x(p) < \deg_x(q)$ for $x := \mathrm{ld}(q)$. A subset S of R is said to be *auto-reduced* if $S \cap K = \emptyset$ and for every $p, q \in S$, $p \neq q$, the differential polynomial p is reduced with respect to q.

Remarks A.3.6. Let $p \in R$ and $q \in R - K$.

a) If p is not partially reduced with respect to q, then Euclidean pseudo-division transforms p into a differential polynomial p' which has this property. There exist a jet variable v which occurs in p and a monomial $\theta \in \mathrm{Mon}(\Delta)$, $\theta \neq 1$, such that $v = \theta\,\mathrm{ld}(q)$. By the defining property b) of a ranking (cf. Def. A.3.1), the leader of θq is v. The rules of differentiation imply that the degree of θq as a polynomial in v is one, and the coefficient of v in θq equals the partial derivative of q with respect to $\mathrm{ld}(q)$, which is called the *separant* of q, denoted by $\mathrm{sep}(q)$. Let d be the degree of v in p and c the coefficient of v^d in p. Then

$$\mathrm{sep}(q) \cdot p - c \cdot v^{d-1} \cdot \theta q$$

eliminates v^d from p. Iteration of this pseudo-division yields, after finitely many steps, a differential polynomial p' which is partially reduced with respect to q.

b) If p is partially reduced, but not reduced with respect to q, then pseudo-division can also be used to produce a differential polynomial p' which is reduced with respect to q. Suppose that $v := \mathrm{ld}(p) = \mathrm{ld}(q)$, and let $d := \deg_v(p)$, $d' := \deg_v(q)$. Then we have $d \geq d'$. The *initial* of p is defined to be the coefficient of v^d in p and denoted by $\mathrm{init}(p)$. Similarly, $\mathrm{init}(q)$ is the coefficient of $v^{d'}$ in q. Then

$$\mathrm{init}(q) \cdot p - \mathrm{init}(p) \cdot v^{d-d'} \cdot q$$

eliminates v^d from p. Again, iteration of this pseudo-division yields, after finitely many steps, a differential polynomial p' which is reduced with respect to q. The same method can be applied recursively to each coefficient of p, if p is not reduced with respect to q and $\mathrm{ld}(p) \neq \mathrm{ld}(q)$.

c) Let $S \subset R - K$ be finite. We apply differential and algebraic reductions (as in a) and b)) to pairs (p,q) of distinct elements of S and after each reduction, we replace p with p' in S if $p' \neq 0$ and remove p from S otherwise. Each reduction step transforms a polynomial p into another one called p' which either has smaller degree in $\mathrm{ld}(p)$ if $\mathrm{ld}(p') = \mathrm{ld}(p)$ or has a leader which is smaller than $\mathrm{ld}(p)$ with respect to the ranking $>$. Since degrees can decrease only finitely many times and since $>$ is a well-ordering, we obtain after finitely many steps either a subset of R which contains a non-zero constant or an auto-reduced subset of R.

d) Both the differential and the algebraic reduction perform a pseudo-division in the sense that p is multiplied by a possibly non-constant polynomial. If p originates from a set of differential polynomials representing a system of partial differential equations, then replacing p with the pseudo-remainder p' in this set may lead to an inequivalent system. If $\mathrm{init}(q)$ and $\mathrm{sep}(q)$ are non-zero when evaluated at any solution of the system, then the replacement of p with the differential polynomial p' computed by the above methods does not change the solution set. Assuming that this condition holds for every initial and separant which is used for pseudo-division, the process in c) computes for a given finite system of partial differential equations an equivalent one which either is recognized to be inconsistent (because of a non-zero constant left hand side) or is auto-reduced.

Remark A.3.7. Every auto-reduced subset of R is finite. This is due to the facts that the elements of an auto-reduced subset of R have pairwise distinct leaders and that every sequence of such leaders in which no element is a derivative of a previous one is finite.

Ritt introduced the following binary relation on the set of all auto-reduced subsets of R.

Definition A.3.8. Let p, $q \in R - K$. The differential polynomial p is said to have *higher rank* than q if either $\mathrm{ld}(p) > \mathrm{ld}(q)$, or if we have $\mathrm{ld}(p) = \mathrm{ld}(q) =: x$ and $\deg_x(p) > \deg_x(q)$. If $\mathrm{ld}(p) = \mathrm{ld}(q) =: x$ and $\deg_x(p) = \deg_x(q)$, then p and q are said to have *the same rank*.

Definition A.3.9. Let $A = \{p_1,\ldots,p_s\}$, $B = \{q_1,\ldots,q_t\} \subseteq R - K$ be auto-reduced sets. We assume that p_{i+1} has higher rank than p_i for all $i = 1, \ldots, s-1$ and that q_{i+1} has higher rank than q_i for all $i = 1, \ldots, t-1$. Then A is said to have *higher rank* than B if one of the following two conditions holds.

a) There exists $j \in \{1,\ldots,\min(s,t)\}$ such that p_i and q_i have the same rank for all $i = 1, \ldots, j-1$ and p_j has higher rank than q_j.
b) We have $s < t$, and p_i and q_i have the same rank for all $i = 1, \ldots, s$.

Remark A.3.10. In every non-empty set $\mathscr{A}$ of auto-reduced subsets of R there exists one which does not have higher rank than any other auto-reduced set in $\mathscr{A}$. Each such auto-reduced set is also said to be of *lowest rank* among those in $\mathscr{A}$. Assuming that each auto-reduced set is sorted as in the previous definition, the existence follows from Remark A.3.7 by considering first those sets in $\mathscr{A}$ whose first elements do not have higher rank than any other first element of sets in $\mathscr{A}$, considering then, if necessary, among these sets the ones whose second elements do not have higher rank than any other second element of these sets and so on.

Definition A.3.11. Let I be a differential ideal of R, $I \neq R$, and define $\mathscr{A}$ to be the set of auto-reduced subsets A of I which satisfy that the separant of each element of A is not an element of I. An auto-reduced subset A of I in $\mathscr{A}$ of lowest rank is called a *characteristic set* of I.

Remark A.3.12. Let I be a differential ideal of R, $I \neq R$. By Remark A.3.10, a characteristic set A of I exists. The definition of a characteristic set implies that the only element of I which is reduced with respect to every element of A is the zero polynomial.

We denote by $\langle A \rangle$ the differential ideal of R which is generated by A, and we define the product q of the initials and separants of all elements of A. Moreover, the *saturation* of $\langle A \rangle$ with respect to q is defined by

$$\langle A \rangle : q^\infty := \{\, p \in R \mid q^r \cdot p \in \langle A \rangle \text{ for some } r \in \mathbb{Z}_{\geq 0}\,\}.$$

Then we have

$$\langle A \rangle \subseteq I \subseteq \langle A \rangle : q^\infty.$$

If A is a characteristic set of $\langle A \rangle : q^\infty$, then not only is zero the unique element of $\langle A \rangle : q^\infty$ which is reduced with respect to every element of A, but membership to the ideal $\langle A \rangle : q^\infty$ can be decided by applying A in the pseudo-reduction process discussed in Remarks A.3.6.

Let P be a prime differential ideal of R and A a characteristic set of P (with respect to the chosen ranking on R). Then we have $P = \langle A \rangle : q^\infty$, where q is the product of the initials and separants of all elements of A. In particular, no initial and no separant of any element of A is an element of P.

It is not known how one could decide effectively whether a prime differential ideal is contained in another one or not (cf. also [Kol73, Sect. IV.9]).

A.3.3 Homogeneous Differential Polynomials

In this subsection we collect a few results about differential ideals which are generated by differential polynomials of a special kind, namely homogeneous differential polynomials, i.e., those for which each term has the same degree, and isobaric differential polynomials, i.e., those for which each term has the same total number of differentiations.

Let Ω be an open and connected subset of $\mathbb{C}^n$ with coordinates $z_1, \ldots, z_n$, and let K be the differential field of meromorphic functions on Ω with commuting derivations $\partial_{z_1}, \ldots, \partial_{z_n}$ that are defined by partial differentiation with respect to the coordinates $z_1, \ldots, z_n$, respectively.

Let $R := K\{u_1, \ldots, u_m\}$ be the differential polynomial ring in the differential indeterminates $u_1, \ldots, u_m$ with commuting derivations which restrict to the derivations $\partial_{z_1}, \ldots, \partial_{z_n}$ on K and which we again denote by $\partial_{z_1}, \ldots, \partial_{z_n}$. We endow R with the standard grading, i.e., each jet variable $(u_k)_J$, $k = 1, \ldots, m$, $J \in (\mathbb{Z}_{\geq 0})^n$, is homogeneous of degree 1. Of course, every homogeneous component of $K\{u_1, \ldots, u_m\}$ of degree greater than zero is an infinite dimensional K-vector space.

For notational convenience we restrict our attention now to differential polynomial rings in one differential indeterminate u, i.e., $R = K\{u\}$; the results of this subsection can easily be generalized to several differential indeterminates.

For any set S of analytic functions on Ω we define the *vanishing ideal of S* by

$$\mathscr{I}_R(S) := \{\, p \in R \mid p(f) = 0 \text{ for all } f \in S \,\},$$

where $p(f)$ is obtained from p by substitution of f for u and of the partial derivatives of f for the corresponding jet variables in u. For any differential ideal I of $K\{u\}$ we denote by $\mathrm{Sol}_\Omega(I)$ the set of analytic functions on Ω that are solutions of the system of partial differential equations $\{\, p = 0 \mid p \in I \,\}$.

Definition A.3.13. A differential ideal of $K\{u\}$ is said to be *homogeneous*, if it is generated by homogeneous differential polynomials, i.e., differential polynomials that are homogeneous with respect to the standard grading of $K\{u\}$ (not necessarily of the same degree).

Lemma A.3.14. *Let S be a set of analytic functions on Ω having the property that for each f in S the function $c \cdot f$ also is in S for all c in some infinite subset of $\mathbb{C}$. Then the differential ideal $\mathscr{I}_R(S)$ of $R = K\{u\}$ is homogeneous. Conversely, if I is a homogeneous differential ideal of $K\{u\}$, then for every $f \in \mathrm{Sol}_\Omega(I)$ we have $c \cdot f \in \mathrm{Sol}_\Omega(I)$ for all $c \in \mathbb{C}$.*

Proof. For every $c \in \mathbb{C}$ and every analytic function f on Ω, the result of substituting $c \cdot f$ into any non-constant differential monomial in u differs from the one for f exactly by the factor c^d, where d is the total degree of the monomial. Moreover, for every $p \in K\{u\}$ we may consider $p(c \cdot u)$ as a polynomial in c with coefficients in $K\{u\}$. Then, for any fixed analytic function f on Ω, the equality $p(c \cdot f) = 0$ for infinitely many $c \in \mathbb{C}$ implies that $p(c \cdot f)$ is the zero polynomial. Therefore, the assumption on S implies that the homogeneous components of a differential polynomial which annihilates S are annihilating polynomials as well. The statement claiming the converse also follows easily. $\square$

Example A.3.15. Let $S = \{c \cdot \exp(z) \mid c \in \mathbb{C}\}$. Then $\mathscr{I}_R(S)$ is generated by $u_z - u$ and hence is homogeneous.

Remark A.3.16. The notion of homogeneity depends on the chosen coordinate system. Expressed in terms of jet coordinates, homogeneity is not invariant under coordinate changes of the dependent variables. If we introduce in the previous example a coordinate v which is related to u by $u = v + 1$, then $u_z - u$ is transformed into the non-homogeneous differential polynomial $v_z - v - 1$, which annihilates exactly the set of analytic functions $\{c \cdot \exp(z) - 1 \mid c \in \mathbb{C}\}$.

Definition A.3.17. We denote by $L = \mathbb{C}((z_1, \ldots, z_n))$ the field of formal Laurent series in $z_1, \ldots, z_n$. Let $L\{u\}$ be the differential polynomial ring with commuting derivations $\partial_{z_1}, \ldots, \partial_{z_n}$ that act on L by partial differentiation.

a) A differential polynomial $p \in L\{u\}$ is said to be[1] *isobaric of weight d* if $p = 0$ or if it can be written as $p = \sum_{i=1}^{r} a_i c_i m_i$ for some $r \in \mathbb{N}$, where a_i are non-zero complex numbers, c_i are Laurent monomials in $z_1, \ldots, z_n$, and m_i are non-constant differential monomials in $L\{u\}$, such that $\mathrm{ord}(m_i) - \deg(c_i)$ equals d for all $i = 1, \ldots, r$. (In this case, each coefficient of p in L is a homogeneous Laurent polynomial.)

b) Let K be the field of meromorphic functions on Ω as above and let $w \in \Omega$. A differential polynomial $p \in K\{u\}$ is said to be *isobaric of weight d around w* if the coefficient-wise expansion around w of its representation as sum of terms is isobaric of weight d as differential polynomial in $\mathbb{C}((y_1, \ldots, y_n))\{u\}$, where $y_i := z_i - w_i$, $i = 1, \ldots, n$.

A differential ideal of $L\{u\}$ (or $K\{u\}$) is said to be *isobaric (around $w \in \Omega$)* if it is generated by differential polynomials which are isobaric (around w). (The generators need not be isobaric of the same weight.)

Example A.3.18. Every differential monomial m of $K\{u\}$ is an isobaric differential polynomial of weight $\mathrm{ord}(m)$ (around any point). The differential polynomials $z_2 u_{z_1, z_2} + u_{z_1}$ and $u_{z_1} + (\frac{1}{z_1} + \frac{1}{z_2}) u$ in $\mathbb{C}((z_1, z_2))\{u\}$ are isobaric of weight 1.

Remark A.3.19. Under a change of coordinates of Ω an isobaric differential polynomial is not necessarily transformed to an isobaric one. For instance, let Ω be a simply connected open subset of $\mathbb{C} - \{0\}$ with coordinate z. Under the coordinate transformation defined by $\exp(\tilde{z}) = z$, the differential polynomial u_z, which is isobaric of weight 1 around any point in Ω, is mapped to $\exp(-\tilde{z}) u_{\tilde{z}}$, which is not isobaric. However, as another example, if $\Omega = \mathbb{C}$ is chosen with coordinate z, then the

---

[1] The notion of isobaric differential polynomial differs from that defined in [Kol73, Sect. I.7] (cf. also [GMO91]) inasmuch as degrees of formal Laurent series coefficients are taken into account.

translation $\tilde{z}+1 = z$ transforms the differential polynomial $z^d\,\partial_z^k u$, $d \in \mathbb{Z}$, $k \in \mathbb{Z}_{\geq 0}$, which is isobaric of weight $k - d$ around 0, into $(\tilde{z}+1)^d\,\partial_{\tilde{z}}^k u$, which is isobaric of the same weight around -1.

We define
$$X(\Omega) := \{c \in \mathbb{C} \mid c \cdot z \in \Omega \text{ for all } z \in \Omega\}.$$

For any analytic function f on Ω and any $c \in X(\Omega)$ we denote by $f(c \cdot -)$ the analytic function on Ω defined by

$$f(c \cdot -)(z_1,\dots,z_n) := f(c \cdot z_1,\dots,c \cdot z_n), \qquad (z_1,\dots,z_n) \in \Omega.$$

Lemma A.3.20. *Assume that $X(\Omega)$ is an infinite set and $0 \in \Omega$. Let S be a set of analytic functions on Ω having the property that for each f in S the function $f(c \cdot -)$ also is in S for all c in some infinite subset of $X(\Omega)$. Then the differential ideal $\mathscr{I}_R(S)$ of $R = K\{u\}$ is isobaric around 0. Conversely, if I is a differential ideal of $K\{u\}$ that is isobaric around 0, then for every $f \in \mathrm{Sol}_\Omega(I)$ we have $f(c \cdot -) \in \mathrm{Sol}_\Omega(I)$ for all $c \in X(\Omega)$.*

Proof. The chain rule implies that for every jet variable $u_J \in K\{u\}$, $J \in (\mathbb{Z}_{\geq 0})^n$, and every analytic function f on Ω we have $u_J(f(c \cdot -)) = c^{|J|} \cdot u_J(f)(c \cdot -)$ for all $c \in X(\Omega)$. We may now argue in the same way as in the proof of Lemma A.3.14. A differential polynomial $p \in K\{u\}$ is isobaric of weight d around 0 if and only if for every analytic function f on Ω we have $p(f(c \cdot -)) = c^d \cdot p(f)(c \cdot -)$ for infinitely many $c \in X(\Omega)$. $\qquad\square$

Examples A.3.21. Let x, y be coordinates of $\Omega = \mathbb{C}^2$, K the field of meromorphic functions on Ω, and $R = K\{u\}$ as above. Then we have $X(\Omega) = \mathbb{C}$.

a) Let $S = \{f_1(x^2) + f_2(xy) \mid f_1, f_2 \text{ analytic}\}$. Then the vanishing ideal $\mathscr{I}_R(S)$ is isobaric (around 0) because $\mathscr{I}_R(S)$ is generated by the differential polynomial

$$x u_{x,y} - y u_{y,y} - u_y,$$

which is isobaric (around 0) of weight 1. Note that $\mathscr{I}_R(S)$ is also generated by

$$u_{x,y} - \frac{y}{x} u_{y,y} - \frac{1}{x} u_y,$$

which is an isobaric differential polynomial of weight 2. (We refer to Sect. 3.2 for more details on methods which determine the vanishing ideal of such a set S.)

b) Let $S = \{f_1(x+2y) + f_2(x+y) \cdot e^x \mid f_1, f_2 \text{ analytic}\}$. Then the vanishing ideal $\mathscr{I}_R(S)$ is not isobaric (around 0) because it is generated by the differential polynomial

$$2 u_{x,x} - 3 u_{x,y} + u_{y,y} - 2 u_x + u_y,$$

and the zero polynomial is therefore the only isobaric element of $\mathscr{I}_R(S)$.

A.3.4 Basis Theorem and Nullstellensatz

We recall two fundamental theorems of differential algebra, which are analogs of Hilbert's Basis Theorem and Hilbert's Nullstellensatz in commutative algebra. For more details, we refer to [Rit34, §§ 77–85], [Rau34], [Rit50, Sects. I.12–14, II.7–11, IX.27], [Sei56], [Kol73, Sect. III.4], [Kap76, § 27], [Kol99, pp. 572–583].

Let K be a differential field of characteristic zero and $R := K\{u_1,\ldots,u_m\}$ the differential polynomial ring in the differential indeterminates $u_1, \ldots, u_m$ with commuting derivations $\partial_1, \ldots, \partial_n$.

There are differential ideals of R which are not finitely generated, as the example of the differential ideal generated by the infinitely many differential monomials

$$(\partial u)(\partial^2 u), \quad (\partial^2 u)(\partial^3 u), \quad \ldots, \quad (\partial^k u)(\partial^{k+1} u), \quad \ldots$$

shows, where $n = 1, m = 1, u = u_1, \partial = \partial_1$; cf. [Rit34, p. 12]. However, the following theorem shows that radical differential ideals admit a representation in terms of a finite generating set.

Theorem A.3.22 (Basis Theorem of Ritt-Raudenbush). *Every radical differential ideal of R is finitely generated.*

For the sake of completeness, we include a proof following [Kap76, § 27].

Proof. The theorem is proved by induction on m. The statement is trivial for $m = 0$. We assume now that $m > 0$ and that the statement is true for smaller values of m, and we consider $K\{u_1,\ldots,u_{m-1}\}$ as a differential subring of $R = K\{u_1,\ldots,u_m\}$.

Let us assume that there exists a radical differential ideal of R which is not finitely generated and derive a contradiction. The non-empty set of such differential ideals is partially ordered by set inclusion and every totally ordered subset has an upper bound given by the union of its elements. Therefore, by Zorn's Lemma, there exists a maximal element I of that set.

We show that I is prime. Assuming the contrary, there exist $p, q \in R$ such that $pq \in I$, but $p \notin I$ and $q \notin I$. Then the radical differential ideals I_1 and I_2 of R which are generated by I and p and by I and q, respectively, are finitely generated, say, by $G_1 = \{p,p_1,\ldots,p_s\}$ and $G_2 = \{q,q_1,\ldots,q_t\}$, respectively, where we may assume that $p_1, \ldots, p_s, q_1, \ldots, q_t \in I$. Then $I_1 \cdot I_2$ is contained in the radical differential ideal which is generated by $\{g_1 \cdot g_2 \mid g_1 \in G_1, g_2 \in G_2\}$. On the other hand, the latter ideal contains I because $I \subseteq I_1$ and $I \subseteq I_2$ imply that it contains the square of every element of I and it is a radical ideal. Since G_1 and G_2 are finite, we conclude that I is finitely generated as a radical differential ideal, which is a contradiction.

Let J be the radical differential ideal of $R = K\{u_1,\ldots,u_m\}$ which is generated by $I \cap K\{u_1,\ldots,u_{m-1}\}$. Since the radical differential ideal $I \cap K\{u_1,\ldots,u_{m-1}\}$ of $K\{u_1,\ldots,u_{m-1}\}$ is finitely generated by the induction hypothesis, J is also finitely generated. By the assumption on I, the ideal J is a proper subset of I.

We choose an arbitrary ranking $>$ on R (cf. Def. A.3.1). Among the differential polynomials in $I - J$ with least leader v with respect to $>$ let r be one of least degree

d in v. Then we have $\mathrm{init}(r) \notin J$ and $\mathrm{sep}(r) \notin J$ because otherwise $r - \mathrm{init}(r)\,v^d$ or $r - \frac{1}{d}\,\mathrm{sep}(r)\,v$ would be either zero or in $I - J$, both possibilities being contradictions to the choice of r. Since $\mathrm{init}(r)$ and $\mathrm{sep}(r)$ have leader smaller than v with respect to $>$ or degree in v less than d and since both differential polynomials are not in J, we also have $\mathrm{init}(r) \notin I$ and $\mathrm{sep}(r) \notin I$. This implies $\mathrm{init}(r) \cdot \mathrm{sep}(r) \notin I$ because I is a prime ideal. By the choice of I, the radical differential ideal H of R which is generated by I and $\mathrm{init}(r) \cdot \mathrm{sep}(r)$ is finitely generated, say, by the differential polynomials $\mathrm{init}(r) \cdot \mathrm{sep}(r), h_1, \ldots, h_k$, where we may assume that $h_1, \ldots, h_k \in I$.

We claim that $\mathrm{init}(r) \cdot \mathrm{sep}(r) \cdot I$ is contained in the radical differential ideal L of R which is generated by J and r. In fact, for every element $a \in I$, the pseudo-reduction process described in Remarks A.3.6 yields an element $b \in I$ of the form $b = \mathrm{init}(r)^i \cdot \mathrm{sep}(r)^j \cdot a - c \cdot r$ for some $i, j \in \mathbb{Z}_{\geq 0}$ and $c \in R$ such that b is reduced with respect to r. Then we have $b \in J$ and $\mathrm{init}(r)^{\max(i,j)} \cdot \mathrm{sep}(r)^{\max(i,j)} \cdot a^{\max(i,j)} \in L$ and therefore $\mathrm{init}(r) \cdot \mathrm{sep}(r) \cdot a \in L$, which proves the claim.

Finally, we obtain a contradiction to the choice of I by showing that I is equal to the radical differential ideal of R which is generated by L and $h_1, \ldots, h_k$. The latter ideal is finitely generated as a radical differential ideal because J and L are so. Clearly it is contained in I. Conversely, the square of every element of I is contained in $H \cdot I$ by the definition of H, hence in the radical differential ideal of R which is generated by $\mathrm{init}(r) \cdot \mathrm{sep}(r) \cdot I, h_1 \cdot I, \ldots, h_k \cdot I$, and therefore in the one generated by L and $h_1, \ldots, h_k$. This proves the reverse inclusion. $\square$

The following theorem can be found, e.g., in [Kap76, § 29], [Rit50, Sect. II.3], [Rau34, Thms. 5 and 6].

Theorem A.3.23. *Every radical differential ideal of R is an intersection of finitely many prime differential ideals. A minimal representation as such an intersection (i.e., one in which none of the prime differential ideals is contained in another one) is uniquely determined up to reordering of the components.*

Proof. Let us assume that there exists a radical differential ideal I of R which has no representation as intersection of finitely many prime differential ideals and derive a contradiction. By the Basis Theorem of Ritt-Raudenbush (Theorem A.3.22), every ascending chain of radical differential ideals of R terminates. Hence, we may assume that I is chosen to be maximal among those radical differential ideals of R having the above property.

Since I is not prime, there exist $p, q \in R$ such that $pq \in I$, but $p \notin I$ and $q \notin I$. Then the radical differential ideals I_1 and I_2 of R which are generated by I and p and by I and q, respectively, are intersections of finitely many prime differential ideals. Moreover, $I_1 \cdot I_2$ is contained in the radical differential ideal which is generated by I and pq, which is equal to I. Therefore, I contains the square of every element of $I_1 \cap I_2$. Since I is radical, this implies $I_1 \cap I_2 \subseteq I$, and then we conclude from $I \subseteq I_1$ and $I \subseteq I_2$ that we have $I = I_1 \cap I_2$. Hence, I has a representation as intersection of finitely many prime differential ideals, which is a contradiction.

Let $P_1 \cap \ldots \cap P_r = Q_1 \cap \ldots \cap Q_s$ be two minimal representations of a radical differential ideal of R as intersection of prime differential ideals. Then we have

$P_1 \cap \ldots \cap P_r \subseteq Q_1$. Since Q_1 is prime, there exists $i \in \{1, \ldots, r\}$ such that $P_i \subseteq Q_1$. On the other hand, we have $Q_1 \cap \ldots \cap Q_s \subseteq P_i$ and, therefore, $Q_j \subseteq P_i$ for some $j \in \{1, \ldots, s\}$. By the minimality of the representation $Q_1 \cap \ldots \cap Q_s$, we have $j = 1$. An iteration of this argument shows that we have $\{P_1, \ldots, P_r\} = \{Q_1, \ldots, Q_s\}$. $\quad\square$

Let Ω be a connected open subset of $\mathbb{C}^n$ and denote by K the differential field of meromorphic functions on Ω. Analytic solutions of systems of differential equations with coefficients in K may have domains of definition which are properly contained in Ω. For more details, we refer to [Rit50, Sects. II.7–11, IX.27], [Rau34, Thm. 9].

Theorem A.3.24 (Nullstellensatz for Analytic Functions). *Let $p_1, \ldots, p_s \in R$ and I the differential ideal of R they generate. Moreover, let $q \in R$ be a differential polynomial which vanishes for all analytic solutions of I. Then some power of q is an element of I.*

Proof (Sketch). The assertion is equivalent to the following statement. If $q \in R$ is not an element of the radical differential ideal $\sqrt{I}$ which is generated by $p_1, \ldots, p_s$, then there exists an analytic solution of $p_1 = 0, \ldots, p_s = 0$ which is not a solution of $q = 0$.

Let $\sqrt{I} = P_1 \cap \ldots \cap P_r$ be a representation of $\sqrt{I}$ as intersection of prime differential ideals (cf. Thm. A.3.23). Since we assume that $q \notin \sqrt{I}$, there exists $j \in \{1, \ldots, r\}$ such that $q \notin P_j$. Let A be a characteristic set of P_j (with respect to a chosen Riquier ranking on R) defining a passive differential system (i.e., one which incorporates all integrability conditions, cf. also the introduction to Sect. 2.1). It is enough to show the existence of an analytic solution of this differential system for which q does not vanish.

Note that no initial and no separant of any element of A is contained in P_j. We consider A as a system of algebraic equations (in finitely many variables) for analytic functions, neglecting the differential relationships among the jet variables. By a version of Hilbert's Nullstellensatz for analytic functions (cf. [Rit50, Sects. IV.13–14]), a solution of this system exists for which neither q nor any initial or separant of elements of A vanishes. A point $w \in \Omega$ may be chosen such that all coefficients of q and all coefficients of all equations defined by A are analytic around w and such that w is not a zero of the evaluation of q at the above solution or those of the initials and separants of elements of A. The equations given by A are solved for their leaders and now considered as differential equations with boundary values at $z = w$ determined by evaluating the above (algebraic) solution at w. If a differential indeterminate occurs in an equation, but none of its derivatives is a leader of an equation, then it is replaced with an arbitrary analytic function taking the value at w which is prescribed by the above (algebraic) solution. A solution of this boundary value problem exists by Riquier's Existence Theorem. By construction, it is not a solution of q. $\quad\square$

References

[ALMM99] Aubry, P., Lazard, D., Moreno Maza, M.: On the theories of triangular sets. J. Symbolic Comput. **28**(1-2), 105–124 (1999). Polynomial elimination—algorithms and applications

[Ama90] Amasaki, M.: Application of the generalized Weierstrass preparation theorem to the study of homogeneous ideals. Trans. Amer. Math. Soc. **317**(1), 1–43 (1990)

[Ape98] Apel, J.: The theory of involutive divisions and an application to Hilbert function computations. J. Symbolic Comput. **25**(6), 683–704 (1998)

[Arn63] Arnol'd, V.I.: Representation of continuous functions of three variables by the superposition of continuous functions of two variables. Amer. Math. Soc. Transl. (2) **28**, 61–147 (1963)

[Bäc14] Bächler, T.: Counting Solutions of Algebraic Systems via Triangular Decomposition. Ph.D. thesis, RWTH Aachen University, Germany (2014). Available online at http://darwin.bth.rwth-aachen.de/opus/volltexte/2014/5104

[Bak10] Baker, H.F.: Principles of geometry. Volume 3. Solid geometry. Cambridge Library Collection. Cambridge University Press, Cambridge (2010). Reprint of the 1923 original

[Bar01] Barakat, M.: Jets. A MAPLE-package for formal differential geometry. In: V.G. Ganzha, E.W. Mayr, E.V. Vorozhtsov (eds.) Computer algebra in scientific computing (Konstanz, 2001), pp. 1–12. Springer, Berlin (2001)

[BC80] Bullough, R.K., Caudrey, P.J.: The soliton and its history. In: R.K. Bullough, P.J. Caudrey (eds.) Solitons, *Topics in Current Physics*, vol. 17, pp. 1–64. Springer (1980)

[BCA10] Bluman, G.W., Cheviakov, A.F., Anco, S.C.: Applications of symmetry methods to partial differential equations, *Applied Mathematical Sciences*, vol. 168. Springer, New York (2010)

[BCG+03a] Blinkov, Y.A., Cid, C.F., Gerdt, V.P., Plesken, W., Robertz, D.: The MAPLE Package "Janet": I. Polynomial Systems. In: V.G. Ganzha, E.W. Mayr, E.V. Vorozhtsov (eds.) Proceedings of the 6th International Workshop on Computer Algebra in Scientific Computing, Passau (Germany), pp. 31–40 (2003). http://wwwb.math.rwth-aachen.de/Janet

[BCG+03b] Blinkov, Y.A., Cid, C.F., Gerdt, V.P., Plesken, W., Robertz, D.: The MAPLE Package "Janet": II. Linear Partial Differential Equations. In: V.G. Ganzha, E.W. Mayr, E.V. Vorozhtsov (eds.) Proceedings of the 6th International Workshop on Computer Algebra in Scientific Computing, Passau (Germany), pp. 41–54 (2003). http://wwwb.math.rwth-aachen.de/Janet

[BCP97] Bosma, W., Cannon, J., Playoust, C.: The Magma algebra system. I. The user language. J. Symbolic Comput. **24**(3-4), 235–265 (1997). Computational algebra and number theory (London, 1993)

[Bem06] Bemelmans, J.: Exakte Lösungen der Navier-Stokes-Gleichungen. Lecture Notes, RWTH Aachen University, Germany (SS 2006)

[Ber78] Bergman, G.M.: The diamond lemma for ring theory. Adv. in Math. **29**(2), 178–218 (1978)

[BG08] Blinkov, Y.A., Gerdt, V.P.: The specialized computer algebra system GINV. Programmirovanie **34**(2), 67–80 (2008). http://invo.jinr.ru

[BG94] Bachmair, L., Ganzinger, H.: Buchberger's algorithm: a constraint-based completion procedure. In: J.P. Jouannaud (ed.) Constraints in computational logics (Munich, 1994), *Lecture Notes in Comput. Sci.*, vol. 845, pp. 285–301. Springer, Berlin (1994)

[BGL+10] Bächler, T., Gerdt, V.P., Lange-Hegermann, M., Robertz, D.: Thomas Decomposition of Algebraic and Differential Systems. In: V.P. Gerdt, W. Koepf, E.W. Mayr, E.H. Vorozhtsov (eds.) Computer Algebra in Scientific Computing, 12th International Workshop, CASC 2010, Tsakhkadzor, Armenia, *Lecture Notes in Comput. Sci.*, vol. 6244, pp. 31–54. Springer (2010)

[BGL+12] Bächler, T., Gerdt, V.P., Lange-Hegermann, M., Robertz, D.: Algorithmic Thomas decomposition of algebraic and differential systems. J. Symbolic Comput. **47**(10), 1233–1266 (2012)

[Bjö79] Björk, J.E.: Rings of differential operators, *North-Holland Mathematical Library*, vol. 21. North-Holland Publishing Co., Amsterdam (1979)

[BK00] Berrick, A.J., Keating, M.E.: An introduction to rings and modules with K-theory in view, *Cambridge Studies in Advanced Mathematics*, vol. 65. Cambridge University Press, Cambridge (2000)

[BKRM01] Bouziane, D., Kandri Rody, A., Maârouf, H.: Unmixed-dimensional decomposition of a finitely generated perfect differential ideal. J. Symbolic Comput. **31**(6), 631–649 (2001)

[BLH] Bächler, T., Lange-Hegermann, M.: AlgebraicThomas and DifferentialThomas: Thomas decomposition of algebraic and differential systems. Available at http://wwwb.math.rwth-aachen.de/thomasdecomposition

[BLMM10] Boulier, F., Lemaire, F., Moreno Maza, M.: Computing differential characteristic sets by change of ordering. J. Symbolic Comput. **45**(1), 124–149 (2010)

[BLOP09] Boulier, F., Lazard, D., Ollivier, F., Petitot, M.: Computing representations for radicals of finitely generated differential ideals. Appl. Algebra Engrg. Comm. Comput. **20**(1), 73–121 (2009)

[BLOP95] Boulier, F., Lazard, D., Ollivier, F., Petitot, M.: Representation for the radical of a finitely generated differential ideal. In: A.H.M. Levelt (ed.) Proceedings of ISSAC'95, pp. 158–166. ACM, New York, NY, USA (1995)

[Bou] Boulier, F.: BLAD – Bibliothèques Lilloises d'Algèbre Différentielle. Available online at http://www.lifl.fr/ boulier/

[Bou07] Bourbaki, N.: Éléments de mathématique. Algèbre. Chapitre 10. Algèbre homologique. Springer-Verlag, Berlin (2007). Reprint of the 1980 original

[Bou98a] Bourbaki, N.: Algebra I. Chapters 1–3. Elements of Mathematics (Berlin). Springer-Verlag, Berlin (1998). Translated from the French. Reprint of the 1989 English translation

[Bou98b] Bourbaki, N.: Commutative algebra. Chapters 1–7. Elements of Mathematics (Berlin). Springer-Verlag, Berlin (1998). Translated from the French. Reprint of the 1989 English translation

[BR08] Barakat, M., Robertz, D.: homalg: a meta-package for homological algebra. J. Algebra Appl. **7**(3), 299–317 (2008). http://wwwb.math.rwth-aachen.de/homalg

[Bro87] Brownawell, W.D.: Bounds for the degrees in the Nullstellensatz. Ann. of Math. (2) **126**(3), 577–591 (1987)

[BS87] Bayer, D., Stillman, M.: A criterion for detecting m-regularity. Invent. Math. **87**(1), 1–11 (1987)

[Buc06] Buchberger, B.: An algorithm for finding the basis elements of the residue class ring of a zero dimensional polynomial ideal. J. Symbolic Comput. **41**(3-4), 475–511 (2006). Translated from the 1965 German original by M. P. Abramson

[Buc79] Buchberger, B.: A criterion for detecting unnecessary reductions in the construction of Gröbner-bases. In: E.W. Ng (ed.) Symbolic and algebraic computation (EUROSAM '79, Internat. Sympos., Marseille, 1979), *Lecture Notes in Comput. Sci.*, vol. 72, pp. 3–21. Springer, Berlin (1979)

[Buc87] Buchberger, B.: History and basic features of the critical-pair/completion procedure. J. Symbolic Comput. **3**(1-2), 3–38 (1987). Rewriting techniques and applications (Dijon, 1985)

[Car14] Cartan, E.: Sur l'équivalence absolue de certains systèmes d'équations différentielles et sur certaines familles de courbes. Bull. Soc. Math. France **42**, 12–48 (1914). URL http://www.numdam.org/item?id=BSMF_1914__42__12_1

[CF07] Carrà Ferro, G.: A survey on differential Gröbner bases. In: Gröbner bases in symbolic analysis, *Radon Ser. Comput. Appl. Math.*, vol. 2, pp. 77–108. Walter de Gruyter, Berlin (2007)

[CF97] Carrà Ferro, G.: A resultant theory for ordinary algebraic differential equations. In: T. Mora, H. Mattson (eds.) Applied algebra, algebraic algorithms and error-correcting codes (Toulouse, 1997), *Lecture Notes in Comput. Sci.*, vol. 1255, pp. 55–65. Springer, Berlin (1997)

[CGO04] Creutzig, C., Gehrs, K., Oevel, W.: Das MuPAD Tutorium, third edn. Springer, Berlin (2004)

[Chy98] Chyzak, F.: Fonctions holonomes en calcul formel. Ph.D. thesis, Ecole Polytechnique, Palaiseau, France (1998)

[CJ84] Castro-Jiménez, F.J.: Théorème de division pour les opérateurs differentiels et calcul des multiplicités. Ph.D. thesis, Université Paris VII, France (1984)

[CJMF03] Castro-Jiménez, F.J., Moreno-Frías, M.A.: Janet bases, δ-bases and Gröbner bases in $A_n(k)$. In: Comptes rendus de la première recontre maroco-andalouse sur les algèbres et leurs applications (Tétouan, 2001), pp. 108–116. Univ. Abdelmalek Essaâdi. Fac. Sci. Tétouan, Tétouan (2003)

[CKM97] Collart, S., Kalkbrener, M., Mall, D.: Converting bases with the Gröbner walk. J. Symbolic Comput. **24**(3-4), 465–469 (1997). Computational algebra and number theory (London, 1993)

[CLO07] Cox, D., Little, J., O'Shea, D.: Ideals, varieties, and algorithms, third edn. Undergraduate Texts in Mathematics. Springer, New York (2007). An introduction to computational algebraic geometry and commutative algebra

[CoC] CoCoATeam: CoCoA: a system for doing Computations in Commutative Algebra. Available online at http://cocoa.dima.unige.it

[Coh71] Cohn, P.M.: Free rings and their relations. Academic Press, London (1971). London Mathematical Society Monographs, No. 2

[Coh91] Cohn, P.M.: Algebra. Vol. 3, second edn. John Wiley & Sons Ltd., Chichester (1991)

[Cou95] Coutinho, S.C.: A primer of algebraic D-modules, *London Mathematical Society Student Texts*, vol. 33. Cambridge University Press, Cambridge (1995)

[CQ08] Cluzeau, T., Quadrat, A.: Factoring and decomposing a class of linear functional systems. Linear Algebra Appl. **428**(1), 324–381 (2008)

[CQ09] Cluzeau, T., Quadrat, A.: OreMorphisms: a homological algebraic package for factoring, reducing and decomposing linear functional systems. In: J.J. Loiseau, W. Michiels, S.I. Niculescu, R. Sipahi (eds.) Topics in time delay systems, *Lecture Notes in Control and Inform. Sci.*, vol. 388, pp. 179–194. Springer, Berlin (2009). Cf. also http://www-sop.inria.fr/members/Alban.Quadrat/OreMorphisms or http://perso.ensil.unilim.fr/ cluzeau/OreMorphisms

[CQR05] Chyzak, F., Quadrat, A., Robertz, D.: Effective algorithms for parametrizing linear control systems over Ore algebras. Appl. Algebra Engrg. Comm. Comput. **16**(5), 319–376 (2005)

[CQR07] Chyzak, F., Quadrat, A., Robertz, D.: OreModules: a symbolic package for the study of multidimensional linear systems. In: J. Chiasson, J.J. Loiseau (eds.) Applications of time delay systems, *Lecture Notes in Control and Inform. Sci.*, vol. 352, pp. 233–264. Springer, Berlin (2007). http://wwwb.math.rwth-aachen.de/OreModules

[ČŠ90] Čadek, M., Šimša, J.: Decomposable functions of several variables. Aequationes Math. **40**(1), 8–25 (1990)

[ČŠ91] Čadek, M., Šimša, J.: Decomposition of smooth functions of two multidimensional variables. Czechoslovak Math. J. **41(116)**(2), 342–358 (1991)

[CS98] Chyzak, F., Salvy, B.: Non-commutative elimination in Ore algebras proves multivariate identities. J. Symbolic Comput. **26**(2), 187–227 (1998)

[CTvB95] Cheb-Terrab, E.S., von Bülow, K.: A computational approach for the analytic solving of partial differential equations. Computer Physics Communications **90**, 102–116 (1995)

[d'A47] d'Alembert, J.: Recherches sur la courbe que forme une corde tendüe mise en vibration. Histoire de l'académie royale des sciences et belles lettres de Berlin **3**, 214–249 (1747)

[Dar73] Darboux, G.: Sur les solutions singulières des équations aux dérivées ordinaires du premier ordre. Bulletin des sciences mathématiques et astronomiques **4**, 158–176 (1873). URL http://www.numdam.org/item?id=BSMA_1873__4__158_0

[DGP99] Decker, W., Greuel, G.M., Pfister, G.: Primary decomposition: algorithms and comparisons. In: B.H. Matzat, G.M. Greuel, G. Hiss (eds.) Algorithmic algebra and number theory (Heidelberg, 1997), pp. 187–220. Springer, Berlin (1999)

[DGPS12] Decker, W., Greuel, G.M., Pfister, G., Schönemann, H.: SINGULAR 3-1-6 — A computer algebra system for polynomial computations (2012). http://www.singular.uni-kl.de

[Die69] Dieudonné, J.: Foundations of modern analysis. Academic Press, New York (1969). Enlarged and corrected printing, Pure and Applied Mathematics, Vol. 10-I

[Dio92] Diop, S.: Differential-algebraic decision methods and some applications to system theory. Theoret. Comput. Sci. **98**(1), 137–161 (1992). Second Workshop on Algebraic and Computer-theoretic Aspects of Formal Power Series (Paris, 1990)

[DJS14] D'Alfonso, L., Jeronimo, G., Solernó, P.: Effective differential Nullstellensatz for ordinary DAE systems with constant coefficients. J. Complexity **30**(5), 588–603 (2014)

[Dra01] Draisma, J.: Recognizing the symmetry type of O.D.E.s. J. Pure Appl. Algebra **164**(1-2), 109–128 (2001). Effective methods in algebraic geometry (Bath, 2000)

[Dub90] Dubé, T.W.: The structure of polynomial ideals and Gröbner bases. SIAM J. Comput. **19**(4), 750–775 (1990)

[Ehr70] Ehrenpreis, L.: Fourier analysis in several complex variables. Pure and Applied Mathematics, Vol. XVII. Wiley-Interscience Publishers A Division of John Wiley & Sons, New York-London-Sydney (1970)

[EHV92] Eisenbud, D., Huneke, C., Vasconcelos, W.: Direct methods for primary decomposition. Invent. Math. **110**(2), 207–235 (1992)

[Eis05] Eisenbud, D.: The geometry of syzygies, *Graduate Texts in Mathematics*, vol. 229. Springer-Verlag, New York (2005). A second course in commutative algebra and algebraic geometry

[Eis95] Eisenbud, D.: Commutative algebra, *Graduate Texts in Mathematics*, vol. 150. Springer-Verlag, New York (1995). With a view toward algebraic geometry

[EM99] Emiris, I.Z., Mourrain, B.: Matrices in elimination theory. J. Symbolic Comput. **28**(1-2), 3–44 (1999). Polynomial elimination—algorithms and applications

[Eva10] Evans, L.C.: Partial differential equations, *Graduate Studies in Mathematics*, vol. 19, second edn. American Mathematical Society, Providence, RI (2010)

[EW07] Evans, G.A., Wensley, C.D.: Complete involutive rewriting systems. J. Symbolic Comput. **42**(11-12), 1034–1051 (2007)

[Fab09] Fabiańska, A.W.: Algorithmic analysis of presentations of groups and modules. Ph.D. thesis, RWTH Aachen University, Germany (2009). Available online at http://darwin.bth.rwth-aachen.de/opus3/volltexte/2009/2950/

[Fau99] Faugère, J.C.: A new efficient algorithm for computing Gröbner bases (F_4). J. Pure Appl. Algebra **139**(1-3), 61–88 (1999). Effective methods in algebraic geometry (Saint-Malo, 1998)

[FLMR95] Fliess, M., Lévine, J., Martin, P., Rouchon, P.: Flatness and defect of non-linear systems: introductory theory and examples. Internat. J. Control **61**(6), 1327–1361 (1995)

[FO98] Fröhler, S., Oberst, U.: Continuous time-varying linear systems. Systems Control Lett. **35**(2), 97–110 (1998)

[FQ07] Fabiańska, A., Quadrat, A.: Applications of the Quillen-Suslin theorem to multidimensional systems theory. In: H. Park, G. Regensburger (eds.) Gröbner bases in control theory and signal processing, *Radon Ser. Comput. Appl. Math.*, vol. 3, pp. 23–106. Walter de Gruyter, Berlin (2007)

[Gal85] Galligo, A.: Some algorithmic questions on ideals of differential operators. In: B.F. Caviness (ed.) EUROCAL '85, Vol. 2 (Linz, 1985), *Lecture Notes in Comput. Sci.*, vol. 204, pp. 413–421. Springer, Berlin (1985)

[Gao03] Gao, X.S.: Implicitization of differential rational parametric equations. J. Symbolic Comput. **36**(5), 811–824 (2003)

[GAP] The GAP Group: GAP – Groups, Algorithms, and Programming, Version 4.7.4 (2014). URL http://www.gap-system.org

[GB05a] Gerdt, V.P., Blinkov, Y.A.: Janet-like monomial division. In: V.G. Ganzha, E.W. Mayr, E.V. Vorozhtsov (eds.) Computer algebra in scientific computing, *Lecture Notes in Comput. Sci.*, vol. 3718, pp. 174–183. Springer, Berlin (2005)

[GB05b] Gerdt, V.P., Blinkov, Y.A.: Janet-like Gröbner bases. In: V.G. Ganzha, E.W. Mayr, E.V. Vorozhtsov (eds.) Computer algebra in scientific computing, *Lecture Notes in Comput. Sci.*, vol. 3718, pp. 184–195. Springer, Berlin (2005)

[GB11] Gerdt, V.P., Blinkov, Y.A.: Involutive Division Generated by an Antigraded Monomial Ordering. In: V.P. Gerdt, W. Koepf, E.W. Mayr, E.V. Vorozhtsov (eds.) Computer Algebra in Scientific Computing, 13th International Workshop, CASC 2011, Kassel, Germany, *Lecture Notes in Comput. Sci.*, vol. 6885, pp. 158–174. Springer (2011)

[GB98a] Gerdt, V.P., Blinkov, Y.A.: Involutive bases of polynomial ideals. Math. Comput. Simulation **45**(5-6), 519–541 (1998). Simplification of systems of algebraic and differential equations with applications

[GB98b] Gerdt, V.P., Blinkov, Y.A.: Minimal involutive bases. Math. Comput. Simulation **45**(5-6), 543–560 (1998). Simplification of systems of algebraic and differential equations with applications

[GBC98] Gerdt, V.P., Berth, M., Czichowski, G.: Involutive divisions in mathematica: Implementation and some applications. In: J. Calmet (ed.) Proceedings of the 6th Rhein Workshop on Computer Algebra (Sankt-Augustin), pp. 74–91. Institute for Algorithms and Scientific Computing, GMD-SCAI, Sankt-Augustin, Germany (1998)

[GC08] Grigor'ev, D.Y., Chistov, A.L.: Complexity of the standard basis of a *D*-module. Algebra i Analiz **20**(5), 41–82 (2008)

[Ger05] Gerdt, V.P.: Involutive algorithms for computing Gröbner bases. In: S. Cojocaru, G. Pfister, V. Ufnarovski (eds.) Computational commutative and non-commutative algebraic geometry, *NATO Sci. Ser. III Comput. Syst. Sci.*, vol. 196, pp. 199–225. IOS, Amsterdam (2005)

[Ger08] Gerdt, V.P.: On decomposition of algebraic PDE systems into simple subsystems. Acta Appl. Math. **101**(1-3), 39–51 (2008)

[Ger09] Gerdt, V.P.: Algebraically simple involutive differential systems and Cauchy problem. Zap. Nauchn. Sem. S.-Peterburg. Otdel. Mat. Inst. Steklov. (POMI) **373**(Teoriya Predstavlenii, Dinamicheskie Sistemy, Kombinatornye Metody. XVII), 94–103, 347 (2009)

[GKOS09] Golubitsky, O., Kondratieva, M., Ovchinnikov, A., Szanto, A.: A bound for orders in differential Nullstellensatz. J. Algebra **322**(11), 3852–3877 (2009)

[GKZ94] Gel'fand, I.M., Kapranov, M.M., Zelevinsky, A.V.: Discriminants, resultants, and multidimensional determinants. Mathematics: Theory & Applications. Birkhäuser Boston Inc., Boston, MA (1994)

[GL11] Gallego, C., Lezama, O.: Gröbner bases for ideals of σ-*PBW* extensions. Comm. Algebra **39**(1), 50–75 (2011)

[Gla01] Glasby, S.P.: On the tensor product of polynomials over a ring. J. Aust. Math. Soc. **71**(3), 307–324 (2001)

[GMO91] Gallo, G., Mishra, B., Ollivier, F.: Some constructions in rings of differential polynomials. In: H.F. Mattson, T. Mora, T.R.N. Rao (eds.) Applied algebra, algebraic algorithms and error-correcting codes (New Orleans, LA, 1991), *Lecture Notes in Comput. Sci.*, vol. 539, pp. 171–182. Springer, Berlin (1991)

[Gou30] Goursat, E.: Sur une généralisation du problème de Monge. Ann. Fac. Sci. Toulouse Sci. Math. Sci. Phys. (3) **22**, 249–295 (1930). URL http://www.numdam.org/item?id=AFST_1930_3_22__249_0

[GR06] Gerdt, V.P., Robertz, D.: A Maple Package for Computing Gröbner Bases for Linear Recurrence Relations. Nuclear Instruments and Methods in Physics Research, A: Accelerators, Spectrometers, Detectors and Associated Equipment **559**(1), 215–219 (2006)

[GR10] Gerdt, V.P., Robertz, D.: Consistency of Finite Difference Approximations for Linear PDE Systems and its Algorithmic Verification. In: S.M. Watt (ed.) Proceedings of the 2010 International Symposium on Symbolic and Algebraic Computation, TU München, Germany, pp. 53–59 (2010)

[GR12] Gerdt, V.P., Robertz, D.: Computation of Difference Gröbner Bases. Computer Science Journal of Moldova **20**(2 (59)), 203–226 (2012)

[GR89] Gauchman, H., Rubel, L.A.: Sums of products of functions of x times functions of y. Linear Algebra Appl. **125**, 19–63 (1989)

[Gri05] Grigor'ev, D.Y.: Weak Bézout inequality for D-modules. J. Complexity **21**(4), 532–542 (2005)

[Gri09] Grigor'ev, D.Y.: Analogue of Newton-Puiseux series for non-holonomic D-modules and factoring. Mosc. Math. J. **9**(4), 775–800, 934 (2009)

[Gri89] Grigor'ev, D.Y.: Complexity of quantifier elimination in the theory of ordinary differentially closed fields. Zap. Nauchn. Sem. Leningrad. Otdel. Mat. Inst. Steklov. (LOMI) **176**(Teor. Slozhn. Vychisl. 4), 53–67, 152 (1989)

[Gri91] Grigor'ev, D.Y.: Complexity of the solution of systems of linear equations in rings of differential operators. Zap. Nauchn. Sem. Leningrad. Otdel. Mat. Inst. Steklov. (LOMI) **192**(Teor. Slozhn. Vychisl. 5), 47–59, 174 (1991)

[Gri96] Grigor'ev, D.Y.: NC solving of a system of linear ordinary differential equations in several unknowns. Theoret. Comput. Sci. **157**(1), 79–90 (1996). Algorithmic complexity of algebraic and geometric models (Creteil, 1994)

[Grö70] Gröbner, W.: Algebraische Geometrie. 2. Teil: Arithmetische Theorie der Polynomringe. Bibliographisches Institut, Mannheim (1970). B. I. Hochschultaschenbücher, 737/737a*

[GS] Grayson, D.R., Stillman, M.E.: Macaulay2, a software system for research in algebraic geometry. Available at http://www.math.uiuc.edu/Macaulay2

[GS04] Grigori'ev, D.Y., Schwarz, F.: Factoring and solving linear partial differential equations. Computing **73**(2), 179–197 (2004)

[GTZ88] Gianni, P., Trager, B., Zacharias, G.: Gröbner bases and primary decomposition of polynomial ideals. J. Symbolic Comput. **6**(2-3), 149–167 (1988). Computational aspects of commutative algebra

[Ham93] Hamburger, M.: Über die singulären Lösungen der algebraischen Differentialgleichungen erster Ordnung. J. Reine und Angew. Math. **112**, 205–246 (1893)

[Har95] Harris, J.: Algebraic geometry, *Graduate Texts in Mathematics*, vol. 133. Springer-Verlag, New York (1995). A first course. Corrected reprint of the 1992 original

[HE71] Harrison, B.K., Estabrook, F.B.: Geometric approach to invariance groups and solution of partial differential systems. J. Mathematical Phys. **12**, 653–666 (1971)

[Hea99] Hearn, A.C.: REDUCE User's and Contributed Packages Manual. Santa Monica (CA) and Codemist Ltd. (1999)

[Her26] Hermann, G.: Die Frage der endlich vielen Schritte in der Theorie der Polynomideale. Math. Ann. **95**(1), 736–788 (1926)

[Her97] Hereman, W.: Review of symbolic software for Lie symmetry analysis. Math. Comput. Modelling **25**(8-9), 115–132 (1997). Algorithms and software for symbolic analysis of nonlinear systems

[Hil12] Hilbert, D.: Über den Begriff der Klasse von Differentialgleichungen. Math. Ann. **73**(1), 95–108 (1912)

[Hil90] Hilbert, D.: Über die Theorie der algebraischen Formen. Math. Ann. **36**(4), 473–534 (1890)

[HS01] Hillebrand, A., Schmale, W.: Towards an effective version of a theorem of Stafford. J. Symbolic Comput. **32**(6), 699–716 (2001). Effective methods in rings of differential operators

[HS02] Hausdorf, M., Seiler, W.M.: Involutive Bases in MuPAD I: Involutive Divisions. mathPAD **11**, 51–56 (2002)

[HSS02] Hausdorf, M., Seiler, W.M., Steinwandt, R.: Involutive bases in the Weyl algebra. J. Symbolic Comput. **34**(3), 181–198 (2002)

[Hub00] Hubert, E.: Factorization-free decomposition algorithms in differential algebra. J. Symbolic Comput. **29**(4-5), 641–662 (2000). Symbolic computation in algebra, analysis, and geometry (Berkeley, CA, 1998). www-sop.inria.fr/members/Evelyne.Hubert/diffalg

[Hub03a] Hubert, E.: Notes on triangular sets and triangulation-decomposition algorithms. I. Polynomial systems. In: F. Winkler, U. Langer (eds.) Symbolic and numerical scientific computation (Hagenberg, 2001), *Lecture Notes in Comput. Sci.*, vol. 2630, pp. 1–39. Springer, Berlin (2003)

[Hub03b] Hubert, E.: Notes on triangular sets and triangulation-decomposition algorithms. II. Differential systems. In: F. Winkler, U. Langer (eds.) Symbolic and numerical scientific computation (Hagenberg, 2001), *Lecture Notes in Comput. Sci.*, vol. 2630, pp. 40–87. Springer, Berlin (2003)

[Hub97] Hubert, E.: Algebra and algorithms for singularities of implicit differential equations. Ph.D. thesis, Institute National Polytechnique de Grenoble, France (1997)

[Hub99] Hubert, E.: Essential components of an algebraic differential equation. J. Symbolic Comput. **28**(4-5), 657–680 (1999). Differential algebra and differential equations

[Inc56] Ince, E.L.: Ordinary Differential Equations. Dover Publications, New York (1956)

[IP98] Insa, M., Pauer, F.: Gröbner bases in rings of differential operators. In: B. Buchberger, F. Winkler (eds.) Gröbner bases and applications (Linz, 1998), *London Math. Soc. Lecture Note Ser.*, vol. 251, pp. 367–380. Cambridge Univ. Press, Cambridge (1998)

[Jam11] Jambor, S.: Computing minimal associated primes in polynomial rings over the integers. J. Symbolic Comput. **46**(10), 1098–1104 (2011)

[Jan20] Janet, M.: Sur les systèmes d'équations aux dérivées partielles. Thèses françaises de l'entre-deux-guerres. Gauthiers-Villars, Paris (1920). URL http://www.numdam.org/item?id=THESE_1920__19__1_0

[Jan29] Janet, M.: Leçons sur les systèmes d'équations aux dérivées partielles. Cahiers Scientifiques IV. Gauthiers-Villars, Paris (1929)

[Jan71] Janet, M.: P. Zervos et le problème de Monge. Bull. Sci. Math. (2) **95**, 15–26 (1971)

[Joh69a] Johnson, J.: Differential dimension polynomials and a fundamental theorem on differential modules. Amer. J. Math. **91**, 239–248 (1969)

[Joh69b] Johnson, J.: Kähler differentials and differential algebra. Ann. of Math. (2) **89**, 92–98 (1969)

[Jou91] Jouanolou, J.P.: Le formalisme du résultant. Adv. Math. **90**(2), 117–263 (1991)

[JRR95] Johnson, J., Reinhart, G.M., Rubel, L.A.: Some counterexamples to separation of variables. J. Differential Equations **121**(1), 42–66 (1995)

[JS92] Jenks, R.D., Sutor, R.S.: AXIOM. Numerical Algorithms Group Ltd., Oxford (1992). The scientific computation system. With a foreword by D. V. Chudnovsky and G. V. Chudnovsky

[Kap76] Kaplansky, I.: An introduction to differential algebra, second edn. Hermann, Paris (1976). Actualités Scientifiques et Industrielles, No. 1251, Publications de l'Institut de Mathématique de l'Université de Nancago, No. V

[Kas03] Kashiwara, M.: *D*-modules and microlocal calculus, *Translations of Mathematical Monographs*, vol. 217. American Mathematical Society, Providence, RI (2003). Translated from the 2000 Japanese original by Mutsumi Saito, Iwanami Series in Modern Mathematics

[KB70] Knuth, D.E., Bendix, P.B.: Simple word problems in universal algebras. In: J.W. Leech (ed.) Computational Problems in Abstract Algebra (Proc. Conf., Oxford, 1967), pp. 263–297. Pergamon, Oxford (1970)

[KMS93] Kolář, I., Michor, P.W., Slovák, J.: Natural operations in differential geometry. Springer-Verlag, Berlin (1993)

[Knu97] Knuth, D.E.: The art of computer programming. Vol. 1, third edn. Addison-Wesley, Reading, MA (1997). Fundamental algorithms

[Kol63] Kolmogorov, A.N.: On the representation of continuous functions of many variables by superposition of continuous functions of one variable and addition. Amer. Math. Soc. Transl. (2) **28**, 55–59 (1963)

[Kol64] Kolchin, E.R.: The notion of dimension in the theory of algebraic differential equations. Bull. Amer. Math. Soc. **70**, 570–573 (1964)

[Kol73] Kolchin, E.R.: Differential algebra and algebraic groups, *Pure and Applied Mathematics*, vol. 54. Academic Press, New York-London (1973)

[Kol99] Kolchin, E.R.: Selected works of Ellis Kolchin with commentary. American Mathematical Society, Providence, RI (1999). Commentaries by A. Borel, M. F. Singer, B. Poizat, A. Buium and P. J. Cassidy, edited and with a preface by H. Bass, A. Buium and P. J. Cassidy

[KP02] Krantz, S.G., Parks, H.R.: A primer of real analytic functions, second edn. Birkhäuser Advanced Texts: Basler Lehrbücher. Birkhäuser Boston, Inc., Boston, MA (2002)

[Kre93] Kredel, H.: Solvable Polynomial Rings. Shaker-Verlag, Aachen (1993)

[KRW90] Kandri-Rody, A., Weispfenning, V.: Noncommutative Gröbner bases in algebras of solvable type. J. Symbolic Comput. **9**(1), 1–26 (1990)

[Lam06] Lam, T.Y.: Serre's problem on projective modules. Springer Monographs in Mathematics. Springer-Verlag, Berlin (2006)

[Lam99] Lam, T.Y.: Lectures on modules and rings, *Graduate Texts in Mathematics*, vol. 189. Springer-Verlag, New York (1999)

[Lem02] Lemaire, F.: Contribution à l'algorithmique en algèbre différentielle. Ph.D. thesis, Université des Sciences et Technologies de Lille, France (2002). Available online at http://tel.archives-ouvertes.fr/tel-00001363/fr/

[Lev05] Levandovskyy, V.: Non-commutative Computer Algebra for polynomial algebras: Gröbner bases, applications and implementation. Ph.D. thesis, Universität Kaiserslautern, Germany (2005). Available online at http://kluedo.ub.uni-kl.de/volltexte/2005/1883

[Lev10] Levin, A.B.: Dimension polynomials of intermediate fields and Krull-type dimension of finitely generated differential field extensions. Math. Comput. Sci. **4**(2-3), 143–150 (2010)

[Ley04] Leykin, A.: Algorithmic proofs of two theorems of Stafford. J. Symbolic Comput. **38**(6), 1535–1550 (2004)

[LGO97] Leedham-Green, C.R., O'Brien, E.A.: Recognising tensor products of matrix groups. Internat. J. Algebra Comput. **7**(5), 541–559 (1997)

[LH14] Lange-Hegermann, M.: Counting Solutions of Differential Equations. Ph.D. thesis, RWTH Aachen University, Germany (2014). Available online at http://darwin.bth.rwth-aachen.de/opus/volltexte/2014/4993

[LHR13] Lange-Hegermann, M., Robertz, D.: Thomas decompositions of parametric nonlinear control systems. In: Proceedings of the 5th Symposium on System Structure and Control, Grenoble (France), pp. 291–296 (2013)

[LL66] Landau, L.D., Lifschitz, E.M.: Lehrbuch der theoretischen Physik IV. Hydrodynamik. Akademie-Verlag, Berlin (1966)

[LMMX05] Lemaire, F., Moreno Maza, M., Xie, Y.: The RegularChains library in MAPLE. SIGSAM Bull. **39**, 96–97 (2005)

[LMW10] Li, X., Mou, C., Wang, D.: Decomposing polynomial sets into simple sets over finite fields: The zero-dimensional case. Comput. Math. Appl. **60**(11), 2983–2997 (2010)

[LST03] Li, Z., Schwarz, F., Tsarev, S.P.: Factoring systems of linear PDEs with finite-dimensional solution spaces. J. Symbolic Comput. **36**(3-4), 443–471 (2003). International Symposium on Symbolic and Algebraic Computation (ISSAC'2002) (Lille)

[LW99] Li, Z., Wang, D.: Coherent, regular and simple systems in zero decompositions of partial differential systems. System Science and Mathematical Sciences **12**, 43–60 (1999)

[Mal62] Malgrange, B.: Systèmes différentiels à coefficients constants. Séminaire Bourbaki **246**, 11 pages (1962–64). URL http://www.numdam.org/item?id=SB_1962-1964__8__79_0

[Man91] Mansfield, E.L.: Differential Gröbner Bases. Ph.D. thesis, University of Sydney, Australia (1991)

[MAP] Maple. Waterloo Maple Inc. www.maplesoft.com

[Mat86] Matsumura, H.: Commutative ring theory, *Cambridge Studies in Advanced Mathematics*, vol. 8. Cambridge University Press, Cambridge (1986). Translated from the Japanese by M. Reid

[May97] Mayr, E.W.: Some complexity results for polynomial ideals. J. Complexity **13**(3), 303–325 (1997)

[Mér80] Méray, C.: Démonstration générale de l'existence des intégrales des équations aux dérivées partielles. Journal de mathématiques pures et appliquées, 3e série **tome VI**, 235–265 (1880)

[Mil77] Miller Jr., W.: Symmetry and separation of variables, *Encyclopedia of Mathematics and its Applications*, vol. 4. Addison-Wesley Publishing Co., Reading, Mass.-London-Amsterdam (1977). With a foreword by R. Askey

[Mis93] Mishra, B.: Algorithmic algebra. Texts and Monographs in Computer Science. Springer-Verlag, New York (1993)

[MLW13] Mou, C., Li, X., Wang, D.: Decomposing polynomial sets into simple sets over finite fields: The positive-dimensional case. Theoret. Comput. Sci. **468**, 102–113 (2013)

[MM82] Mayr, E.W., Meyer, A.R.: The complexity of the word problems for commutative semigroups and polynomial ideals. Adv. in Math. **46**(3), 305–329 (1982)

[Mor94] Mora, T.: An introduction to commutative and noncommutative Gröbner bases. Theoret. Comput. Sci. **134**(1), 131–173 (1994). Second International Colloquium on Words, Languages and Combinatorics (Kyoto, 1992)

[MR01] McConnell, J.C., Robson, J.C.: Noncommutative Noetherian rings, *Graduate Studies in Mathematics*, vol. 30, revised edn. American Mathematical Society, Providence, RI (2001). With the cooperation of L. W. Small

[MR88] Mora, T., Robbiano, L.: The Gröbner fan of an ideal. J. Symbolic Comput. **6**(2-3), 183–208 (1988). Computational aspects of commutative algebra

[MRC98] Mansfield, E.L., Reid, G.J., Clarkson, P.A.: Nonclassical reductions of a $(3+1)$-cubic nonlinear Schrödinger system. Comput. Phys. Comm. **115**(2-3), 460–488 (1998)

[Neu82] Neuman, F.: Factorizations of matrices and functions of two variables. Czechoslovak Math. J. **32(107)**(4), 582–588 (1982)

[Neu90] Neuman, F.: Finite sums of products of functions in single variables. Linear Algebra Appl. **134**, 153–164 (1990)

[NR91] Neuman, F., Rassias, T.M.: Functions decomposable into finite sums of products (old and new results, problems and trends). In: Constantin Carathéodory: an international tribute, Vol. I, II, pp. 956–963. World Sci. Publ., Teaneck, NJ (1991)

[NS20] Noether, E., Schmeidler, W.: Moduln in nichtkommutativen Bereichen, insbesondere aus Differential- und Differenzenausdrücken. Math. Z. **8**(1-2), 1–35 (1920)

[O'B06] O'Brien, E.A.: Towards effective algorithms for linear groups. In: Finite geometries, groups, and computation, pp. 163–190. Walter de Gruyter, Berlin (2006)

[Obe90] Oberst, U.: Multidimensional constant linear systems. Acta Appl. Math. **20**(1-2), 1–175 (1990)

[Oll91] Ollivier, F.: Standard bases of differential ideals. In: S. Sakata (ed.) Applied algebra, algebraic algorithms and error-correcting codes (Tokyo, 1990), *Lecture Notes in Comput. Sci.*, vol. 508, pp. 304–321. Springer, Berlin (1991)

[Olv93] Olver, P.J.: Applications of Lie groups to differential equations, *Graduate Texts in Mathematics*, vol. 107, second edn. Springer-Verlag, New York (1993)

[OP01] Oberst, U., Pauer, F.: The constructive solution of linear systems of partial difference and differential equations with constant coefficients. Multidimens. Systems Signal Process. **12**(3-4), 253–308 (2001). Special issue: Applications of Gröbner bases to multidimensional systems and signal processing

[OR86] Olver, P.J., Rosenau, P.: The construction of special solutions to partial differential equations. Phys. Lett. A **114**(3), 107–112 (1986)

[Ore33] Ore, O.: Theory of non-commutative polynomials. Ann. of Math. (2) **34**(3), 480–508 (1933)

[Pal70] Palamodov, V.P.: Linear differential operators with constant coefficients. Translated from the Russian by A. A. Brown. Die Grundlehren der mathematischen Wissenschaften, Band 168. Springer-Verlag, New York (1970)

[Pan89] Pankrat'ev, E.V.: Computations in differential and difference modules. Acta Appl. Math. **16**(2), 167–189 (1989). Symmetries of partial differential equations, Part III

[PB14] Plesken, W., Bächler, T.: Counting Polynomials for Linear Codes, Hyperplane Arrangements, and Matroids. Doc. Math. **19**, 285–312 (2014)

[PG97] Péladan-Germa, A.: Tests effectifs de nullité dans des extensions d'anneaux différentiels. Ph.D. thesis, Ecole Polytechnique, Palaiseau, France (1997)

[Ple09a] Plesken, W.: Counting solutions of polynomial systems via iterated fibrations. Arch. Math. (Basel) **92**(1), 44–56 (2009)

[Ple09b] Plesken, W.: Gauss-Bruhat decomposition as an example of Thomas decomposition. Arch. Math. (Basel) **92**(2), 111–118 (2009)

[Pom01] Pommaret, J.F.: Partial differential control theory, *Mathematics and its Applications*, vol. 530. Kluwer Academic Publishers Group, Dordrecht (2001). Vol. I. Mathematical tools; Vol. II. Control systems. With a foreword by J. C. Willems

[Pom78] Pommaret, J.F.: Systems of partial differential equations and Lie pseudogroups, *Mathematics and its Applications*, vol. 14. Gordon & Breach Science Publishers, New York (1978). With a preface by André Lichnerowicz

[Pom94] Pommaret, J.F.: Partial differential equations and group theory, *Mathematics and its Applications*, vol. 293. Kluwer Academic Publishers Group, Dordrecht (1994). New perspectives for applications

[PQ99] Pommaret, J.F., Quadrat, A.: Localization and parametrization of linear multidimensional control systems. Systems Control Lett. **37**(4), 247–260 (1999)

[PR05] Plesken, W., Robertz, D.: Janet's approach to presentations and resolutions for polynomials and linear PDEs. Arch. Math. (Basel) **84**(1), 22–37 (2005)

[PR06] Plesken, W., Robertz, D.: Representations, commutative algebra, and Hurwitz groups. J. Algebra **300**(1), 223–247 (2006)

[PR07] Plesken, W., Robertz, D.: Some Elimination Problems for Matrices. In: V.G. Ganzha, E.W. Mayr, E.V. Vorozhtsov (eds.) Computer Algebra in Scientific Computing, 10th International Workshop, CASC 2007, Bonn, Germany, *Lecture Notes in Comput. Sci.*, vol. 4770, pp. 350–359. Springer (2007). Cf. also http://wwwb.math.rwth-aachen.de/elimination

[PR08] Plesken, W., Robertz, D.: Elimination for coefficients of special characteristic polynomials. Experiment. Math. **17**(4), 499–510 (2008). Cf. also http://wwwb.math.rwth-aachen.de/charpoly

[PR10] Plesken, W., Robertz, D.: Linear differential elimination for analytic functions. Math. Comput. Sci. **4**(2-3), 231–242 (2010)

[QR05a] Quadrat, A., Robertz, D.: On the blowing-up of stably free behaviours. In: Proceedings of the 44th IEEE Conference on Decision and Control and European Control Conference ECC 2005, Seville (Spain), pp. 1541–1546 (2005)

[QR05b] Quadrat, A., Robertz, D.: Parametrizing all solutions of uncontrollable multidimensional linear systems. In: Proceedings of the 16th IFAC World Congress, Prague (Czech Republic) (2005)

[QR06] Quadrat, A., Robertz, D.: On the Monge problem and multidimensional optimal control. In: Proceedings of the 17th International Symposium on Mathematical Theory of Networks and Systems (MTNS 2006), Kyoto (Japan), pp. 596–605 (2006)

[QR07] Quadrat, A., Robertz, D.: Computation of bases of free modules over the Weyl algebras. J. Symbolic Comput. **42**(11-12), 1113–1141 (2007)

[QR08] Quadrat, A., Robertz, D.: Baer's extension problem for multidimensional linear systems. In: Proceedings of the 18th International Symposium on Mathematical Theory of Networks and Systems (MTNS 2008), Virginia Tech, Blacksburg, Virginia (USA) (2008)

[QR10] Quadrat, A., Robertz, D.: Controllability and differential flatness of linear analytic ordinary differential systems. In: Proceedings of the 19th International Symposium on Mathematical Theory of Networks and Systems (MTNS 2010), Budapest, Hungary (2010)

[QR13] Quadrat, A., Robertz, D.: Stafford's reduction of linear partial differential systems. In: Proceedings of the 5th Symposium on System Structure and Control, Grenoble (France), pp. 309–314 (2013)

[Qua10a] Quadrat, A.: An introduction to constructive algebraic analysis and its applications. In: Les cours du CIRM, tome 1, numéro 2: Journées Nationales de Calcul Formel, pp. 281–471 (2010). Available online at http://ccirm.cedram.org/item?id=CCIRM_2010__1_2_281_0

[Qua10b] Quadrat, A.: Systèmes et Structures: Une approche de la théorie mathématique des systèmes par l'analyse algébrique constructive. Habilitation thesis, Université de Nice (Sophia Antipolis), France (2010)

[Qua99] Quadrat, A.: Analyse algébrique des systèmes de contrôle linéaires multidimensionnels. Ph.D. thesis, Ecole Nationale des Ponts et Chaussées, Marne-la-Vallée, France (1999)

[Ras86] Rassias, T.M.: A criterion for a function to be represented as a sum of products of factors. Bull. Inst. Math. Acad. Sinica **14**(4), 377–382 (1986)

[Rau34] Raudenbush Jr., H.W.: Ideal theory and algebraic differential equations. Trans. Amer. Math. Soc. **36**(2), 361–368 (1934)

[Riq10] Riquier, C.: Les systèmes d'équations aux dérivées partielles. Gauthiers-Villars, Paris (1910)

[Rit34] Ritt, J.F.: Differential Equations from the Algebraic Standpoint. American Mathematical Society Colloquium Publications, Vol. XIV. American Mathematical Society, New York, N. Y. (1934)

[Rit36] Ritt, J.F.: On the singular solutions of algebraic differential equations. Ann. of Math. (2) **37**(3), 552–617 (1936)

[Rit50] Ritt, J.F.: Differential Algebra. American Mathematical Society Colloquium Publications, Vol. XXXIII. American Mathematical Society, New York, N. Y. (1950)

[Rob] Robertz, D.: InvolutiveBases – Methods for Janet bases and Pommaret bases in Macaulay2. Available online at http://www.math.uiuc.edu/Macaulay2/Packages/

[Rob06] Robertz, D.: Formal Computational Methods for Control Theory. Ph.D. thesis, RWTH Aachen University, Germany (2006). Available online at http://darwin.bth.rwth-aachen.de/opus/volltexte/2006/1586

[Rob07] Robertz, D.: Janet bases and applications. In: M. Rosenkranz, D. Wang (eds.) Gröbner bases in symbolic analysis, *Radon Ser. Comput. Appl. Math.*, vol. 2, pp. 139–168. Walter de Gruyter, Berlin (2007)

[Rob09] Robertz, D.: Noether normalization guided by monomial cone decompositions. J. Symbolic Comput. **44**(10), 1359–1373 (2009)

[Rob14] Robertz, D.: Recent progress in an algebraic analysis approach to linear systems. To appear in Multidimens. Systems Signal Process. (2014)

[Ros59] Rosenfeld, A.: Specializations in differential algebra. Trans. Amer. Math. Soc. **90**, 394–407 (1959)

[Rot09] Rotman, J.J.: An introduction to homological algebra, second edn. Universitext. Springer, New York (2009)

[RR04] Renardy, M., Rogers, R.C.: An introduction to partial differential equations, *Texts in Applied Mathematics*, vol. 13, second edn. Springer-Verlag, New York (2004)

[RR92] Rochberg, R., Rubel, L.A.: A functional equation. Indiana Univ. Math. J. **41**(2), 363–376 (1992)

[RS10] Rueda, S.L., Sendra, J.R.: Linear complete differential resultants and the implicitization of linear DPPEs. J. Symbolic Comput. **45**(3), 324–341 (2010)

[RŠ95] Rassias, T.M., Šimša, J.: Finite sums decompositions in mathematical analysis. Pure and Applied Mathematics (New York). John Wiley & Sons, Chichester (1995)

[RWB96] Reid, G.J., Wittkopf, A.D., Boulton, A.: Reduction of systems of nonlinear partial differential equations to simplified involutive forms. European J. Appl. Math. **7**(6), 635–666 (1996)

[Sad00] Sadik, B.: Une note sur les algorithmes de décomposition en algèbre différentielle. C. R. Acad. Sci. Paris Sér. I Math. **330**(8), 641–646 (2000)

[Sch08a] Schwarz, F.: Algorithmic Lie theory for solving ordinary differential equations, *Pure and Applied Mathematics (Boca Raton)*, vol. 291. Chapman & Hall/CRC, Boca Raton, FL (2008)

[Sch08b] Schwarz, F.: ALLTYPES in the Web. ACM Communications in Computer Algebra **42**(3), 185–187 (2008). URL http://www.alltypes.de. ISSAC 2008 Software Exhibition Abstracts

[Sch80] Schreyer, F.O.: Die Berechnung von Syzygien mit dem verallgemeinerten Weierstraßschen Divisionssatz und eine Anwendung auf analytische Cohen-Macaulay-Stellenalgebren minimaler Multiplizität. Diploma thesis, Universität Hamburg, Germany (1980)

[Sch84] Schwarz, F.: The Riquier-Janet theory and its application to nonlinear evolution equations. Phys. D **11**(1-2), 243–251 (1984)

[Sch99] Schwingel, R.: The tensor product of polynomials. Experiment. Math. **8**(4), 395–397 (1999)

[Sei10] Seiler, W.M.: Involution, *Algorithms and Computation in Mathematics*, vol. 24. Springer-Verlag, Berlin (2010). The formal theory of differential equations and its applications in computer algebra

[Sei56] Seidenberg, A.: An elimination theory for differential algebra. Univ. California Publ. Math. (N.S.) **3**, 31–65 (1956)

[Sha01] Shankar, S.: The lattice structure of behaviors. SIAM J. Control Optim. **39**(6), 1817–1832 (electronic) (2001)

[SST00] Saito, M., Sturmfels, B., Takayama, N.: Gröbner deformations of hypergeometric differential equations, *Algorithms and Computation in Mathematics*, vol. 6. Springer-Verlag, Berlin (2000)

[Sta78] Stafford, J.T.: Module structure of Weyl algebras. J. London Math. Soc. (2) **18**(3), 429–442 (1978)

[Sta96] Stanley, R.P.: Combinatorics and commutative algebra, *Progress in Mathematics*, vol. 41, second edn. Birkhäuser Boston Inc., Boston, MA (1996)

[Stu96] Sturmfels, B.: Gröbner bases and convex polytopes, *University Lecture Series*, vol. 8. American Mathematical Society, Providence, RI (1996)

[SW91] Sturmfels, B., White, N.: Computing combinatorial decompositions of rings. Combinatorica **11**(3), 275–293 (1991)

[SWPD08] Sendra, J.R., Winkler, F., Pérez-Díaz, S.: Rational algebraic curves, *Algorithms and Computation in Mathematics*, vol. 22. Springer, Berlin (2008). A computer algebra approach

[SY96] Shimoyama, T., Yokoyama, K.: Localization and primary decomposition of polynomial ideals. J. Symbolic Comput. **22**(3), 247–277 (1996)

[Tho28] Thomas, J.M.: Riquier's existence theorems. Ann. of Math. (2) **30**(1-4), 285–310 (1928/29)

[Tho34] Thomas, J.M.: Riquier's existence theorems. Ann. of Math. (2) **35**(2), 306–311 (1934)

[Tho37] Thomas, J.M.: Differential Systems. American Mathematical Society Colloquium Publications, Vol. XXI. American Mathematical Society, New York, N. Y. (1937)

[Tho62] Thomas, J.M.: Systems and Roots. William Byrd Press, Richmond, VA (1962)

[Top89] Topunov, V.L.: Reducing systems of linear differential equations to a passive form. Acta Appl. Math. **16**(2), 191–206 (1989). Symmetries of partial differential equations, Part III

[Vas98] Vasconcelos, W.V.: Computational methods in commutative algebra and algebraic geometry, *Algorithms and Computation in Mathematics*, vol. 2. Springer-Verlag, Berlin (1998). With chapters by D. Eisenbud, D. R. Grayson, J. Herzog and M. Stillman

[vdPS03] van der Put, M., Singer, M.F.: Galois theory of linear differential equations, *Grundlehren der Mathematischen Wissenschaften*, vol. 328. Springer-Verlag, Berlin (2003)

[Vin84] Vinogradov, A.M.: Local symmetries and conservation laws. Acta Appl. Math. **2**(1), 21–78 (1984)

[vzGG03] von zur Gathen, J., Gerhard, J.: Modern computer algebra, third edn. Cambridge University Press, Cambridge (2013)

[Wan01] Wang, D.: Elimination methods. Texts and Monographs in Symbolic Computation. Springer-Verlag, Vienna (2001)

[Wan04] Wang, D.: Elimination practice. Imperial College Press, London (2004). Software tools and applications. With 1 CD-ROM (UNIX/LINUX, Windows). http://www-salsa.lip6.fr/ wang/epsilon

[Wan98] Wang, D.: Decomposing polynomial systems into simple systems. J. Symbolic Comput. **25**(3), 295–314 (1998)

[WBM99] Wolf, T., Brand, A., Mohammadzadeh, M.: Computer algebra algorithms and routines for the computation of conservation laws and fixing of gauge in differential expressions. J. Symbolic Comput. **27**(2), 221–238 (1999)

[Win84] Winkler, F.: On the complexity of the Gröbner-bases algorithm over $K[x, y, z]$. In: J. Fitch (ed.) EUROSAM 84 (Cambridge, 1984), *Lecture Notes in Comput. Sci.*, vol. 174, pp. 184–194. Springer, Berlin (1984)

[Wol04] Wolf, T.: Applications of CRACK in the classification of integrable systems. In: Superintegrability in classical and quantum systems, *CRM Proc. Lecture Notes*, vol. 37, pp. 283–300. Amer. Math. Soc., Providence, RI (2004). http://lie.math.brocku.ca/Crack_demo.html

[Wol99] Wolfram, S.: The Mathematica® book, fourth edn. Wolfram Media, Inc., Champaign, IL (1999)

[Woo00] Wood, J.: Modules and behaviours in nD systems theory. Multidimens. Systems Signal Process. **11**(1-2), 11–48 (2000). Recent progress in multidimensional control theory and applications

[Wu00] Wu, W.t.: Mathematics mechanization, *Mathematics and its Applications*, vol. 489. Kluwer Academic Publishers Group, Dordrecht; Science Press, Beijing (2000). Mechanical geometry theorem-proving, mechanical geometry problem-solving and polynomial equations-solving

[Wu89] Wu, W.t.: On the foundation of algebraic differential geometry. Systems Sci. Math. Sci. **2**(4), 289–312 (1989)

[Wu91] Wu, W.t.: On the construction of Groebner basis of a polynomial ideal based on Riquier-Janet theory. Systems Sci. Math. Sci. **4**(3), 193–207 (1991). Also in: Dongming Wang, Zhiming Zheng (eds.) Differential Equations with Symbolic Computation, *Trends Math.*, pp. 351–368. Birkhäuser, Basel, 2005.

[Zas66] Zassenhaus, H.: The sum-intersection method (1966). Manuscript, Ohio State University, Columbus

[ZB96] Zharkov, A.Y., Blinkov, Y.A.: Involution approach to investigating polynomial systems. Math. Comput. Simulation **42**(4-6), 323–332 (1996). Symbolic computation, new trends and developments (Lille, 1993)

[Zei90] Zeilberger, D.: A holonomic systems approach to special functions identities. J. Comput. Appl. Math. **32**(3), 321–368 (1990)

[Zer00] Zerz, E.: Topics in multidimensional linear systems theory, *Lecture Notes in Control and Information Sciences*, vol. 256. Springer-Verlag London Ltd., London (2000)

[Zer06] Zerz, E.: An algebraic analysis approach to linear time-varying systems. IMA J. Math. Control Inform. **23**(1), 113–126 (2006)

[Zer32] Zervos, P.: Le problème de Monge. Mémorial des Sciences Mathématiques, fasc. LIII. Gauthier-Villars, Paris (1932)

[ZL04] Zhang, S.q., Li, Z.b.: An implementation for the algorithm of Janet bases of linear differential ideals in the Maple system. Acta Math. Appl. Sin. Engl. Ser. **20**(4), 605–616 (2004)

List of Algorithms

© Springer International Publishing Switzerland 2014
D. Robertz, *Formal Algorithmic Elimination for PDEs*,
Lecture Notes in Mathematics 2121, DOI 10.1007/978-3-319-11445-3

List of Examples

Some families of analytic functions dealt with in the text are referenced below.

© Springer International Publishing Switzerland 2014
D. Robertz, *Formal Algorithmic Elimination for PDEs*,
Lecture Notes in Mathematics 2121, DOI 10.1007/978-3-319-11445-3

Index of Notation

$\mathbb{N} = \{1,2,3,\ldots\}$
$\mathbb{Z}$ (integers)
$\mathbb{Q}$ (rational numbers)
$\mathbb{R}$ (real numbers)
$\mathbb{C}$ (complex numbers)

$A - B$ (set difference)
$A \uplus B$ (disjoint union)

Symbols

$>$ (a term ordering)   22
$>$ (a total ordering on $\{x_1,\ldots,x_n\}$)   11, 60
$>$ (a ranking)   249
$[]$ (multiple-closed set of monomials)   24, 92
$\langle\,\rangle$ (differential ideal)   88
$_D\langle\,\rangle$ (left D-module)   25
$|\cdot|$ (length of a multi-index)   9
$|$ (divisibility relation)   9

A

$\mathbb{A}^n$ (affine space)   128
$A_n(K)$ (Weyl algebra)   19

B

$B_n(K)$ (algebra of differential operators with rational function coefficients)   19

D

deg (total degree)   18, 22, 44
$\deg_{y_i}$ (exponent of a variable in a monomial)   9
Δ (set of derivations)   88, 249
$\Delta_\omega(\alpha,g)$ 166

$\Delta_{\omega,J}$ 168
$\Delta_{\omega,J}^{(i,j)}$ 168
$\Delta_\omega(p)$ 209
$\mathrm{der}_F(A,M)$ 207
$D_h, h \in \mathbb{R}$ 20
disc (discriminant)   60, 88

E

$E : q^\infty$ (saturation)   57

F

$f(c \cdot -)$ (function with scaled arguments)   255
$(\overline{F_i})_L$ (name for a residue class $\theta_i F_i + Z$)   209

G

gld (global dimension)   240

H

$H_{M,\Gamma}$ (Hilbert series of a graded module)   42
$H_{M,\Phi}$ (Hilbert series of a filtered module)   42
H_M (generalized Hilbert series for simple differential systems)   112
$H_{M,i}$ (component of a generalized Hilbert series)   112
H_S (Hilbert series for simple differential systems)   113
H_S (generalized Hilbert series for Ore algebras)   40
$H_{S,k}$ (component of a generalized Hilbert series)   40

© Springer International Publishing Switzerland 2014
D. Robertz, *Formal Algorithmic Elimination for PDEs*,
Lecture Notes in Mathematics 2121, DOI 10.1007/978-3-319-11445-3

Index

© Springer International Publishing Switzerland 2014
D. Robertz, *Formal Algorithmic Elimination for PDEs*,
Lecture Notes in Mathematics 2121, DOI 10.1007/978-3-319-11445-3

LECTURE NOTES IN MATHEMATICS   🐎 Springer

Edited by J.-M. Morel, B. Teissier; P.K. Maini

Editorial Policy (for the publication of monographs)

1. Lecture Notes aim to report new developments in all areas of mathematics and their applications - quickly, informally and at a high level. Mathematical texts analysing new developments in modelling and numerical simulation are welcome.

   Monograph manuscripts should be reasonably self-contained and rounded off. Thus they may, and often will, present not only results of the author but also related work by other people. They may be based on specialised lecture courses. Furthermore, the manuscripts should provide sufficient motivation, examples and applications. This clearly distinguishes Lecture Notes from journal articles or technical reports which normally are very concise. Articles intended for a journal but too long to be accepted by most journals, usually do not have this "lecture notes" character. For similar reasons it is unusual for doctoral theses to be accepted for the Lecture Notes series, though habilitation theses may be appropriate.

2. Manuscripts should be submitted either online at www.editorialmanager.com/lnm to Springer's mathematics editorial in Heidelberg, or to one of the series editors. In general, manuscripts will be sent out to 2 external referees for evaluation. If a decision cannot yet be reached on the basis of the first 2 reports, further referees may be contacted: The author will be informed of this. A final decision to publish can be made only on the basis of the complete manuscript, however a refereeing process leading to a preliminary decision can be based on a pre-final or incomplete manuscript. The strict minimum amount of material that will be considered should include a detailed outline describing the planned contents of each chapter, a bibliography and several sample chapters.

   Authors should be aware that incomplete or insufficiently close to final manuscripts almost always result in longer refereeing times and nevertheless unclear referees' recommendations, making further refereeing of a final draft necessary.

   Authors should also be aware that parallel submission of their manuscript to another publisher while under consideration for LNM will in general lead to immediate rejection.

3. Manuscripts should in general be submitted in English. Final manuscripts should contain at least 100 pages of mathematical text and should always include

   - a table of contents;
   - an informative introduction, with adequate motivation and perhaps some historical remarks: it should be accessible to a reader not intimately familiar with the topic treated;
   - a subject index: as a rule this is genuinely helpful for the reader.

   For evaluation purposes, manuscripts may be submitted in print or electronic form (print form is still preferred by most referees), in the latter case preferably as pdf- or zipped ps-files. Lecture Notes volumes are, as a rule, printed digitally from the authors' files. To ensure best results, authors are asked to use the LaTeX2e style files available from Springer's web-server at:

   ftp://ftp.springer.de/pub/tex/latex/svmonot1/ (for monographs) and
   ftp://ftp.springer.de/pub/tex/latex/svmultt1/ (for summer schools/tutorials).

Additional technical instructions, if necessary, are available on request from lnm@springer.com.

4. Careful preparation of the manuscripts will help keep production time short besides ensuring satisfactory appearance of the finished book in print and online. After acceptance of the manuscript authors will be asked to prepare the final LaTeX source files and also the corresponding dvi-, pdf- or zipped ps-file. The LaTeX source files are essential for producing the full-text online version of the book (see http://www.springerlink.com/openurl.asp?genre=journal&issn=0075-8434 for the existing online volumes of LNM). The actual production of a Lecture Notes volume takes approximately 12 weeks.

5. Authors receive a total of 50 free copies of their volume, but no royalties. They are entitled to a discount of 33.3 % on the price of Springer books purchased for their personal use, if ordering directly from Springer.

6. Commitment to publish is made by letter of intent rather than by signing a formal contract. Springer-Verlag secures the copyright for each volume. Authors are free to reuse material contained in their LNM volumes in later publications: a brief written (or e-mail) request for formal permission is sufficient.

Addresses:
Professor J.-M. Morel, CMLA,
École Normale Supérieure de Cachan,
61 Avenue du Président Wilson, 94235 Cachan Cedex, France
E-mail: morel@cmla.ens-cachan.fr

Professor B. Teissier, Institut Mathématique de Jussieu,
UMR 7586 du CNRS, Équipe "Géométrie et Dynamique",
175 rue du Chevaleret
75013 Paris, France
E-mail: teissier@math.jussieu.fr

For the "Mathematical Biosciences Subseries" of LNM:

Professor P. K. Maini, Center for Mathematical Biology,
Mathematical Institute, 24-29 St Giles,
Oxford OX1 3LP, UK
E-mail: maini@maths.ox.ac.uk

Springer, Mathematics Editorial, Tiergartenstr. 17,
69121 Heidelberg, Germany,
Tel.: +49 (6221) 4876-8259

Fax: +49 (6221) 4876-8259
E-mail: lnm@springer.com